Liapunov Functionals for Integral Equations

T. A. Burton

Department of Mathematics
Southern Illinois University
Carbondale, Illinois
taburton@olypen.com

©Theodore Allen Burton
732 Caroline St.
Port Angeles, WA 98362
October 2, 2009
All rights reserved.

Preface

Many investigators apply Liapunov's direct method to an integral equation by first differentiating it. This is the first book which also applies the method directly to the given integral equation. This substantially broadens the application, as we discuss in some detail in the first chapter. Moreover, it allows us to study integral equations at a more fundamental level than can be done after conversion to differential equations.

There is hardly an area of science which does not use differential or integral equations either directly or indirectly. Moreover, many of the problems require very exact qualitative results. In particular:

1. The functions cannot be replaced by convenient approximations.
2. It is critical that we know the behavior of all solutions.
3. We must know the behavior of solutions for arbitrarily large time.

Even with the most advanced computer methods, it is difficult to meet all three criteria. Yet, more than one hundred years ago a Russian mathematician, A. M. Liapunov, advanced a simple method which did, indeed, meet those criteria for ordinary differential equations.

Nothing comes to us free of charge and Liapunov's method is no exception. To implement the method for a given equation one must first construct a measuring device called a Liapunov function. That construction is an art, rather than a science. But in the 1950s a number of mathematicians, including notably Barbashin and Krasovskii, constructed a considerable collection upon which all of us working in the area continue to feed. Included in their work was a very perceptive idea by Krasovskii showing how to add a functional to a known Liapunov function so that the work was advanced to the study of a functional differential equation.

Liapunov's method was directed at ordinary differential equations, but it expanded seamlessly to partial differential equations and functional differential equations of both ordinary and partial types. We come now to the topic at hand. If an integral equation could be differentiated to produce an integrodifferential equation, then the method could be applied there as well. Unfortunately, this often turns things around. Some of the most important integral equations arise in the process of inverting a perturbed differential operator which has proved intractable by other methods, including Liapunov's direct method. In any event, there are substantial reasons why we cannot or should not differentiate the integral equation. This brings us to the year 1992, exactly one hundred years after the publication of Liapunov's work. It was then noted that the aforementioned functional of Krasovskii was actually a Liapunov functional for an associated integral equation and that the method of Liapunov could be extended to integral equations without differentiating the equation.

This book is an introduction to integral equations from the ground up focusing on the use of Liapunov functionals. It contains a variety of basic existence theorems, results on the resolvent, and several variation of parameters formulae. It also contains a collection of Liapunov functionals and techniques which enable the investigator to obtain very concrete qualitative information about solutions of a wide variety of integral equations of the type expected in many scientific investigations. We hope that this collection can serve in the same way as did the collection of Barbashin and Krasovskii.

At the same time, we emphasize that the intersection of the contents of this book with those of so many excellent classical treatments is very small. The path given here is totally different than that of any other book. Not only are the methods different, but the new methods yield results which are of types not previously seen in the theory of integral equations. Chapter 2 contains several surprises along that line.

It has long been our view that fixed point theory and Liapunov theory work hand-in-glove and we have sought throughout this work to tie them together. Chapters 3 through 6 represent a close weaving of these two areas.

In spite of the length of this book, we consider it to be a mere introduction. It is our hope that it will spur research in this very important area. The outstanding feature is its great simplicity, as was the case of Liapunov's original contribution. It is highly accessible to advanced undergraduate students in mathematics, as well as researchers in biology, chemistry, economics, engineering, mathematics, physics, and so many other areas of science.

Thanks go to Leigh Becker for furnishing much of the material in Chapter 1 concerning his construction of the resolvent equation for integrodifferential equations. Thanks go to Tetsuo Furumochi who came with his family from Shimane University in Japan to Southern Illinois University for ten months in 1993 and 1994 for collaboration on the development of Liapunov's direct method for integral equations. Thanks to Leigh Becker, Tetsuo Furumochi, Geza Makay, Ioannis Purnaras, Cristian Vladimirescu, and Bo Zhang for carefully reading portions or all of the material which went into the manuscript. Thanks to Charles Gibson for furnishing an excellent template for the word processing.

A special thanks goes to the University of Memphis for a visiting professorship in Spring 2009 to offer a graduate course over the material in this book. Thanks also to Professors David Dwiggins, John Haddock, and Alistair Windsor who attended all the lectures, participated in many of them, and worked through much of the material for the book.

T. A. Burton
Northwest Research Institute
Port Angeles, Washington
October 2, 2009
taburton @olypen.com

Contents

Preface		**vi**
Contents		**ix**
0	**Introduction and Overview**	**1**
	0.1 An Idea with many Applications	1
	0.2 An Introductory Lecture	6
	0.3 Notation	17
1	**Linear Equations, Existence, and Resolvents**	**18**
	1.1 A Qualitative Approach	18
	1.2 Existence and Resolvents	24
2	**The Remarkable Resolvent**	**52**
	2.1 Perfect Harmony	52
	2.2 A Strong Resolvent: the scalar case	70
	2.3 An Unusual Scalar Equation	89
	2.4 A Mathematical David & Goliath: Periodicity	95
	2.5 Floquet Theory	101
	2.6 Liapunov Functionals for Resolvents	111
	2.7 Integrodifferential Equations	135
	2.8 A Nonlinear Application	152
	2.9 Notes	164
3	**Existence Properties**	**166**
	3.1 A Global Existence Theorem	166
	3.2 Classical Theory	174
	3.3 Equations of Chapter 2	180
	3.4 Resolvents and Nonlinearities	196
	3.5 Notes	198

CONTENTS

4 Applications and Origins — **200**
- 4.0 Introduction — 200
- 4.1 Differential Equations — 201
- 4.2 Classical Problems — 214
- 4.3 The Nature of the Kernel — 216

5 Infinite Delay: Schaefer's Theorem — **236**
- 5.1 Introduction — 236
- 5.2 Qualitative Theory — 257
- 5.3 A Refined Liapunov Functional — 267
- 5.4 *A Priori* Bounds — 280

6 The Krasnoselskii-Schaefer Theorem — **289**
- 6.1 Method and Problems — 289
- 6.2 A Linear Truncated Equation — 295
- 6.3 A Nonlinear Truncated Equation — 299
- 6.4 Neutral Integral Equations — 303

7 Stability: Convex Memory — **320**
- 7.1 Introduction — 320
- 7.2 Near Equilibria — 324

8 Appendix: Preparing the Kernel — **341**
- 8.1 Introduction — 341
- 8.2 Adding x and x' — 342
- 8.3 Uniform Boundedness — 345
- 8.4 Another Unusual Equation — 347

References — **350**

Author Index — **358**

Subject Index — **360**

Chapter 0

Introduction and Overview

0.1 An Idea with many Applications

History

In 1892 Liapunov (1992) published a major work in which he advanced an idea for studying qualitative properties of solutions of an ordinary differential equation,

$$x' = F(t, x)$$

where $' = \frac{d}{dt}$ and $F : [0, \infty) \times \Re^n \to \Re^n$ is usually at least continuous. Thus, for each $(t_0, x_0) \in [0, \infty) \times \Re^n$ there is at least one solution $x(t) := x(t, t_0, x_0)$ through (t_0, x_0) defined on some interval $[t_0, t_0 + \alpha)$ for some $\alpha > 0$. His theory ranged from utter simplicity to mathematically deep and intricate properties, typified by the span from simple boundedness to periodicity of solutions.

While it started as a technique to study ordinary differential equations, it progressed smoothly to functional differential equations, control theory, partial differential equations, and difference equations. In many ways it has provided a unifying link between these seemingly different subjects.

Here is its simplest form. We mainly follow Yoshizawa (1966; p. 2) and say that a *Liapunov function* is a differentiable scalar function. Suppose we construct a differentiable scalar function $V : [0, \infty) \times \Re^n \to [0, \infty)$ with the property that $V(t, x) \to \infty$ as $|x| \to \infty$ uniformly for $0 \leq t < \infty$. Now $V(t, x)$ and $x(t) = x(t, t_0, x_0)$ are totally unrelated, but if we write $V(t, x(t))$ then they are related and by the chain rule, for example, we can compute

$$\frac{dV(t, x(t))}{dt} = \frac{\partial V}{\partial t} + \frac{\partial V}{\partial x_1}\frac{dx_1}{dt} + \cdots + \frac{\partial V}{\partial x_n}\frac{dx_n}{dt},$$

where the dx_i/dt are just the components of the KNOWN function F.

Here is the first of two critical points. An existence theorem tells us that the solution $x(t)$ exists; but virtually never can we find that solution. We do not need to know the solution because we can compute the derivative of V DIRECTLY from the differential equation itself. Hence, the name "Liapunov's direct method."

The second critical point is this. That derivative which we just obtained may yield no information at all unless it is of a special form. For example, if we have shrewdly constructed $V(t,x)$ we may find that

$$\frac{dV(t,x(t))}{dt} \leq 0.$$

This will mean that for $t \geq t_0$ then

$$V(t,x(t)) \leq V(t_0,x_0)$$

and so $x(t)$ is bounded since $V(t,x) \to \infty$ as $|x| \to \infty$. This is the crudest qualitative property and much more can be obtained with further study.

We have achieved a great victory and it hinged completely on being able to relate $V(t,x)$ to $x(t)$ by the chain rule.

The task, then, is to learn to construct suitable functions, $V(t,x)$, called Liapunov functions. Early on Barbashin (1968) constructed an immense set of Liapunov functions for a great variety of problems. It turned out that investigators were able to modify those functions to make them applicable to more problems, and that process continues to this day. From thousands of pages of literature on such constructions and intricate theorems we select a few more examples. Perhaps the oldest is the Liapunov function consisting of the sum of the kinetic and potential energy of a Liénard equation. This was advanced to a delay equation by Krasovskii using a simple device which converted a functional to a function. Krasovskii's work showed us how to generally advance our Liapunov functions for differential equations to Liapunov functionals for many functional differential equations.

That sum of kinetic and potential energy was also advanced to the control problem of Lurie and then on to the damped wave equation. That single Liapunov functional spanned and brought some unification to three distinct areas of mathematics. It unified areas in the sense that an investigator moved from one area to the other with the same tools and applicable experience.

Much of the work in constructing Liapunov functions was *ad hoc*, but from earliest investigations there was general theory of linear equations of the form $x' = Ax$, A being a constant $n \times n$ matrix, using a Liapunov function of the form $V = x^T B x$ and a scheme is given for constructing

0.1. AN IDEA WITH MANY APPLICATIONS 3

B. In recent years Bo Zhang (2001, 2005) presented formal theory for constructing Liapunov functions for systems of linear ordinary differential equations, delay equations, and partial differential equations. He was joined by Graef and Qian (2004) in advancing the theory to difference equations.

Two of the most prominent classical works with transition from ordinary differential equations to functional differential equations are Krasovskii (1963) and Yoshizawa (1966). The transition of the aforementioned sum of kinetic and potential energy from a Liénard equation to a delay equation, on to the damped wave equation, and then to a Lurie control problem can be seen in Barbashin (1968; p. 1102), Krasovskii (1968; p. 173), Burton (1991), and Burton and Somolinos (2007). An introductory treatment showing a smooth transition of Liapunov functions from ordinary to partial differential equations is found in Henry (1981).

The Present Problem

We reiterate that the success of Liapunov's direct method is tied to our ability to unite the differential equation $x' = F(t,x)$ to a totally independent function, $V(t,x)$.

In the same general category of equations under discussion above is the integral equation

$$x(t) = f(t) + \int_0^t g(t,s,x(s))ds.$$

Investigators have long desired to advance Liapunov's direct method to such, but the difficulty is that we do not know how to unite a Liapunov function, $V(t,x)$, to this equation.

The book by R. K. Miller (1971a) presents an excellent introduction to integral equations from a classical point of view. On p. 337 he begins Chapter VI as follows: "If a system of integral equations of the form

$$x(t) = f(t) + \int_0^t g(t,s,x(s))ds$$

can be written in differentiated form

$$x'(t) = f'(t) + g(t,t,x(t)) + \int_0^t g_t(t,s,x(s))ds, x(0) = f(0),$$

then it may be possible to analyze the behavior of the solution $x(t)$ by means of Lyapunov's second method." Miller has plainly stated the prevailing view.

In large measure the present book begins at this point. Frequently we do find it profitable to differentiate an integral equation, as we will do many times in this book. However, there is a long list of reasons why we do not want to differentiate the integral equation. That list includes:

1. The functions may not be differentiable, so it is impossible.
2. Integration smooths, but differentiation produces roughness; thus, differentiation can introduce complications and great disorder.
3. If the functions are differentiable, then we can prove existence and uniqueness of solutions. But the derived equation may have functions which are, at most, continuous. Thus, the derived equation may have nonunique solutions and we may have introduced extraneous solutions. This means that the properties we prove for the derived equation are actually absent in the original equation.
4. The problem arises as an integral equation and it would seem to be a matter of integrity to deal with it in that category.
5. There is a vast category of problems in the form

$$x(t) = f(t) + \int_0^t [C(t,s)x(s) + g(t,s,x(s))]ds$$

in which there is no hope for differentiation. But it is possible to extract the resolvent problem

$$R(t,s) = C(t,s) - \int_s^t R(t,u)C(u,s)du$$

which we may treat with Liapunov functionals and reformulate the original equation with $R(t,s)$ about which we have very precise information. Virtually all of our lengthy Chapter 2 involves this analysis by direct application of Liapunov's method.

This book offers a variety of attacks on integral equations using Liapunov's direct method, frequently without differentiating the equation. Again, taking our cue from Yoshizawa we will say that a *Liapunov functional* is a differentiable scalar functional. Our task in this book is to show how the Liapunov functional can be united with the integral equation in a way parallel to the union achieved through the use of the chain rule described above.

Outline of Coverage

Section 1.1 contains a collection of equations and results which are enlarged upon in the first two chapters. It lists the basic linear integral

0.1. AN IDEA WITH MANY APPLICATIONS

equation, the resolvent, and the variation of parameters formula. It mentions the great challenge which is ever before us to understand why the equation

$$x(t) = t + \sin t + (t+1)^{1/2} \sin(t+1)^{1/3} - \int_0^t C(t,s)x(s)ds$$

has a solution almost indistinguishable from that of

$$y(t) = \sin t - \int_0^t C(t,s)y(s)ds$$

when $C(t,s)$ is a "nice" function. Finally, it offers the contraction mapping theorem and a brief introduction to Liapunov theory, together with examples, as the main tools used in the first half of this book.

Section 1.2 is long and has many subsections. It offers theorems and proofs concerning existence, uniqueness, resolvents, and variation of parameters formulae for linear equations. It is largely the work of Leigh Becker.

Chapter 2 focuses on the many wonders of the resolvent, $R(t,s)$, and the variation of parameters formula

$$x(t) = a(t) - \int_0^t R(t,s)a(s)ds$$

for the solution of

$$x(t) = a(t) - \int_0^t C(t,s)x(s)ds.$$

It explains that, under general conditions, there are large vector spaces containing bounded and unbounded functions, $a(t)$, such as $(t+1)^{1/2}$, for which

$$\int_0^t R(t,s)a(s)ds$$

represents almost an exact copy of $a(t)$ for large t, while there are vector spaces of small functions, such as $\sin t$, for which that integral is totally incapable of duplicating $a(t)$. We try to bring this into perspective early on by offering in Section 0.2 an introductory lecture which gives a sample of the things we try to accomplish in Chapter 2. It is a lecture we have given in several countries during the last few years as we develop the material.

Section 2.3 presents an unusual integro-differential equation derived from the integral equation by other than the usual differentiation process and which is very useful in establishing qualitative properties.

Sections 2.4 and 2.5 deal with periodicity and offer a Floquet theory. We obtain a partial answer to the challenge mentioned above.

Section 3.1 employs Schaefer's fixed point theorem, together with Liapunov functions, to obtain a global existence result. It introduces us to the marriage of fixed point theory and Liapunov theory which is the main focus of Section 4.1 and Chapters 5 and 6. Section 3.2 offers classical local existence, uniqueness, and continuability results for nonlinear equations. With these properties in hand, we review some of the work from Chapter 2 for nonlinear equations and see how the Liapunov functionals are changed from the linear to nonlinear case.

Chapter 4 introduces Krasnoselskii's fixed point theorem and his working hypothesis which has had such an impact on the direction of integral equations in the last 15 years. We unite Krasnoselskii's theorem with Schaefer's theorem which is our principle link with Liapunov's direct method. We also present brief descriptions of classical applications. The chapter concludes with a discussion of two strikingly different kernels which generate very similar solutions.

Finally, Chapter 7 offers a brief introduction to stability theory for some integral equations.

0.2 An Introductory Lecture

Integral equations confront us with one surprise after another in the most elementary investigations, particularly when the kernel is nonconvolution and not integrable to ∞. They have been studied so much less than differential equations and they offer most exciting opportunities for investigation.

This lecture has been prepared over the last several years as the material in Chapter 2 was being developed and has been presented in several countries. We strive to present elementary techniques for displaying qualitative properties of solutions of integral equations. These techniques are simple and "clean." The investigator is able to see clearly what is happening in a given problem. Moreover, the same techniques are applied to a great variety of problems in Chapter 2 and the last section of Chapter 3. Not only is the work readily accessible to well-prepared third year university students, but it offers enticing problems for research.

In the classical theory, if the kernel is "nice", then the solution of the integral equation

$$x(t) = a(t) - \int_0^t C(t,s)x(s)ds \qquad (0.2.1)$$

"follows" $a(t)$:

0.2. AN INTRODUCTORY LECTURE

If $a(t)$ is bounded, so is $x(t)$.
If $a(t) \to 0$, so does $x(t)$.
If $a(t) \in L^1$, so is $x(t)$.
If $a(t)$ is asymptotically periodic, so is $x(t)$.
PROBLEM: Would we expect the solution of

$$x(t) = \sin t + (t+1)^{1/2} + \int_0^t C(t,s)x(s)ds$$

to follow $\sin t$? Would we ever expect the big function $(t+1)^{1/2}$ to be largely ignored and $\sin t$ to dominate?

In fact, that is exactly what does happen.

There is a resolvent $R(t,s)$ and a variation of parameters formula

$$x(t) = a(t) - \int_0^t R(t,s)a(s)ds.$$

One discovers big vector spaces, V, containing unbounded functions ϕ such that $\phi \in V$ implies that

$$\phi(t) - \int_0^t R(t,s)\phi(s)ds$$

is an L^p-function. That integral creates a copy of ϕ which is correct to within an L^p-function. By contrast, there are big vector spaces of little functions ϕ such that $\int_0^t R(t,s)\phi(s)ds$ is unable to copy any ϕ in the space with a corresponding error bound.

There will be two parts to this lecture. In Part I we will present a variety of results of classical type which lead us to believe that when the kernel, $C(t,s)$, is "nice", then the solution of the scalar equation

$$x(t) = a(t) - \int_0^t C(t,s)x(s)ds$$

follows $a(t)$. In Part II we show that this is vastly premature.

We deal with spaces of functions $\phi: [0, \infty) \to \Re$:

(i) $(BC, \|\cdot\|)$ is the Banach space of bounded continuous functions with supremum norm.

(ii) Continuous functions in $L^p[0, \infty)$

(iii) $(Y, \|\cdot\|)$ is the Banach space of continuous functions written as $\phi = p + q$ where $p \in \mathcal{P}_T$, the set of T-periodic functions, and $q \in Q$, the set of functions tending to zero as $t \to \infty$.

PART I: Perfect Harmony

We will sketch proofs showing that:
(i) If $a \in BC$, then $x \in BC$.
(ii) If a is continuous and in $L^p[0, \infty)$, so is x.
(iii) If $a \in Y$, then $x \in Y$.

Here are some details. Let $a : [0, \infty) \to \Re$ and $C : [0, \infty) \times [0, \infty) \to \Re$ both be continuous and consider the scalar integral equation

$$x(t) = a(t) - \int_0^t C(t,s)x(s)ds. \qquad (0.2.1)$$

Associated with (0.2.1) is the resolvent equation

$$R(t,s) = C(t,s) - \int_s^t R(t,u)C(u,s)du \qquad (0.2.2)$$

(note that $R(t,s)$ depends only on $C(t,s)$) and the variation of parameters formula

$$x(t) = a(t) - \int_0^t R(t,s)a(s)ds. \qquad (0.2.3)$$

One of the interesting parts of the investigation is that, while ideal theory of Ritt (1966) shows that $R(t,s)$ is arbitrarily complicated, the integral in (0.2.3) often strips away all complications and constructs a function closely approximating $a(t)$ for large t.

Theorem 0.2.1. *Let $\int_0^t |C(t,s)|ds \leq \alpha < 1$. If $a \in BC$, so is x.*

Notice that the integration is with respect to the variable s. We have $a \in BC$, $x \in BC$, and the variation of parameters formula is $x(t) = a(t) - \int_0^t R(t,s)a(s)ds$, so $\int_0^t R(t,s)a(s)ds \in BC$. Every function in that group is in the same space.

Proof. We define a mapping P by $\phi \in BC$ implies that

$$(P\phi)(t) := a(t) - \int_0^t C(t,s)\phi(s)ds.$$

Then $P : BC \to BC$ and is a contraction mapping with unique fixed point $\phi \in BC$, the solution of (0.2.1).

Theorem 0.2.2. *Let $\int_0^\infty |C(u+t,t)|du \leq \alpha < 1$. If $a \in L^1$ so is x.*

0.2. AN INTRODUCTORY LECTURE

Notice that the integration is with respect to the variable t. We have $a \in L^1$, $x \in L^1$, and by the variation of parameters formula $\int_0^t R(t,s)a(s)ds \in L^1$. Again, every function in that group is in the same space.

Proof. Much of this book will concern the intricacies of Liapunov functionals for integral equations. But here we will quickly introduce one, show its power, and leave explanations for later.

Note that if $x(t)$ solves (0.2.1) then $|x(t)| \leq |a(t)| + \int_0^t |C(t,s)x(s)|ds$. Define a Liapunov functional by

$$V(t) = \int_0^t \int_{t-s}^\infty |C(u+s,s)|du|x(s)|ds$$

so that

$$V'(t) = \int_0^\infty |C(u+t,t)|du|x(t)| - \int_0^t |C(t,s)x(s)|ds$$

$$\leq \alpha|x(t)| - \int_0^t |C(t,s)x(s)|ds.$$

But our initial inequality now shows that

$$V'(t) \leq \alpha|x(t)| - |x(t)| + |a(t)|$$

which yields

$$0 \leq V(t) \leq V(0) + \int_0^\infty |a(t)|dt - (1-\alpha)\int_0^t |x(s)|ds.$$

Theorem 0.2.3. *Suppose there are constants $\alpha < 1$ and $\beta < 1$ with*

$$\int_0^t |C(t,s)|ds \leq \alpha \quad \text{and} \quad \int_0^\infty |C(u+t,t)|du \leq \beta.$$

If there is an integer $n > 0$ with $a \in L^{2^n}[0,\infty)$ then the solution of (0.2.1) satisfies $x \in L^{2^n}[0,\infty)$.

Notice that we are now integrating with respect to both s and t. Again, we can see that each function in the group is in L^{2^n}.

Proof. Define a Liapunov functional

$$V(t) = \int_0^t \int_{t-s}^\infty |C(u+s,s)| du x^{2^n}(s) ds$$

so that by Leibnitz's rule we have

$$V'(t) = \int_0^\infty |C(u+t,t)| du x^{2^n}(t) - \int_0^t |C(t,s)| x^{2^n}(s) ds$$
$$\leq \beta x^{2^n}(t) - x^{2^n}(t) + M^{2^n-1} a^{2^n}(t)$$

where the last line is obtained by repeatedly squaring our integral equation, using the Schwarz inequality, and choosing a suitably large constant M. The details are found in the proof of Lemma 2.1.2. Thus, an integration yields

$$V(t) \leq V(0) - (1-\beta) \int_0^t x^{2^n}(s) ds + M^{2^n-1} \int_0^\infty a^{2^n}(t) dt.$$

Theorem 0.2.4. *Suppose that for a fixed positive constant T we have $C(t+T, s+T) = C(t,s)$, that $\int_{-\infty}^t |C(t,s)| ds$ is continuous, and that $\int_0^t |C(t,s)| ds \leq \alpha < 1$. In addition, let $\int_{-\infty}^t C(t,s) p(s) ds$ be continuous for $p \in \mathcal{P}_T$ while*

$$(i) \int_{-\infty}^0 |C(t,s)| ds \to 0 \text{ as } t \to \infty$$

and

$$(ii) \int_0^t C(t,s) q(s) ds \to 0 \text{ as } t \to \infty \text{ for } q \in Q.$$

Then $a \in Y$ implies that the solution of (0.2.1) satisfies $x \in Y$.

Notice again that each function in the group is in Y.

0.2. AN INTRODUCTORY LECTURE

Proof. Let $(Y, \|\cdot\|)$ be the Banach space of continuous functions $\phi = p + q$ where $p \in \mathcal{P}_T$ and $q \in Q$ with the supremum norm. Also, let $a = p^* + q^* \in Y$. Define $P : Y \to Y$ by $\phi = p + q \in Y$ implies that

$$(P\phi)(t) = a(t) - \int_0^t C(t,s)[p(s) + q(s)]ds$$
$$= \left[p^*(t) - \int_{-\infty}^t C(t,s)p(s)ds\right]$$
$$+ \left[q^*(t) + \int_{-\infty}^0 C(t,s)p(s)ds - \int_0^t C(t,s)q(s)ds\right]$$
$$=: B\phi + A\phi.$$

This defines operators A and B on Y. Note that $B : Y \to \mathcal{P}_T \subset Y$ and $A : Y \to Q \subset Y$.

But from the first line of this array we see that P is a contraction with unique fixed point $\phi \in Y$ and that proves the result.

Notice that x, a, and $\int_0^t R(t,s)a(s)ds$ are always in the same space. In classical theory we think of x and a being in the same space, without thinking of the integral. We show in the next part that it is a and $\int_0^t R(t,s)a(s)ds$ which are in the same space so that x is always small. Thus, if a is large, then x and a will not be in the same space. The integral faithfully duplicates a so that x must be content with whatever is left over.

Notice also that $C(t,s) \equiv 0$ is allowed in every one of the preceding results. Thus, if x is to be bounded, then a must be bounded. That will now change so that a can be large.

PART II: Great Discord

Here is an overview of what we will see. Let $C(t,s)$ be "nice." Suppose that $0 < \beta < 1$ so that $[(t+1)^\beta]' \in L^p$ for some $p < \infty$. The solution of

$$x(t) = (t+1)^\beta - \int_0^t C(t,s)x(s)ds \in L^p,$$

which means that it does not follow $(t+1)^\beta$, while

$$x(t) = (t+1)^\beta - \int_0^t R(t,s)(s+1)^\beta ds \in L^p.$$

Hence,

$$\int_0^t R(t,s)(s+1)^\beta ds$$

copies $(t+1)^\beta$ correct to within an L^p-function. It is also true at the "boundary" where $\beta = 1$. In fact, if we consider the vector space of functions ϕ with $\phi' \in L^p$ for some $p < \infty$, it is true that $\int_0^t R(t,s)\phi(s)ds$ creates a correspondingly good copy of every ϕ in that space.

The Big Function Vanishes

We continue the example and add to $a(t)$ the function $\sin(t+1)^\beta$ with variation of parameters formula

$$x(t) = \sin(t+1)^\beta + (t+1)^\beta$$
$$- \int_0^t R(t,s)[\sin(s+1)^\beta + (s+1)^\beta]ds \in L^p$$

so that the integral is making only an L^p-error in copying $a(t)$. However, at $\beta = 1$ the process breaks down and the solution becomes $p + q$ where $p \in \mathcal{P}_T, q \in Q$. The large function t is almost totally lost, while the solution follows $\sin t$. The integral $\int_0^t R(t,s)\phi(s)$ has been so successful at copying large functions ϕ, but it is totally unable to copy $\sin(t+1)$. This time R is unable to hold onto its "boundary" function. Thus, at $\beta = 1$, since $x = p + q$, that integral makes such a good copy of t that the error tends to zero.

Here are the details. It will sometimes avoid confusion if we denote the partial derivative of C with respect to t by either C_t or C_1. Assume that a' and $C_t(t,s)$ are continuous and write (0.2.1) as

$$x'(t) = a'(t) - C(t,t)x(t) - \int_0^t C_t(t,s)x(s)ds. \tag{0.2.4}$$

Theorem 0.2.5. *Suppose there is a positive integer n with $a'(t) \in L^{2n}[0,\infty)$, a constant $\alpha > 0$, and a constant $N > 0$ with*

$$\frac{2n-1}{2nN^{\frac{2n}{2n-1}}} - C(t,t) + \frac{2n-1}{2n}\int_0^t |C_t(t,s)|ds$$
$$+ \frac{1}{2n}\int_0^\infty |C_1(u+t,t)|du \leq -\alpha.$$

Then the solution x of (0.2.1) is bounded and $x \in L^{2n}[0,\infty)$.

Notice that we integrate with respect to t and s; a big $C(t,s)$ is required. Also $x(t) = a(t) - \int_0^t R(t,s)a(s)ds \in L^{2n}$.

0.2. AN INTRODUCTORY LECTURE

Proof. For a fixed solution of (0.2.4) we define the function

$$V(t) = \frac{x^{2n}(t)}{2n} + \frac{1}{2n}\int_0^t \int_{t-s}^\infty |C_1(u+s,s)| du \, x^{2n}(s) ds$$

with

$$V'(t) = -C(t,t)x^{2n} - \int_0^t C_t(t,s)x(s)x^{2n-1}(t) ds$$
$$+ x^{2n-1}(t)a'(t) + \frac{1}{2n}\int_0^\infty |C_1(u+t,t)| du \, x^{2n}(t)$$
$$- \frac{1}{2n}\int_0^t |C(t,s)| x^{2n}(s) ds$$

(Use Hölder's inequality on the 2nd & 3rd terms.)

$$\leq \frac{(2n-1)x^{2n}(t)}{2nN^{\frac{2n}{2n-1}}} + \frac{(Na'(t))^{2n}}{2n} - C(t,t)x^{2n}$$
$$+ \int_0^t |C_t(t,s)| \left[\frac{(2n-1)x^{2n}(t)}{2n} + \frac{x^{2n}(s)}{2n}\right] ds$$
$$+ \frac{1}{2n}\int_0^\infty |C_1(u+t,t)| du \, x^{2n}(t)$$
$$- \frac{1}{2n}\int_0^t |C_1(t,s)| x^{2n}(s) ds$$
$$= \frac{(Na'(t))^{2n}}{2n} + x^{2n}(t)\left[\frac{(2n-1)}{2nN^{\frac{2n}{2n-1}}} - C(t,t)\right.$$
$$\left. + \frac{2n-1}{2n}\int_0^t |C_t(t,s)| ds + \frac{1}{2n}\int_0^\infty |C_1(u+t,t)| du\right]$$
$$\leq -\alpha x^{2n}(t) + \frac{N^{2n}}{2n}|a'(t)|^{2n}$$

for large N. It follows that

$$\frac{x^{2n}(t)}{2n} \leq V(t) \leq V(0) - \alpha \int_0^t x^{2n}(s) ds + k \int_0^\infty (a'(s))^{2n} ds$$

for some $k > 0$ from which the result follows.

For $0 < \beta < 1$ this theorem now gives us

$$x(t) = (t+1)^\beta + \sin(t+1)^\beta$$
$$- \int_0^t R(t,s)[\sin(s+1)^\beta + (s+1)^\beta] ds \in L^p.$$

Notice that x and a are not in the same space. The big function $t+1$ vanishes, as does the oscillating function. On average, the integral faithfully copies both functions. We now want a periodic theorem to show that precisely at $\beta = 1$ the process breaks down. The integral still faithfully copies the large function $t+1$, but it can not copy $\sin(t+1)$ which totally rules the equation; that L^p-error is now a thing of the past.

Conclusion

Take courage when encountering a large forcing function which threatens to destabilize an equation. With care you may show that the large function is totally harmless.

The Little Function Rules

Let $a' \in Y$, $C(t,s) = C(t+T, s+T)$, and let $C(t,s)$ and $C_t(t,s)$ be continuous. Write (0.2.1) as

$$x'(t) = a'(t) - C(t,t)x(t) - \int_0^t C_t(t,s)x(s)ds. \tag{0.2.4}$$

This is what will happen. If $a(t) = t + \sin t$ then $a'(t) = 1 + \cos t \in Y$. The big function t will seem to vanish and $\sin t$ will rule the equation.

By the variation of parameters formula

$$x(t) = x(0)e^{-\int_0^t C(s,s)ds}$$
$$+ \int_0^t e^{-\int_u^t C(s,s)ds}\left[a'(u) - \int_0^u C_1(u,s)x(s)ds\right]du. \tag{0.2.5}$$

We suppose that there is a number $c^* > 0$ with

$$C(t,t) \geq c^* \tag{0.2.6}$$

and an $\alpha < 1$ with

$$\int_0^t e^{-\int_u^t C(s,s)ds} \int_0^u |C_1(u,s)|dsdu \leq \alpha. \tag{0.2.7}$$

One may note that this requires a large $C(t,s)$. We remind the reader that $p \in \mathcal{P}_T$ and $q \in Q$ where this means that p is periodic and q tends to zero. Recall also that the solution of interest satisfies $x(0) = a(0)$.

Write $a'(t) = p^*(t) + q^*(t) \in Y$ and define a mapping by $\phi = p + q \in Y$ implies that

$$(P\phi)(t) = a(0)e^{-\int_0^t C(s,s)ds}$$
$$+ \int_0^t e^{-\int_u^t C(s,s)ds}\Big[p^*(u) + q^*(u)$$
$$- \int_0^u C_1(u,s)[p(s) + q(s)]ds\Big]du. \qquad (0.2.8)$$

Let $\int_{-\infty}^0 |C_1(t,s)|ds$ be continuous,

$$\int_{-\infty}^0 |C_1(t,s)|ds \to 0 \text{ as } t \to \infty, \qquad (0.2.9)$$

and for $q \in Q$ let

$$\int_0^t C_1(t,s)q(s)ds \to 0 \text{ as } t \to \infty. \qquad (0.2.10)$$

Theorem 0.2.6. *In (0.2.4) let (0.2.6), (0.2.7), (0.2.9), (0.2.10) hold. Suppose, in addition, that*

$$\int_{-\infty}^t |C_1(t,s)|ds \qquad (0.2.11)$$

is bounded. If $a' \in Y$ so is x, the solution of (0.2.1).

Notice that for $a(t) = \sin(t+1) + (t+1)$ then $a'(t) \in Y$ so $x(t) = a(t) - \int_0^t R(t,s)a(s)ds \in Y$. Thus, $\int_0^t R(t,s)sds$ so accurately copies t that t is effectively removed from the equation.

Proof. Use (0.2.8) to define a mapping $P: Y \to Y$ by $\phi = p + q \in Y$ implies that

$$(P\phi)(t) = a(0)e^{-\int_0^t C(s,s)ds}$$
$$+ \int_0^t e^{-\int_u^t C(s,s)ds}\Big[a'(u) - \int_0^u C_1(u,s)\phi(s)ds\Big]du.$$

By (0.2.7) it is clearly a contraction, but we must show that $P: Y \to Y$. Write $a' = p^* + q^*$ and then

$$(P\phi)(t) = \int_{-\infty}^{t} e^{-\int_{u}^{t} C(s,s)ds} \left[p^*(u) - \int_{-\infty}^{u} C_1(u,s)p(s)ds \right] du$$

$$- \int_{0}^{t} e^{-\int_{u}^{t} C(s,s)ds} \int_{0}^{u} C_1(u,s)q(s)ds\,du + a(0)e^{-\int_{0}^{t} C(s,s)ds}$$

$$- \int_{-\infty}^{0} e^{-\int_{u}^{t} C(s,s)ds} \left[p^*(u) - \int_{-\infty}^{u} C_1(u,s)p(s)ds \right] du$$

$$+ \int_{0}^{t} e^{-\int_{u}^{t} C(s,s)ds} \int_{-\infty}^{0} C_1(u,s)p(s)ds\,du$$

$$+ \int_{0}^{t} e^{-\int_{u}^{t} C(s,s)ds} q^*(u)du.$$

The first term on the right-hand-side is clearly in $\mathcal{P}_T$. In the second term, $\int_{0}^{u} C_1(u,s)q(s)ds \in Q$ by (0.2.10). Hence the second term is in Q by (0.2.6). The third term is in Q by (0.2.6). The fourth term is in Q by (0.2.6), (0.2.11), and the fact that $p^* \in \mathcal{P}_T$ and, hence, is bounded. The next to last term is in Q because of (0.2.9) followed by (0.2.10). The last term is in Q by (0.2.6).

More detail is given with Theorem 2.4.1.

A significant instability can occur at $\beta = 1$. Under the conditions on $C(t,s)$ of Theorem 0.2.5 the integral of that resolvent has been faithfully following $\sin(t+1)^{\beta}$ so that the difference is an L^p function. Suddenly, that relationship breaks completely and the integral with the resolvent seems to "struggle along trying to catch up with $\sin(t+1)$" but always is out of step, lagging by a nontrivial periodic function plus a function tending to zero. The little function, $\sin(t+1)$, dominates everything.

Corollary. *If the conditions of Theorem 0.2.5 hold and if $0 < \beta < 1$ then $\sin(t+1)^{\beta} - \int_{0}^{t} R(t,s)\sin(s+1)^{\beta}ds \in L^s$ for some $s < \infty$. But at $\beta = 1$, under conditions on $C(t,s)$ of Theorem 0.2.6 then $s = \infty$ and that difference approaches a periodic function.*

We can now state the promised result, a corollary of Theorems 0.2.5 and 0.2.6.

Corollary. *Let the conditions on $C(t,s)$ of Theorems 0.2.5 and 0.2.6 hold. For fixed $\beta \in (0,1)$ there is a $p \in \mathcal{P}_T$, $q \in Q$, and $u \in L^s[0,\infty)$ for some $s > 0$ so that the solution of*

$$x(t) = \sin t + (t+1)^{\beta} - \int_{0}^{t} C(t,s)x(s)ds$$

may be written as
$$x(t) = p(t) + q(t) + u(t).$$

Proof. The solution is
$$x(t) = \sin t + (t+1)^\beta - \int_0^t R(t,s)[\sin s + (s+1)^\beta]ds.$$

But
$$(t+1)^\beta - \int_0^t R(t,s)(s+1)^\beta ds =: u(t) \in L^s[0,\infty),$$

while
$$\sin t - \int_0^t R(t,s)\sin s\, ds$$

is the solution described in Theorem 0.2.6 and it has the required form of $p+q$.

See Example 2.4.2 for specific functions satisfying these conditions.

The conclusion is that small functions can totally destabilize an equation.

0.3 Notation

For a functional differential equation we have a Liapunov functional $V(t, x(\cdot))$ where $x(\cdot)$ is drawn from a certain space, X, of functions. The functional V may possess many properties, such as being positive definite, independently of $x \in X$. But fundamental properties of a Liapunov functional for an integral equation, such as being positive definite, may depend very explicitly on x being a solution of the integral equation. To emphasize this, we will write $V(t)$, rather than $V(t, x(\cdot))$ with the explicit understanding that x is a solution of the integral equation. For continuity, we continue that practice even after differentiating the integral equation.

Section 3.1 and Chapter 7 are different and we will revert to the first mentioned notation there.

Chapter 1

Linear Equations, Existence, and Resolvents

1.1 A Qualitative Approach

Much of the first half of this book is concerned with a linear integral equation written

$$x(t) = a(t) - \int_0^t C(t,s)x(s)ds \qquad (1.1.1)$$

where x and a are n-vectors, $n \geq 1$, while C is an $n \times n$ matrix. Usually, we ask that a be at least continuous on $[0, \infty)$ and that C be at least continuous on $[0, \infty) \times [0, \infty)$. The first surprise awaiting the student of differential equations is that the scalar case, $n = 1$, is fundamental since a linear n^{th} order normalized ordinary differential equation can be expressed in that way. By contrast, an integral equation of the form of (1.1.1) can be expressed as a differential equation only under very special conditions.

Along with (1.1.1) we study the resolvent equation

$$R(t,s) = C(t,s) - \int_s^t R(t,u)C(u,s)du \qquad (1.1.2)$$

and the variation of parameters formula

$$x(t) = a(t) - \int_0^t R(t,s)a(s)ds. \qquad (1.1.3)$$

The resolvent, $R(t, s)$, depends only on $C(t, s)$ and not on $a(t)$. Once we have determined certain basic properties of R, the equation (1.1.3) will speak volumes about (1.1.1) for a wide variety of functions a.

1.1. A QUALITATIVE APPROACH

In addition to the aforementioned surprise, two more await us. While $R(t,s)$ depends only on $C(t,s)$, under a variety of conditions on C we can identify vector spaces of functions ϕ such that $\int_0^t R(t,s)\phi(s)ds$ is a good copy of ϕ. We will, for example, give conditions on C so that if $\phi' \in L^p[0,\infty)$ for some $p > 0$, then in (1.1.3) we have

$$x(t) = \phi(t) - \int_0^t R(t,s)\phi(s)ds \in L^p[0,\infty);$$

the function $(t+1)^\beta$ for $0 < \beta < 1$ is included.

This gives a glimpse into the types of conditions and conclusions to be studied. No integrations are involved, no graphs are plotted, and no initial conditions are provided. Instead, a vector space of functions ϕ with $\phi' \in L^p$ is specified and the conclusion is that the solution is in L^p. Thus, we know that the solution resides near 0 much of the time although it may have spikes. Also, there are large functions $a = \phi$ which seem to affect the solution very little; they are harmless perturbations.

In each of the problems we discuss we present three packages: P_1, P_2, P_3. P_1 states conditions on C. P_2 describes the vector space in which a resides. P_3 describes the behavior of all solutions generated by elements of the vector space. These properties are in the way of stability results. Because of uncertainties or even stochastic elements, a may not be known. Yet we find vast vector spaces containing functions a which would yield acceptable solutions. Along the same lines, the vector spaces can be greatly expanded to include functions b which may not satisfy the exact technical conditions such as $b' \in L^p$, but if $a' \in L^p$ and $|a(t) - b(t)|$ is bounded, then b may be included.

As we saw in the introductory lecture, the example which is ever before us can be stated as the following question. If $C(t,s)$ is a "nice" function, what is the difference between the solution of

$$x(t) = t + \sin t + (t+1)^{1/2}\sin(t+1)^{1/3} - \int_0^t C(t,s)x(s)ds \quad (1.1.4)$$

and

$$y(t) = \sin t - \int_0^t C(t,s)y(s)ds? \quad (1.1.5)$$

There would be no surprise and no story to tell unless the answer were that, in spite of the large difference in size of the forcing functions, the difference in solutions is an L^p-function. This is the third surprise. Obviously, much is hidden in the package P_1 which we call "nice" and that, itself, is a long problem to unravel.

Here, we find two curious properties. The first is that, on average,

$$\int_0^t R(t,s)[(s+1)^{1/2}\sin(s+1)^{1/3}]ds$$

is a good copy of $(t+1)^{1/2}\sin(t+1)^{1/3}$ for large t; a surprising fact since the function oscillates unboundedly. By contrast, $\int_0^t R(t,s)\sin s\,ds$ is utterly unable to copy the small and nicely behaved function $\sin t$ and (1.1.5) has an asymptotically periodic solution. Substituting these functions into (1.1.3), together with a bit more unstated work, leaves us with both (1.1.4) and (1.1.5) having an asymptotically periodic solution.

The example leads us into the main work of the first hundred pages of this book and it tells us many important things. First, take courage when encountering a large function $a(t)$ in a delicate physical problem; such a function may offer little cause for worry. On the other hand, never dismiss small terms as they can have absolutely disastrous effects and must be studied with respect. But far more than that is the idea that we are looking for broad qualitative properties. We are more than satisfied with the information that a solution is in $L^p[0,\infty)$ or that it is asymptotically periodic.

Finally, we come to methods. This book is basically elementary. The methods employed in the analysis consist mainly of contraction mappings and Liapunov functionals. These methods are fully accessible to third year university students who are mathematically well-prepared. Before we become immersed in technical details let us look at two examples which set the tone for the book. We will consider (1.1.1) for $n = 1$, the scalar case, and give two simple results. The first uses contraction mappings and is very old. While the proof given here is newer, the result itself is probably one of the first ever obtained for integral equations and it is named accordingly. The second result is obtained by means of a Liapunov functional and it is very new. Taken together these results lead us to understand the richness of the nonconvolution case.

In preparation for the first result and, indeed the entire book, we define a complete metric space and state the contraction mapping principle.

Definition 1.1.1. *A pair $(\mathcal{S},\rho)$ is a metric space if $\mathcal{S}$ is a set and $\rho: \mathcal{S}\times\mathcal{S} \to [0,\infty)$ such that when y, z, and u are in $\mathcal{S}$ then*

(a) $\rho(y,z) \geq 0$, $\rho(y,y) = 0$, and $\rho(y,z) = 0$ implies $y = z$,
(b) $\rho(y,z) = \rho(z,y)$, and
(c) $\rho(y,z) \leq \rho(y,u) + \rho(u,z)$.

The metric space is complete *if every Cauchy sequence in $(\mathcal{S},\rho)$ has a limit in that space.*

Definition 1.1.2. Let (S, ρ) be a metric space and $A : S \to S$. The operator A is a *contraction operator* if there is an $\alpha \in (0,1)$ such that $x \in S$ and $y \in S$ imply

$$\rho[A(x), A(y)] \leq \alpha \rho(x,y).$$

Theorem 1.1.1. *Contraction Mapping Principle* Let (S, ρ) be a complete metric space and $A : S \to S$ a contraction operator. Then there is a unique $\phi \in S$ with $A(\phi) = \phi$. Furthermore, if $\psi \in S$ and if $\{\psi_n\}$ is defined inductively by $\psi_1 = A(\psi)$ and $\psi_{n+1} = A(\psi_n)$, then $\psi_n \to \phi$, the unique fixed point.

Proof. Let $x_0 \in S$ and define a sequence $\{x_n\}$ in S by $x_1 = A(x_0)$, $x_2 = A(x_1) \stackrel{\text{def}}{=} A^2 x_0, \ldots, x_n = A x_{n-1} = A^n x_0$. To see that $\{x_n\}$ is a Cauchy sequence, note that if $m > n$, then

$$\begin{aligned}
\rho(x_n, x_m) &= \rho(A^n x_0, A^m x_0) \\
&\leq \alpha \rho(A^{n-1} x_0, A^{m-1} x_0) \\
&\vdots \\
&\leq \alpha^n \rho(x_0, x_{m-n}) \\
&\leq \alpha^n \{\rho(x_0, x_1) + \rho(x_1, x_2) + \cdots + \rho(x_{m-n-1}, x_{m-n})\} \\
&\leq \alpha^n \{\rho(x_0, x_1) + \alpha \rho(x_0, x_1) + \cdots + \alpha^{m-n-1} \rho(x_0, x_1)\} \\
&= \alpha^n \rho(x_0, x_1) \{1 + \alpha + \cdots + \alpha^{m-n-1}\} \\
&\leq \alpha^n \rho(x_0, x_1) \{1/(1-\alpha)\}.
\end{aligned}$$

Because $\alpha < 1$, the right side tends to zero as $n \to \infty$. Thus, $\{x_n\}$ is a Cauchy sequence, and because (S, ρ) is complete, it has a limit $x \in S$. Now A is certainly continuous, so

$$A(x) = A(\lim_{n \to \infty} x_n) = \lim_{n \to \infty} A(x_n) = \lim_{n \to \infty} x_{n+1} = x,$$

and x is a fixed point. To see that x is unique, consider $A(x) = x$ and $A(y) = y$. Then

$$\rho(x,y) = \rho(A(x), A(y)) \leq \alpha \rho(x,y),$$

and because $\alpha < 1$, we conclude that $\rho(x,y) = 0$, so that $x = y$. This completes the proof.

1. LINEAR EQUATIONS

Frequently the complete metric space will be a Banach space and the metric, ρ, will be the norm. Much of the power of the contraction mapping theorem lies in a careful choice of metric. We begin with perhaps the simplest, the supremum norm. The following result is sometimes called the Adam and Eve theorem since it seems to appear in the very earliest treatments of integral equations.

Theorem 1.1.2. *Suppose there are positive numbers K and α, $\alpha < 1$, such that*

$$|a(t)| \leq K \text{ and } \sup_{t \geq 0} \int_0^t |C(t,s)| ds \leq \alpha.$$

Then there is a unique solution of (1.1.1) and it is bounded and continuous on $[0, \infty)$.

Proof. Let $(X, \|\cdot\|)$ be the Banach space of bounded continuous functions $\phi : [0, \infty) \to \Re^n$ with the supremum norm. Define a mapping $P : X \to X$ by $\phi \in X$ implies

$$(P\phi)(t) = a(t) - \int_0^t C(t,s)\phi(s)ds.$$

Notice that if $\phi, \eta \in X$ then

$$|(P\phi)(t) - (P\eta)(t)| \leq \int_0^t |C(t,s)||\phi(s) - \eta(s)|ds \leq \alpha \|\phi - \eta\|.$$

Hence, P is a contraction and there is a unique $\phi \in X$ with

$$(P\phi)(t) = \phi(t) = a(t) - \int_0^t C(t,s)\phi(s)ds.$$

Notice that we obtain existence, uniqueness, and boundedness all at the same time. Our next example is a contrast in several ways. First, we need an existence theorem to even begin. In the next section we will prove such a result and we do so by means of a contraction mapping with a weighted metric. It is a very interesting construction. The next contrast is that we integrate the first coordinate of C. Finally, the conclusion is that $x \in L^1[0, \infty)$, rather than being bounded. Such a result is always just the first step. Given that the solution is L^1 the enterprising investigator may go back to (1.1.1) and parlay that into boundedness and, often, decay to zero. The reader is welcome to try that on the example below.

1.1. A QUALITATIVE APPROACH

The idea of a Liapunov functional is marvelous. It converts a functional equation into an algebraic equation. Suppose, for example, that we have a linear algebraic equation

$$x = a(t) + b(t)x$$

and we want to show that x is bounded using the facts that $a(t)$ is bounded and $|b(t)| \leq \alpha < 1$. Then we might write

$$|x| \leq |a(t)| + |b(t)||x| \leq |a(t)| + \alpha|x|$$

or

$$|x|(1 - \alpha) \leq |a(t)|$$

so that

$$|x| \leq |a(t)|/(1 - \alpha).$$

Obviously, we can not do that for (1.1.1) since $x(s)$ is in the integral. But a carefully chosen Liapunov functional allows us to do precisely that. Here, we use a rather pedestrian Liapunov functional, but as we proceed through the book we will see more intricate and useful choices.

Theorem 1.1.3. *Suppose that there is an $\alpha < 1$ with*

$$\int_0^\infty |C(u+t,t)|\,du \leq \alpha.$$

If $a \in L^1[0, \infty)$, so is any solution of (1.1.1).

Proof. We begin exactly as we did with our algebraic argument and write

$$|x(t)| \leq |a(t)| + \int_0^t |C(t,s)||x(s)|\,ds.$$

Assume that a solution of (1.1.1) exists on $[0, \infty)$ and contrive to construct a functional having a derivative containing that last term. One simple choice is

$$V(t) = \int_0^t \int_{t-s}^\infty |C(u+s,s)|\,du\,|x(s)|\,ds$$

so that

$$V'(t) = \int_0^\infty |C(u+t,t)|\,du\,|x(t)| - \int_0^t |C(t,s)||x(s)|\,ds.$$

Substituting our first inequality into this expression yields

$$V'(t) \leq \int_0^\infty |C(u+t,t)|du|x(t)| - |x(t)| + |a(t)| \leq (\alpha-1)|x(t)| + |a(t)|.$$

Integrate both sides and obtain

$$0 \leq V(t) \leq V(0) + \int_0^t |a(s)|ds - (1-\alpha)\int_0^t |x(s)|ds$$

from which we obtain

$$\int_0^t |x(s)|ds \leq (1/(1-\alpha))\int_0^t |a(s)|ds,$$

as required.

1.2 Existence and Resolvents

The first four pages of this section will contain the basic relations which will be needed throughout the book. This is the material which the reader will frequently consult. We then give two basic existence theorems and some supplementary remarks. The remainder of the chapter presents a more comprehensive study of the relations and proofs.

We begin with a study of the linear integral equation

$$x(t) = a(t) - \int_0^t C(t,s)x(s)ds \qquad (1.2.1)$$

where $a : [0,\infty) \to \Re^n$ is at least continuous, while C is an $n \times n$ matrix of functions which are at least continuous for $0 \leq s \leq t < \infty$.

Along with (1.2.1) is the resolvent equation

$$R(t,s) = C(t,s) - \int_s^t R(t,u)C(u,s)du. \qquad (1.2.2)$$

There is then the variation of parameters formula

$$x(t) = a(t) - \int_0^t R(t,s)a(s)ds \qquad (1.2.3)$$

so that for x defined by (1.2.3), then x is the unique solution of (1.2.1) on $[0,\infty)$.

1.2. EXISTENCE AND RESOLVENTS

If C, and sometimes a, is differentiable then there are other forms of (1.2.1) which can be very useful. The most obvious is obtained by differentiating (1.2.1) to obtain the integrodifferential equation

$$x'(t) = a'(t) - C(t,t)x(t) - \int_0^t C_t(t,s)x(s)ds \qquad (1.2.4)$$

where C_t or C_1 denotes $\frac{\partial C(t,s)}{\partial t}$.

There is a less-known form obtained as follows. In (1.2.1) we integrate by parts and have

$$x(t) = a(t) - C(t,t)\int_0^t x(u)du + \int_0^t C_s(t,s)\int_0^s x(u)du\,ds. \qquad (1.2.5)$$

If we let $y(t) = \int_0^t x(u)du$, $y(0) = 0$, then (1.2.5) becomes

$$y'(t) = a(t) - C(t,t)y(t) + \int_0^t C_s(t,s)y(s)ds. \qquad (1.2.6)$$

Thus, either (1.2.4) or (1.2.6) leads us to

$$x'(t) = A(t)x(t) + \int_0^t B(t,u)x(u)du + f(t). \qquad (1.2.7)$$

Along with (1.2.7) is Becker's resolvent equation

$$\frac{\partial}{\partial t}Z(t,s) = A(t)Z(t,s) + \int_s^t B(t,u)Z(u,s)du, \quad Z(s,s) = I, \qquad (1.2.8)$$

written less formidably as

$$z'(t) = A(t)z(t) + \int_s^t B(t,u)z(u)du \qquad (1.2.9)$$

where $Z(t,s)$ is the $n \times n$ matrix whose columns are solutions of (1.2.9) on $[s, \infty)$ with $Z(s,s) = I$. We will show that $Z(t,s)$ exists for $0 \leq s \leq t < \infty$ and that for a given constant vector $x(0)$, there is a unique solution of (1.2.7) on $[0, \infty)$ given by the variation of parameters formula

$$x(t) = Z(t,0)x(0) + \int_0^t Z(t,s)f(s)ds. \qquad (1.2.10)$$

A more detailed form is also true. The solution of

$$x'(t) = A(t)x(t) + \int_\tau^t B(t,u)x(u)du + f(t), \quad x(\tau) = x_0, \qquad (1.2.11)$$

is given by

$$x(t) = Z(t,\tau)x_0 + \int_\tau^t Z(t,s)f(s)ds. \qquad (1.2.12)$$

Now (1.2.9) is the resolvent equation of Becker (1979) (2006) and we much prefer it to the classical resolvent equation which is the solution of

$$\frac{\partial H(t,s)}{\partial s} = -H(t,s)A(s) - \int_s^t H(t,u)B(u,s)du, \quad H(t,t) = I. \qquad (1.2.13)$$

We will show that $Z(t,s) = H(t,s)$ and this will be important for the following reason. As the solution of (1.2.1) is also a solution of (1.2.4) and the solution of (1.2.1) is expressed as (1.2.3), while solutions of (1.2.4) are expressed as (1.2.10), it follows that (1.2.10) and (1.2.3) are related. Now if $H = Z$ then it follows that $Z_s(t,s)$ is continuous and that will allow us to integrate that integral in (1.2.10) by parts. We take $f(t) = a'(t)$ and $x(0) = a(0)$ in (1.2.10) and have

$$x(t) = Z(t,0)a(0) + Z(t,s)a(s)\big|_0^t - \int_0^t Z_s(t,s)a(s)ds$$

$$= Z(t,0)a(0) + Z(t,t)a(t) - Z(t,0)a(0) - \int_0^t Z_s(t,s)a(s)ds$$

so that

$$x(t) = a(t) - \int_0^t Z_s(t,s)a(s)ds, \qquad (1.2.14)$$

a variation of parameters formula for (1.2.1) when C_t is continuous.

APPLICATION See Theorems 2.1.3, 2.2.10, and 2.6.2.1 for use of this form. In Burton (2005b; p. 321) we offer a theorem which can be used with many of the Liapunov functionals which we will present in Chapter 2 and exploit (1.2.14) in just this same way.

This is the collection of results which we will need to study (1.2.1). The remainder of this section will be devoted to their properties.

For an $n \times n$ matrix A, we define the operator norm of A as follows. Let $|\cdot|$ denote any vector norm on $\Re^n$ and for any $n \times n$ matrix A let

$$|A| = \sup\{|Ax| : |x| \leq 1, x \in \Re^n\}. \qquad (1.2.15)$$

This is equivalent to

$$|A| = \inf_{0 < M < \infty}\{M : |Ax| \leq M|x|, \text{ for all } x \in \Re^n\}. \qquad (1.2.16)$$

1.2. EXISTENCE AND RESOLVENTS

By (1.2.16),

$$|Ax| \leq |A||x|. \tag{1.2.17}$$

We now show that if A and B are $n \times n$ matrices, then $|AB| \leq |A||B|$. To this end, we note that by (1.2.17) we have

$$|(AB)x| = |A(Bx)| \leq |A||Bx| \leq |A||B||x|.$$

By (1.2.17) again we have

$$|(AB)x| \leq |AB||x|.$$

Finally, by (1.2.16) we then have

$$|AB| \leq |A||B|.$$

For existence results under very weak conditions see Corduneanu (1991), Grippenberg-Londen-Staffans (1990), and Miller (1971a). The solution actually has many continuity properties, as may be seen in Windsor (2009).

Theorem 1.2.1. *Let $a : [0, \infty) \to \Re^n$ be continuous and let $C(t, s)$ be an $n \times n$ matrix of functions continuous for $0 \leq s \leq t < \infty$. Then there is one and only one continuous function $x : [0, \infty) \to \Re^n$ satisfying (1.2.1).*

Proof. Let $b > 0$ and denote by X the vector space of continuous functions $\phi : [0, b] \to \Re^n$. If r is a fixed positive number then we define the norm on X by

$$|\phi|_r := \sup\{|\phi(t)|e^{-rt} : 0 \leq t \leq b\}.$$

Then $(X, |\cdot|_r)$ is a Banach space.

Let b be any fixed positive number. We now show that the unique solution of (1.2.1) exists on [0,b] and, as b is arbitrary, the theorem will follow.

Define $r := \sup_{0 \leq s \leq t \leq b} |C(t, s)| + 1$ and let $P : X \to X$ be defined by $\phi \in X$ implies that

$$(P\phi)(t) = a(t) - \int_0^t C(t, s)\phi(s)ds.$$

Then for $\phi, \eta \in X$ we have

$$|(P\phi)(t) - (P\eta)(t)|e^{-rt} \leq e^{-rt} \int_0^t |C(t,s)||\phi(s) - \eta(s)|ds$$

$$= \int_0^t e^{-r(t-s)}|C(t,s)|e^{-rs}|\phi(s) - \eta(s)|ds$$

$$\leq \int_0^t e^{-r(t-s)}|C(t,s)|ds|\phi - \eta|_r$$

$$\leq \int_0^t e^{-r(t-s)}(r-1)ds|\phi - \eta|_r$$

$$= \frac{r-1}{r}|\phi - \eta|_r.$$

It follows that P is a contraction and there is a unique fixed point satisfying (1.2.1). This completes the proof.

We will give a direct proof of the existence of a continuous function $R(t,s)$ satisfying (1.2.2). One can show by a simple integration argument (Burton and Dwiggins (2009)) that

$$\int_s^t R(t,u)C(u,s)\,du = \int_s^t C(t,u)R(u,s)\,du.$$

It can also be shown by iterations and Miller (1971a, p. 200) shows it by other methods. This means that (1.2.2) may be written as

$$R(t,s) = C(t,s) - \int_s^t C(t,u)R(u,s)\,du$$

and if we replace t by $t+s$, we have

$$R(t+s,s) = C(t+s,s) - \int_s^{t+s} C(t+s,u)R(u,s)\,du$$

or

$$R(t+s,s) = C(t+s,s) - \int_0^t C(t+s,u+s)R(u+s,s)\,du.$$

In this form s is simply a parameter and we may write the last equation as

$$L(t) = D(t) - \int_0^t C(t+s,u+s)L(u)\,du,$$

and the proof of existence and uniqueness may be applied directly to it by treating $L(t)$ one column at a time. Of course, $L(t) = L(t,s)$ and $R(t,s) = L(t-s,s)$.

1.2. EXISTENCE AND RESOLVENTS

We showed in (1.2.14) that there is an alternate form of (1.2.3) obtainable from (1.2.10) when C_t is continuous. If we use Miller's reversal we can get a glimpse into how one might conjecture (1.2.10) from (1.2.2).

Write (1.2.2) as

$$R(t,s) = C(t,s) - \int_s^t C(t,u)R(u,s)du$$

so that

$$R_t(t,s) = C_t(t,s) - Ct,t)R(t,s) - \int_s^t C_t(t,u)R(u,s)du.$$

One can readily argue from our subsequent Theorem 1.2.2.1 that $R(t,s)$ is jointly continuous in (t,s).

For an $n \times n$ matrix D, denote the ith column by D^i. Then $R(t,s)$ is the $n \times n$ matrix with R^i being the solution of

$$x' = -C(t,t)x - \int_s^t C_t(t,u)x(u)du + C_t^i(t,s), \quad x(s) = C^i(s,s). \quad (1.2.18)$$

This is actually a nonhomogeneous form of (1.2.9) when (1.2.7) is obtained from (1.2.4). Moreover, it can be a very effective source for properties of $R(t,s)$.

We return to the undifferentiated R. In order to verify (1.1.3) we will need to know that for fixed t then $R(t,s)$ is continuous in s for $0 \leq s \leq t$. We also need to see that bounds on $R(t,s)$ can be calculated. The next theorem states that $R(t,s)$ exists and is jointly continuous in (t,s). The proof of joint continuity is lengthy and will not be given here since we will see a parallel argument in the proof of Theorem 1.2.2.1 for integrodifferential equations. A proof is also found in Miller (1971a; p. 202). But the proof for the critical knowledge that $R(t,s)$ is continuous in s for fixed t will be given here.

Theorem 1.2.2. *If $C(t,s)$ is continuous for $0 \leq s \leq t < \infty$, then there is a unique continuous function R which satisfies (1.2.2) for $0 \leq s \leq t < \infty$. If $c = \sup_{0 \leq s \leq t} |C(t,s)|$ then $|R(t,s)| \leq nce^{c(t-s)}$.*

Proof. We can transpose our resolvent equation (1.2.2) and write it as

$$R^T(t,s) = C^T(t,s) - \int_s^t C^T(u,s)R^T(t,u)du$$

or

$$Y(t,s) = D(t,s) - \int_s^t D(u,s)Y(t,u)du$$

where

$$R^T(t,s) = Y(t,s), \qquad C^T(t,s) = D(t,s).$$

Using the notation mentioned above, denote the i-th column of an $n \times n$ matrix Q as Q^i so that we have the n integral equations

$$Y^i(t,s) = D^i(t,s) - \int_s^t D(u,s) Y^i(t,u) du.$$

Let t be a fixed positive number and X be the vector space of continuous functions $\phi : [0, t] \to \Re^n$. Thus, $\phi = \phi(t, s)$ where $0 \leq s \leq t$. But t is a fixed and arbitrary positive number, so we will think of it as $\phi(s)$ during our calculations and then revert to $\phi^i(t, s)$ when we have a fixed point ϕ of the i-th integral equation. We take $r = \sup_{0 \leq s \leq t} |C(t,s)| + 1$ and for $\phi \in X$ let

$$|\phi|_r = \sup_{0 \leq s \leq t} |\phi(s)| e^{rs}.$$

(The positive exponent rs is not a misprint.) Then $(X, |\cdot|_r)$ is a Banach space.

Define $P : X \to X$ by $\phi \in X$ implies that

$$(P\phi)(s) = D^i(t,s) - \int_s^t D(u,s)\phi(u) du.$$

Thus, $P\phi$ is a continuous function of s for $0 \leq s \leq t$. If $\phi, \eta \in X$ then

$$|(P\phi)(s) - (P\eta)(s)| e^{rs} \leq e^{rs} \int_s^t |D(u,s)||\phi(u) - \eta(u)| du$$

$$= \int_s^t e^{ru} |\phi(u) - \eta(u)| e^{r(s-u)} |D(u,s)| du$$

$$\leq |\phi - \eta|_r \int_s^t (r-1) e^{r(s-u)} du$$

$$\leq |\phi - \eta|_r \frac{(r-1)}{r}$$

so P is a contraction with unique fixed point $\phi(s)$ which we now designate more fully as $\phi^i(t,s) = D^i(t,s) - \int_s^t D(u,s)\phi^i(t,u) du$. The set of ϕ^i make up the matrix $R(t, s)$ so the existence and uniqueness is proved. We now find a bound for R.

1.2. EXISTENCE AND RESOLVENTS

Let $t > 0$ be fixed and $c = \sup_{0 \le s \le t} |D(t,s)|$. We will find a bound on that fixed point $\phi(s)$ for $0 \le s \le t$. Designating by $\phi(t,s)$ any of the ϕ^i, we have

$$|\phi(t,s)| \le c + \int_s^t |\phi(t,u)| c\, du.$$

Separate variables and integrate from s to t obtaining

$$\int_s^t \frac{-|\phi(t,v)| c\, dv}{c + \int_v^t |\phi(t,u)| c\, du} \ge -c(t-s)$$

and after integration and rearranging we have

$$|\phi(t,s)| \le c + \int_s^t c|\phi(t,u)|\, du \le c e^{c(t-s)}.$$

That is a bound on $\phi(t,s)$ for $0 \le s \le t$. There is one such bound for each i, which represents the i-th row of $R(t,s)$ and so a bound for $R(t,s)$ is $n c e^{c(t-s)}$. If the need arises, one can improve that bound.

As mentioned above, the proof of the joint continuity is parallel to that of Theorem 1.2.2.1.

The work just done in this proof will also prove an old and useful result which we will see throughout the book.

Theorem 1.2.3 (Gronwall's Inequality). *Let $f, g : [0, \alpha] \to [0, \infty)$ be continuous and let c be a nonnegative number. If*

$$f(t) \le c + \int_0^t g(s)f(s)\, ds, \quad 0 \le t < \alpha,$$

then

$$f(t) \le c \exp \int_0^t g(s)\, ds, \quad 0 \le t < \alpha.$$

Proof. Suppose first that $c > 0$. Divide by $c + \int_0^t g(s)f(s)\, ds$ and multiply by $g(t)$ to obtain

$$f(t)g(t) \Big/ \left[c + \int_0^t g(s)f(s)\, ds \right] \le g(t).$$

An integration from 0 to t yields

$$\ln\left\{ \left[c + \int_0^t g(s)f(s)\, ds \right] \Big/ c \right\} \le \int_0^t g(s)\, ds$$

or
$$f(t) \leq c + \int_0^t g(s)f(s)\,ds \leq c\exp\int_0^t g(s)\,ds\,.$$

If $c = 0$, take the limit as $c \to 0$ through positive values. This completes the proof.

We now verify (1.2.3).

Theorem 1.2.4. *Under the continuity conditions of (1.2.1), the unique solution can be expressed as (1.2.3).*

Proof. For fixed $t > 0$, $R(t, u)$ is defined for $0 \leq u \leq t$, while
$$x(u) = a(u) - \int_0^u C(u, s)x(s)\,ds$$
is defined for $0 \leq s \leq u \leq t$.

Left multiply $x(u)$ by $R(t, u)$ and integrate with respect to u from 0 to t. We have
$$\int_0^t R(t,u)x(u)\,du - \int_0^t R(t,u)a(u)\,du$$
$$= -\int_0^t R(t,u)\int_0^u C(u,s)x(s)\,ds\,du$$
$$= -\int_0^t \int_s^t R(t,u)C(u,s)\,du\,x(s)\,ds$$
$$= \int_0^t [R(t,s) - C(t,s)]x(s)\,ds$$

by (1.2.2). Thus
$$-\int_0^t R(t,u)a(u)\,du = -\int_0^t C(t,s)x(s)\,ds\,,$$

which, together with (1.2.1), yields
$$x(t) = a(t) - \int_0^t R(t,u)a(u)\,du\,,$$

as required.

1.2. EXISTENCE AND RESOLVENTS

Remark 1.2.1. *This is a first, and most obscure, clue to a property developed in Section 2.6. If $x(t)$ follows $a(t)$ closely then the equality $\int_0^t R(t,u)a(u)du = \int_0^t C(t,s)x(s)ds$ suggests that $C(t,s)$ might be an L^p-approximation to $R(t,s)$. It is another in our list of surprises which are almost beyond belief.*

We turn now to the integrodifferential equations and first prove that (1.2.11) has a unique solution. That will also yield existence and uniqueness of solutions of (1.2.8) and (1.2.9). Our focus for the remainder of the section will then be on (1.2.9).

Theorem 1.2.5. *Let A and f be continuous on $[\tau, \infty)$ and let B be continuous for $\tau \leq u \leq t < \infty$. Then there is a unique differentiable function $x : [\tau, \infty) \to \Re^n$ with $x(\tau) = x_0$ and satisfying (1.2.11) on $[\tau, \infty)$.*

Proof. If there is such a differentiable function then we can integrate (1.2.11) from τ to $t > \tau$ and obtain

$$x(t) = x_0 + \int_\tau^t A(v)x(v)dv + \int_\tau^t \int_\tau^v B(v,u)x(u)dudv + \int_\tau^t f(v)dv.$$

By interchanging the order of integration we have

$$x(t) = x_0 + \int_\tau^t \left[A(u) + \int_u^t B(v,u)dv \right] x(u)du + \int_\tau^t f(u)du. \quad (1.2.19)$$

Let $|\cdot|$ be any norm on $\Re^n$ and choose the operator norm for matrices. Let $b > \tau$ be arbitrary and pick $r > 1$ by

$$\sup_{\tau \leq u \leq t \leq b} \left[|A(u)| + \int_u^t |B(v,u)|dv \right] = r - 1.$$

Let $(X, |\cdot|_r)$ be the Banach space of continuous functions $\phi : [\tau, b] \to \Re^n$ with $|\phi(t)|_r = \sup_{\tau \leq t \leq b} |\phi(t)| e^{-r(t-\tau)}$. Define $P : X \to X$ by $\phi \in X$ implies that

$$(P\phi)(t) = x_0 + \int_\tau^t \left[A(u) + \int_u^t B(v,u)dv \right] \phi(u)du + \int_\tau^t f(u)du.$$

If $\phi, \eta \in X$ then

$$|(P\phi)(t) - (P\eta)(t)|e^{-r(t-\tau)}$$
$$\leq e^{-r(t-\tau)}\left|\int_\tau^t \left[A(u) + \int_u^t B(v,u)dv\right](\phi(u) - \eta(u))du\right|$$
$$\leq \int_\tau^t (r-1)e^{-r(t-\tau)+r(u-\tau)}|\phi(u) - \eta(u)|e^{-r(u-\tau)}du$$
$$\leq |\phi - \eta|_r \int_\tau^t (r-1)e^{-r(t-u)}du$$
$$\leq |\phi - \eta|_r \frac{r-1}{r}.$$

Thus, P is a contraction with unique fixed point ϕ which satisfies $\phi(\tau) = x_0$. Moreover, as it satisfies the integral equation it inherits differentiability and so solves (1.2.11). As b is arbitrary, this completes the proof.

This also gives existence and uniqueness of solutions of (1.2.8) and of (1.2.9), but there will be interpretation to do on that equation.

Historical Note

The remainder of this section will display work of Leigh C. Becker (1979) concerning the resolvent for integrodifferential equations. Most of that work was part of his doctoral dissertation written in 1979 and available for many years only in the dissertation form. Nevertheless it became widely known and was recognized as one of the most fundamental pieces of work on the subject, forming the basis of a large number of papers by other investigators. In 2005 it was scanned and a pdf file became available on his website. In 2006 it was updated and rewritten for publication in the free Electronic Journal of Qualitative Theory of Differential Equations journal as Becker (2006). We present it here mainly in the form of that paper and is included with permission of both the journal and Becker. That paper contains much more detail and several references not included here. The results are fundamental for the rest of the work in this book.

1.2.1 Resolvent vs Principal Matrix Solution

The variation of parameters formula

$$x(t) = H(t,0)x_0 + \int_0^t H(t,s)f(s)\,ds \qquad (1.2.1.1)$$

1.2. EXISTENCE AND RESOLVENTS

gives the unique solution of the linear nonhomogeneous Volterra vector integro-differential equation

$$x'(t) = A(t)x(t) + \int_0^t B(t,u)x(u)\,du + f(t) \qquad (1.2.1.2)$$

satisfying the initial condition $x(0) = x_0$. Grossman and Miller (1970) defined the matrix function $H(t,s)$, called the *resolvent*, and used it to derive (1.2.1.1). They formally defined $H(t,s)$ by

$$H(t,s) = I + \int_s^t H(t,u)\Psi(u,s)\,du \quad (0 \le s \le t < \infty) \qquad (1.2.1.3)$$

where I is the identity matrix and

$$\Psi(t,s) = A(t) + \int_s^t B(t,v)\,dv. \qquad (1.2.1.4)$$

They proved that $H(t,s)$ exists and is continuous for $0 \le s \le t$ and that it satisfies

$$\frac{\partial}{\partial s}H(t,s) = -H(t,s)A(s) - \int_s^t H(t,u)B(u,s)\,du, \quad H(t,t) = I \quad (1.2.1.5)$$

on the interval $[0,t]$, for each $t > 0$. With this they were able to derive the variation of parameters formula (1.2.1.1) (cf. Grossman and Miller (1970; p. 459)).

Despite the prominence of the resolvent $H(t,s)$ in the literature and its indispensability, its definition (1.2.1.3) is not as conceptually simple as one would like. A "linear system of ODEs" point of view was presented in Becker (1979; Ch. II). There the *principal matrix solution* $Z(t,s)$ of the homogeneous Volterra equation

$$x'(t) = A(t)x(t) + \int_s^t B(t,u)x(u)\,du \qquad (1.2.1.6)$$

was first introduced. Its definition looks exactly like the classical definition of the principal matrix solution of the homogeneous vector differential equation

$$x'(t) = A(t)x(t).$$

Now $Z(t,s)$ is a matrix solution of (1.2.1.6) with columns that are linearly independent such that $Z(s,s) = I$. Using $Z(t,s)$ instead of $H(t,s)$, the variation of parameters formula

$$x(t) = Z(t,0)x_0 + \int_0^t Z(t,s)f(s)\,ds \qquad (1.2.1.7)$$

for (1.2.1.2) is a natural extension of the variation of parameters formula for the nonhomogeneous vector differential equation

$$x'(t) = A(t)x(t) + f(t).$$

The principal matrix version of the resolvent equation (1.2.1.5), namely,

$$\frac{\partial}{\partial t}Z(t,s) = A(t)Z(t,s) + \int_s^t B(t,u)Z(u,s)\,du, \quad Z(s,s) = I \quad (1.2.1.8)$$

has been instrumental in a number of papers for obtaining results that might not have otherwise been obtained with (1.2.1.5) alone.

Not found in Becker (1979) is an alternative to Grossman and Miller's definition of $H(t,s)$. It is this: $H(t,s)$ is the transpose of the principal matrix solution of the adjoint equation

$$y'(s) = -A^T(s)y(s) - \int_s^t B^T(u,s)y(u)\,du \quad (1.2.1.9)$$

for $0 \le s \le t$. The section culminates with the proof that, notwithstanding the difference in their definitions, $Z(t,s)$ and $H(t,s)$ are identical.

1.2.2 Joint Continuity

If we refer back to Theorem 1.2.5, replace τ by s, and delete the forcing function f, then we see that for a given $x_0 \in \Re^n$, the homogeneous equation

$$x'(t) = A(t)x(t) + \int_s^t B(t,u)x(u)\,du \quad (1.2.2.1)$$

has a unique solution x_s satisfying the initial condition $x_s(s) = x_0$. Equivalently, by (1.2.19), x_s is the unique continuous solution of

$$x(t) = x_0 + \int_s^t \Phi(t,u)x(u)\,du \quad (1.2.2.2)$$

where

$$\Phi(t,u) := A(u) + \int_u^t B(v,u)\,dv. \quad (1.2.2.3)$$

Up to now the value of s has been fixed. But with that restriction removed, the totality of values $x_s(t)$ defines a function, x say, on the set

$$\Omega := \{\,(t,s) : 0 \le s \le t < \infty\,\}$$

whose value at $(t_1, s_1) \in \Omega$ is the value of the solution x_{s_1} at $t = t_1$.

1.2. EXISTENCE AND RESOLVENTS

Definition 1.2.2.1. For a given $x_0 \in \Re^n$, let x denote the function with domain Ω whose value at (t, s) is

$$x(t, s) := x_s(t) \tag{1.2.2.4}$$

where x_s is the unique solution of (1.2.2.1) on $[s, \infty)$ satisfying the initial condition $x_s(s) = x_0$.

Since $x(t, s)$ is continuous in t for a fixed s, it is natural to ask if it is also continuous in s for a fixed t—and if so, is it jointly continuous in t and s? The next theorem answers both of these in the affirmative. This will play an essential role in the proof of the variation of parameters formula for (1.2.11) that is given in Section 1.2.4.

Theorem 1.2.2.1. *The function $x(t, s)$ defined by (1.2.2.4) is continuous for $0 \leq s \leq t < \infty$.*

Proof. First extend the domain Ω of the function x to the entire plane by defining $x(t, s) = x_0$ for $s > t$. For any $T > 0$, consider $x(t, s)$ on $[0, T] \times [0, T]$. We will prove $x(t, s)$ is continuous in s uniformly for $t \in [0, T]$, which means that for every $\epsilon > 0$, there exists a $\delta > 0$ such that $|s_1 - s_2| < \delta$ implies that

$$|x(t, s_1) - x(t, s_2)| < \epsilon \tag{1.2.2.5}$$

for all $s_1, s_2 \in [0, T]$ and all $t \in [0, T]$. This and the continuity of $x(t, s)$ in t for each fixed s would establish that $x(t, s)$ is jointly continuous in both variables on the set $[0, T] \times [0, T]$ by the Moore-Osgood theorem (cf. Graves (1946; Thm. 5, p. 102), Hurewicz (1958; p. 13), or Olmsted (1959; Ex. 31, p. 310)).

Proving (1.2.2.5) will require bounds for $x(t, s)$. For a fixed $s \in [0, T]$ and for $t \in [s, T]$, we see from (1.2.2.2) that

$$|x(t, s)| \leq |x_0| + \int_s^t |\Phi(t, u)| |x(u, s)| \, du$$

$$\leq |x_0| + \int_s^t \left[|A(u)| + \int_u^t |B(v, u)| \, dv \right] |x(u, s)| \, du$$

$$\leq |x_0| + \int_s^t k |x(u, s)| \, du, \tag{1.2.2.6}$$

where k is a constant chosen so that

$$|\Phi(t, u)| \leq |A(u)| + \int_u^t |B(v, u)| \, dv \leq k \tag{1.2.2.7}$$

for all $(t, u) \in [0, T] \times [0, T]$. By Gronwall's inequality (Theorem 1.2.3),
$$|x(t, s)| \leq |x_0| e^{\int_s^t k\, du} = |x_0| e^{k(t-s)}$$
for $0 \leq s \leq t \leq T$. Since $|x(t, s)| = |x_0|$ for $s > t$, we have
$$|x(t, s)| \leq |x_0| e^{kT} \qquad (1.2.2.8)$$
for all $(t, s) \in [0, T] \times [0, T]$.

With the aid of (1.2.2.8) we now prove (1.2.2.5). For definiteness, suppose $s_2 > s_1$. For $t \in [0, s_1]$,
$$|x(t, s_1) - x(t, s_2)| = 0 \qquad (1.2.2.9)$$
as $x(t, s) = x_0$ for $t \leq s$.

For $t \in (s_1, s_2]$, it follows from (1.2.2.2) and (1.2.2.7) that
$$|x(t, s_1) - x(t, s_2)| = |x(t, s_1) - x_0| \leq \int_{s_1}^t |\Phi(t, u)|\, |x(u, s_1)|\, du$$
$$\leq \int_{s_1}^{s_2} |\Phi(t, u)|\, |x(u, s_1)|\, du \leq \int_{s_1}^{s_2} k |x(u, s_1)|\, du.$$
Then by (1.2.2.8)
$$|x(t, s_1) - x(t, s_2)| \leq \int_{s_1}^{s_2} k |x_0| e^{kT}\, du = k|x_0| e^{kT} (s_2 - s_1). \qquad (1.2.2.10)$$
For $t \in (s_2, T]$, we have
$$|x(t, s_1) - x(t, s_2)| = \left| \int_{s_1}^t \Phi(t, u) x(u, s_1)\, du - \int_{s_2}^t \Phi(t, u) x(u, s_2)\, du \right|$$
$$= \left| \int_{s_1}^t \Phi(t, u) x(u, s_1)\, du - \int_{s_2}^t \Phi(t, u) x(u, s_1)\, du \right.$$
$$\left. + \int_{s_2}^t \Phi(t, u) x(u, s_1)\, du - \int_{s_2}^t \Phi(t, u) x(u, s_2)\, du \right|$$
$$\leq \int_{s_1}^{s_2} |\Phi(t, u)|\, |x(u, s_1)|\, du + \int_{s_2}^t |\Phi(t, u)|\, |x(u, s_1) - x(u, s_2)|\, du$$
$$\leq \int_{s_1}^{s_2} k\, |x(u, s_1)|\, du + \int_{s_2}^t k\, |x(u, s_1) - x(u, s_2)|\, du.$$
Applying (1.2.2.8) again,
$$|x(t, s_1) - x(t, s_2)| \leq k|x_0| e^{kT} (s_2 - s_1) + \int_{s_2}^t k\, |x(u, s_1) - x(u, s_2)|\, du.$$

1.2. EXISTENCE AND RESOLVENTS

By (1.2.2.10), this holds at $t = s_2$ as well. Therefore, for $t \in [s_2, T]$,

$$|x(t, s_1) - x(t, s_2)| \leq k|x_0|e^{kT}(s_2 - s_1)e^{k(t-s_2)} \quad (1.2.2.11)$$

by Gronwall's inequality.

It follows from (1.2.2.9) - (1.2.2.11) that

$$|x(t, s_1) - x(t, s_2)| \leq k|x_0|e^{2kT}(s_2 - s_1) \quad (1.2.2.12)$$

for all $t \in [0, T]$ and $s_2 > s_1$. Of course, it is also true for $s_2 = s_1$.

We conclude

$$|x(t, s_1) - x(t, s_2)| \leq k|x_0|e^{2kT}|s_1 - s_2| \quad (1.2.2.13)$$

for all $s_1, s_2 \in [0, T]$ and $t \in [0, T]$, which implies (1.2.2.5). Therefore, $x(t, s)$ is continuous on $[0, T] \times [0, T]$. Since T is arbitrary, $x(t, s)$ is continuous on $[0, \infty) \times [0, \infty)$, a fortiori, for $0 \leq s \leq t < \infty$.

1.2.3 Principal Matrix Solution

For a fixed $s \geq 0$, let S denote the set of all solutions of (1.2.2.1) on the interval $[s, \infty)$ that correspond to initial vectors. Let $x(t, s)$ and $\tilde{x}(t, s)$ be two such solutions satisfying the initial conditions $x(s, s) = x_0$ and $\tilde{x}(s, s) = x_1$, respectively. Linearity of (1.2.2.1) implies the *principle of superposition*, namely, that the linear combination $c_1 x(t, s) + c_2 \tilde{x}(t, s)$ is a solution of (1.2.2.1) on $[s, \infty)$ for any $c_1, c_2 \in \Re$. Consequently, the set S is a vector space. Note that S comprises all solutions that have their initial values specified at $t = s$, but not those for which an initial function is specified on an initial interval $[s, t_0]$ for some $t_0 > s$.

Theorem 1.2.3.1. *For a fixed $s \in [0, \infty)$, let S be the set of all solutions of (1.2.2.1) on the interval $[s, \infty)$ corresponding to initial vectors. Then S is an n-dimensional vector space.*

Proof. We have already established that S is a vector space. To complete the proof, we must find n linearly independent solutions spanning S. To this end, let $e^1, \ldots, e^n$ be the standard basis for $\Re^n$, where e^i is the vector whose ith component is 1 and whose other components are 0. By Theorem 1.2.5, there are n unique solutions $x^i(t, s)$ of (1.2.2.1) on $[s, \infty)$ with $x^i(s, s) = e^i$ ($i = 1, \ldots, n$). By the usual argument, these solutions are linearly independent.

To show they span S, choose any $x(t,s) \in S$. Suppose its value at $t = s$ is the vector x_0. Let $\xi_1, \ldots, \xi_n$ be the unique scalars such that $x_0 = \xi_1 e^1 + \cdots + \xi_n e^n$. By the principle of superposition, the linear combination

$$\xi_1 x^1(t,s) + \cdots + \xi_n x^n(t,s) = \sum_{i=1}^n \xi_i x^i(t,s) \tag{1.2.3.1}$$

is a solution of (1.2.2.1). Since its value at $t = s$ is x_0, the uniqueness part of Theorem 1.2.5 implies

$$x(t,s) = \sum_{i=1}^n \xi_i x^i(t,s). \tag{1.2.3.2}$$

Hence, the n solutions $x^1(t,s), \ldots, x^n(t,s)$ span S. This and their linear independence make them a basis for S.

If we define an $n \times n$ matrix function $Z(t,s)$ by

$$Z(t,s) := \begin{bmatrix} x^1(t,s) & x^2(t,s) & \cdots & x^n(t,s) \end{bmatrix}, \tag{1.2.3.3}$$

where the columns $x^1(t,s), \ldots, x^n(t,s)$ are the basis for S defined in the proof of Theorem 1.2.3.1, then (1.2.3.2) can be written as

$$x(t,s) = Z(t,s) x_0. \tag{1.2.3.4}$$

Since $x^i(s,s) = e^i$,

$$Z(s,s) = I, \tag{1.2.3.5}$$

the $n \times n$ identity matrix.

If B is the zero matrix, then the columns of $Z(t,s)$ become linearly independent solutions of the ordinary vector differential equation

$$x'(t) = A(t)x(t). \tag{1.2.3.6}$$

This makes $Z(t,s)$ a *fundamental matrix solution* of (1.2.3.6). In fact, because $Z(s,s) = I$, it is the so-called *principal matrix solution*. These terms are also used for the integro-differential equation (1.2.2.1).

Definition 1.2.3.1. The principal matrix solution of (1.2.2.1) is the $n \times n$ matrix function $Z(t,s)$ defined by (1.2.3.3). In other words, $Z(t,s)$ is a matrix with n columns that are linearly independent solutions of (1.2.2.1) and whose value at $t = s$ is the identity matrix I.

1.2. EXISTENCE AND RESOLVENTS

An alternative term from integral equations is *resolvent*, an apt term in view of (1.2.3.2), which states that every solution of (1.2.2.1) can be resolved into the n columns constituting $Z(t,s)$.

Theorem 1.2.2.1 implies that each of the columns $x^i(t,s)$ of $Z(t,s)$ are continuous for $0 \leq s \leq t < \infty$. Consequently, we have the following.

Theorem 1.2.3.2. *$Z(t,s)$, the principal matrix solution of equation (1.2.2.1), is continuous for $0 \leq s \leq t < \infty$.*

Since the ith column of $Z(t,s)$ is the unique solution of (1.2.2.1) whose value at $t = s$ is e^i, $Z(t,s)$ is the unique matrix solution of the initial value problem

$$\frac{\partial}{\partial t} Z(t,s) = A(t)Z(t,s) + \int_s^t B(t,u)Z(u,s)\,du, \quad Z(s,s) = I \quad (1.2.3.7)$$

for $0 \leq s \leq t < \infty$. Equivalently, it is the unique matrix solution of

$$Z(t,s) = I + \int_s^t \left[A(u) + \int_u^t B(v,u)\,dv \right] Z(u,s)\,du \quad (1.2.3.8)$$

by (1.2.2.2) and (1.2.2.3). Note that this is the principal matrix counterpart of Grossman and Miller's resolvent equation (1.2.1.3).

1.2.4 Variation of Parameters Formula

Let $X(t)$ be any fundamental matrix solution of the homogeneous differential equation

$$x'(t) = A(t)x(t). \quad (1.2.4.1)$$

By definition, the columns of a fundamental matrix solution $X(t)$ are linearly independent solutions of (1.2.4.1). So for $c \in \Re^n$, $x(t) = X(t)c$ is a solution of (1.2.4.1) by the principle of superposition. If $x(\tau) = x_0$, then $X(\tau)c = x_0$. Since $X(\tau)$ is nonsingular (cf. Corduneanu (1977; p. 62)), the unique solution $x(t)$ of (1.2.4.1) satisfying $x(\tau) = x_0$ is

$$x(t) = X(t)X^{-1}(\tau)x_0. \quad (1.2.4.2)$$

Now compare (1.2.4.2) to the unique solution of the nonhomogeneous equation

$$x'(t) = A(t)x(t) + f(t) \quad (1.2.4.3)$$

satisfying $x(\tau) = x_0$. The method of variation of parameters applied to (1.2.4.3) (cf. Corduneanu (1977; p. 65)) yields the following well-known formula for the solution

$$x(t) = X(t)X^{-1}(\tau)x_0 + \int_\tau^t X(t)X^{-1}(s)f(s)\,ds. \qquad (1.2.4.4)$$

Of course, (1.2.4.4) reduces to (1.2.4.2) if $f \equiv 0$.

As for the integro-differential equation (1.2.2.1), the counterpart of (1.2.4.2) is (1.2.3.4), which is stated next as a lemma.

Lemma 1.2.4.1. *The solution of*

$$x'(t) = A(t)x(t) + \int_\tau^t B(t,u)x(u)\,du \quad (\tau \geq 0) \qquad (1.2.4.5)$$

on $[\tau, \infty)$ satisfying the initial condition $x(\tau) = x_0$ is

$$x(t) = Z(t,\tau)x_0, \qquad (1.2.4.6)$$

where $Z(t,\tau)$ is the principal matrix solution of (1.2.4.5).

Suppose $B \equiv 0$ (zero matrix). Then (1.2.4.2), (1.2.4.6), and uniqueness of solutions imply that

$$Z(t,\tau) = X(t)X^{-1}(\tau).$$

In that case, the variation of parameters formula (1.2.4.4) simplifies to

$$x(t) = Z(t,\tau)x_0 + \int_\tau^t Z(t,s)f(s)\,ds. \qquad (1.2.4.7)$$

Lemma 1.2.4.1 extends a classical result for the homogeneous differential equation (1.2.4.1) to the homogeneous integro-differential equation (1.2.2.1). This suggests that a variation of parameters formula similar to (1.2.4.7) may also hold for the nonhomogeneous integro-differential equation (1.2.11).

The essential element in the derivation of the variation of parameters formula (1.2.4.4) is the nonsingularity of $X(t)$ for each t. If the same were true of the principal matrix solution $Z(t,s)$ of (1.2.2.1), then (1.2.4.7) would hold for (1.2.2.1) as well. In fact, as Theorem 1.2.4.1 shows, there are examples of (1.2.2.1) other than (1.2.4.1) for which $\det Z(t,s)$ is never zero.

1.2. EXISTENCE AND RESOLVENTS

Theorem 1.2.4.1. *(Becker (2007; Cor. 3.4)) Assume $a, b\colon [0, \infty) \to \Re$ are continuous functions and $b(t) \geq 0$ on $[0, \infty)$. Let $x(t)$ be the unique solution of the scalar equation*

$$x'(t) = -a(t)x(t) + \int_s^t b(t-u)x(u)\,du \quad (s \geq 0) \tag{1.2.4.8}$$

on $[s, \infty)$ satisfying the initial condition $x(s) = x_0$. If $x_0 \geq 0$, then

$$x_0 e^{-\int_s^t a(u)\,du} \leq x(t) \leq x_0 e^{-\int_s^t p(u)\,du} \tag{1.2.4.9}$$

for all $t \geq s$, where

$$p(u) := a(u) - \int_0^{u-s} e^{\int_{u-v}^u a(r)\,dr} b(v)\,dv. \tag{1.2.4.10}$$

It follows that the *principal solution* $x(t, s)$ of (1.2.4.8) (i.e., the solution whose value at $t = s$ is 1) is always positive. In our notation, $Z(t, s)$ is the 1×1 matrix $[x(t, s)]$ and so

$$\det Z(t, s) = x(t, s) > 0$$

for all $t \geq s \geq 0$.

However, unlike differential equations, the principal matrix solution of an integro-differential equation (1.2.2.1) may be singular at points as the next theorem found in Burton (2005b; p. 86) shows.

Theorem 1.2.4.2. *Assume $a \geq 0$ and $b\colon [0, \infty) \to \Re$ is continuous, where $b(t) \leq 0$ on $[0, \infty)$. Let $x(t)$ be the unique solution of*

$$x'(t) = -ax(t) + \int_0^t b(t-u)x(u)\,du \tag{1.2.4.11}$$

satisfying the initial condition $x(0) = 1$. If there exists a $t_1 > 0$ such that

$$\int_{t_1}^t \int_0^{t_1} b(v-u)\,du\,dv \to -\infty \tag{1.2.4.12}$$

as $t \to \infty$, then there exists a $t_2 > 0$ such that $x(t_2) = 0$.

Theorem 1.2.4.2 clearly establishes that the determinant of the principal matrix solution $Z(t, s)$ may vanish. Consequently, we cannot derive (1.2.4.7) in general for the integro-differential equation (1.2.11) by applying the method of variation of parameters to it. However, as Theorem 1.2.4.3 shows, (1.2.4.7) satisfies (1.2.11) irrespective of the values of $\det Z(t, s)$.

1. LINEAR EQUATIONS

Theorem 1.2.4.3. (Variation of Parameters) *The solution of*

$$x'(t) = A(t)x(t) + \int_\tau^t B(t,u)x(u)\,du + f(t) \quad (\tau \geq 0) \tag{1.2.4.13}$$

on $[\tau, \infty)$ satisfying the initial condition $x(\tau) = x_0$ is

$$x(t) = Z(t,\tau)x_0 + \int_\tau^t Z(t,s)f(s)\,ds, \tag{1.2.4.14}$$

where $Z(t,s)$ is the principal matrix solution of

$$x'(t) = A(t)x(t) + \int_s^t B(t,u)x(u)\,du.$$

Proof. By Theorem 1.2.5, there is a unique solution $x(t)$ of (1.2.4.13) on $[\tau, \infty)$ such that $x(\tau) = x_0$. Let us show that

$$\varphi(t) := Z(t,\tau)x_0 + \int_\tau^t Z(t,s)f(s)\,ds \tag{1.2.4.15}$$

is also a solution of (1.2.4.13) by differentiating it. To this end, define $Z(t,s) = I$ for $s > t$. Then $Z(t,s)$ is continuous on $[0, \infty) \times [0, \infty)$ by Theorem 1.2.3.2. This and (1.2.3.7) imply the same is true of its partial derivative $Z_t(t,s)$. Consequently, the integral function in (1.2.4.15) is differentiable by Leibniz's rule. Differentiating $\varphi(t)$, we obtain

$$\varphi'(t) = \left[A(t)Z(t,\tau) + \int_\tau^t B(t,u)Z(u,\tau)\,du \right]x_0$$

$$+ Z(t,t)f(t) + \int_\tau^t \frac{\partial}{\partial t}Z(t,s)f(s)\,ds$$

by (1.2.3.7) and Leibniz's rule. Applying (1.2.3.7) again, we have

$$\varphi'(t) = A(t)Z(t,\tau)x_0 + \int_\tau^t B(t,u)Z(u,\tau)x_0\,du + If(t)$$

$$+ \int_\tau^t \left[A(t)Z(t,s) + \int_s^t B(t,u)Z(u,s)\,du \right] f(s)\,ds$$

$$= f(t) + A(t)\left[Z(t,\tau)x_0 + \int_\tau^t Z(t,s)f(s)\,ds \right]$$

$$+ \int_\tau^t B(t,u)Z(u,\tau)x_0\,du + \int_\tau^t \int_s^t B(t,u)Z(u,s)f(s)\,du\,ds.$$

1.2. EXISTENCE AND RESOLVENTS

An interchange in the order of integration yields

$$\varphi'(t) = f(t) + A(t)\varphi(t) + \int_\tau^t B(t,u)Z(u,\tau)x_0\, du$$
$$+ \int_\tau^t \int_\tau^u B(t,u)Z(u,s)f(s)\, ds\, du,$$

which simplifies to

$$\varphi'(t) = f(t) + A(t)\varphi(t)$$
$$+ \int_\tau^t B(t,u)\left[Z(u,\tau)x_0 + \int_\tau^u Z(u,s)f(s)\, ds\right] du$$
$$= f(t) + A(t)\varphi(t) + \int_\tau^t B(t,u)\varphi(u)\, du.$$

Thus, $\varphi(t)$ is a solution on $[\tau, \infty)$. By (1.2.4.15), $\varphi(\tau) = x_0$. Therefore, $x(t) \equiv \varphi(t)$ on $[\tau, \infty)$ by uniqueness of solutions.

Note that (1.2.4.14) reduces to (1.2.4.6) when $f \equiv 0$.

Corollary. *Let $\varphi \in C[0, \tau]$ for any $\tau > 0$. The solution of*

$$x'(t) = A(t)x(t) + \int_0^t B(t,u)x(u)\, du + f(t) \tag{1.2.4.16}$$

on $[\tau, \infty)$ satisfying the condition $x(t) = \varphi(t)$ for $0 \le t \le \tau$ is

$$x(t) = Z(t,\tau)\varphi(\tau) + \int_\tau^t Z(t,s)f(s)\, ds$$
$$+ \int_\tau^t Z(t,s)\left\{\int_0^\tau B(s,u)\varphi(u)\, du\right\} ds. \tag{1.2.4.17}$$

Proof. Since $x(t) \equiv \varphi(t)$ on $[0, \tau]$, we can rewrite (1.2.4.16) as follows:

$$x'(t) = A(t)x(t) + \int_\tau^t B(t,u)x(u)\, du + g(t) \tag{1.2.4.18}$$

where

$$g(t) := f(t) + \int_0^\tau B(t,u)\varphi(u)\, du. \tag{1.2.4.19}$$

By Theorem 1.2.5, equation (1.2.4.18) has a unique solution on $[\tau, \infty)$ such that $x(\tau) = \varphi(\tau)$. By the variation of parameters formula (1.2.4.14), the solution is

$$x(t) = Z(t, \tau)\varphi(\tau) + \int_\tau^t Z(t, s)g(s)\,ds,$$

which is (1.2.4.17).

1.2.5 The Adjoint Equation

The differential equation

$$y'(t) = -A^T(t)y(t), \qquad (1.2.5.1)$$

where A^T is the transpose of A, is the so-called adjoint to (1.2.4.1). The associated nonhomogeneous adjoint equation (cf. Hartman (1964; p. 62)) is

$$y'(t) = -A^T(t)y(t) - g(t). \qquad (1.2.5.2)$$

Let us extend this definition to the integro-differential equation (1.2.4.13).

Definition 1.2.5.1. *The adjoint to (1.2.4.13) is*

$$y'(s) = -A^T(s)y(s) - \int_s^t B^T(u, s)y(u)\,du - g(s) \qquad (1.2.5.3)$$

where $s \in [0, t]$.

The next theorem establishes that solutions of (1.2.5.3) do exist and are unique.

Theorem 1.2.5.1. *For a fixed $t > 0$ and a given $y_0 \in \Re^n$, there is a unique solution $y(s)$ of*

$$y'(s) = -A^T(s)y(s) - \int_s^t B^T(u, s)y(u)\,du - g(s)$$

on the interval $[0, t]$ satisfying the condition $y(t) = y_0$.

1.2. EXISTENCE AND RESOLVENTS

Proof. The objective, as it was in proving Theorem 1.2.5, is to find a suitable contraction mapping. To this end, integrate (1.2.5.3) from s to t:

$$y(t) - y(s) = -\int_s^t A^T(v)y(v)\,dv$$
$$-\int_s^t \int_v^t B^T(u,v)y(u)\,du\,dv - \int_s^t g(v)\,dv.$$

Replacing $y(t)$ with y_0 and interchanging the order of integration yields

$$y(s) = y_0 + \int_s^t A^T(v)y(v)\,dv + \int_s^t \int_s^u B^T(u,v)y(u)\,dv\,du + \int_s^t g(v)\,dv$$

or

$$y(s) = y_0 + \int_s^t \left[A^T(u) + \int_s^u B^T(u,v)\,dv \right] y(u)\,du + \int_s^t g(u)\,du. \quad (1.2.5.4)$$

Clearly, the appropriate set of functions on which to define a mapping is

$$C_{y_0}[0,t] := \{\phi \in C[0,t] : \phi(t) = y_0\}.$$

Now define the mapping $\tilde{P}$ by

$$(\tilde{P}\phi)(s) := y_0 + \int_s^t \left[A^T(u) + \int_s^u B^T(u,v)\,dv \right] \phi(u)\,du + \int_s^t g(u)\,du$$

for all $\phi \in C_{y_0}[0,t]$. For a given $\phi \in C_{y_0}[0,t]$, it is apparent that $\tilde{P}\phi$ is continuous on $[0,t]$ and that $(\tilde{P}\phi)(t) = y_0$. Thus, $\tilde{P}\colon C_{y_0}[0,t] \to C_{y_0}[0,t]$.

For an arbitrary pair of functions $\phi, \eta \in C_{y_0}[0,t]$,

$$|(\tilde{P}\phi)(s) - (\tilde{P}\eta)(s)|$$
$$= \left| \int_s^t \left[A^T(u) + \int_s^u B^T(u,v)\,dv \right] (\phi(u) - \eta(u))\,du \right|$$
$$\leq \int_s^t \left[|A^T(u)| + \int_s^u |B^T(u,v)|\,dv \right] |\phi(u) - \eta(u)|\,du.$$

So if $r > 1$ is chosen so that

$$|A^T(u)| + \int_s^u |B^T(u,v)|\,dv \leq r - 1$$

for $0 \leq s \leq u \leq t$, then

$$|(\tilde{P}\phi)(s) - (\tilde{P}\eta)(s)| \leq \int_s^t (r-1)|\phi(u) - \eta(u)|\,du. \quad (1.2.5.5)$$

The proof of Theorem 1.2.5 for τ replaced by s takes place in a Banach space with s fixed and t varying. But now that s varies and t is fixed, let us

1. LINEAR EQUATIONS

alter the norm slightly in order to show that $\tilde{P}$ is a contraction mapping: replacing $-r$ with r yields the norm

$$|\phi|_r = \sup_{0 \le s \le t} |\phi(s)| e^{rs}.$$

What remains then is to show that $\tilde{P}$ is a contraction. Returning to (1.2.5.5), we have

$$|(\tilde{P}\phi)(s) - (\tilde{P}\eta)(s)|e^{rs} \le \int_s^t (r-1)e^{rs-ru}|\phi(u) - \eta(u)|e^{ru}\, du$$
$$\le |\phi - \eta|_r$$
$$\le |\phi - \eta|\frac{r-1}{r}.$$

Therefore, $\tilde{P}$ has a unique fixed point, which translates to the existence of a unique solution of (1.2.5.4) on the interval $[0,t]$.

Definition 1.2.5.2. *The principal matrix solution of*

$$y'(s) = -A^T(s)y(s) - \int_s^t B^T(u,s)y(u)\, du \qquad (1.2.5.6)$$

is the $n \times n$ matrix function

$$Q(t,s) := \begin{bmatrix} y^1(t,s) & y^2(t,s) & \cdots & y^n(t,s) \end{bmatrix}, \qquad (1.2.5.7)$$

where $y^i(t,s)$ (t fixed) is the unique solution of (1.2.5.6) on $[0,t]$ that satisfies the condition $y^i(t,t) = e^i$.

By virtue of this definition, $Q(t,s)$ is the unique matrix solution of

$$\frac{\partial}{\partial s}Q(t,s) = -A^T(s)Q(t,s) - \int_s^t B^T(u,s)Q(t,u)\, du, \ Q(t,t) = I \quad (1.2.5.8)$$

on the interval $[0,t]$. Reasoning as in the proof of Theorem 1.2.3.1, we conclude that for a given $y_0 \in \Re^n$, the unique solution of (1.2.5.6) satisfying the condition $y(t) = y_0$ is

$$y(s) = Q(t,s)y_0 \qquad (1.2.5.9)$$

for $0 \le s \le t$.

1.2. EXISTENCE AND RESOLVENTS

Taking the transpose of (1.2.5.6) and letting $r(s)$ be the row vector $y^T(s)$, we obtain

$$r'(s) = -r(s)A(s) - \int_s^t r(u)B(u,s)\,du.$$

The solution satisfying the condition $r(t) = y_0^T =: r_0$ is the transpose of (1.2.5.9), namely,

$$y^T(s) = y_0^T Q^T(t,s) \tag{1.2.5.10}$$

or

$$r(s) = r_0 H(t,s)$$

where

$$H(t,s) := Q^T(t,s). \tag{1.2.5.11}$$

Consequently, $H(t,s)$ is the principal matrix solution of the transposed equation. As a result, Lemma 1.2.4.1 has the following adjoint counterpart.

Lemma 1.2.5.1. *The solution of*

$$r'(s) = -r(s)A(s) - \int_s^t r(u)B(u,s)\,du \tag{1.2.5.12}$$

on $[0,t]$ satisfying the condition $r(t) = r_0$ is

$$r(s) = r_0 H(t,s), \tag{1.2.5.13}$$

where $H(t,s)$ is the principal matrix solution of (1.2.5.12).

It follows from (1.2.5.8) that $H(t,s)$ is the unique matrix solution of

$$\frac{\partial}{\partial s}H(t,s) = -H(t,s)A(s) - \int_s^t H(t,u)B(u,s)\,du, \quad H(t,t) = I \tag{1.2.5.14}$$

on the interval $[0,t]$. Moreover, it is the unique matrix solution of

$$R(t,s) = I + \int_s^t R(t,u)\left[A(u) + \int_s^u B(u,v)\,dv\right]du \tag{1.2.5.15}$$

for $0 \le s \le t < \infty$, which is derived by integrating (1.2.5.14) from s to t and then interchanging the order of integration.

Now it becomes apparent from comparing (1.2.5.14) and (1.2.5.15) to (1.2.1.5) and (1.2.1.3), respectively, that the principal matrix solution of the adjoint equation (1.2.5.12) is identical to Grossman and Miller's resolvent.

1. LINEAR EQUATIONS

1.2.6 Equivalence of $H(t,s)$ and $Z(t,s)$

The solutions of (1.2.2.1) and its adjoint

$$r'(s) = -r(s)A(s) - \int_s^t r(u)B(u,s)\,du$$

are related via the equation

$$\frac{\partial}{\partial u}[r(u)Z(u,s)] = r(u)\frac{\partial}{\partial u}Z(u,s) + r'(u)Z(u,s) \qquad (1.2.6.1)$$

for $0 \le s \le u \le t$. We exploit this to prove that the principal matrix solution and Grossman and Miller's resolvent are one and the same.

Theorem 1.2.6.1. $H(t,s) \equiv Z(t,s)$.

Proof. Select any $t > 0$. For a given row n-vector r_0, let $r(s)$ be the unique solution of (1.2.5.12) on $[0,t]$ such that $r(t) = r_0$. Now integrate both sides of (1.2.6.1) from s to t:

$$r(t)Z(t,s) - r(s)Z(s,s) = \int_s^t [r(u)Z_u(u,s) + r'(u)Z(u,s)]\,du.$$

By (1.2.5.12), we have

$$r_0 Z(t,s) - r(s) = \int_s^t \Big[r(u)Z_u(u,s) - r(u)A(u)Z(u,s)$$
$$- \Big(\int_u^t r(v)B(v,u)\,dv\Big) Z(u,s)\Big]\,du. \qquad (1.2.6.2)$$

With an interchange in the order of integration, the iterated integral becomes

$$\int_s^t \Big(\int_u^t r(v)B(v,u)\,dv\Big) Z(u,s)\,du$$
$$= \int_s^t r(v)\Big(\int_s^v B(v,u)Z(u,s)\,du\Big)\,dv$$
$$= \int_s^t r(u)\Big(\int_s^u B(u,v)Z(v,s)\,dv\Big)\,du. \qquad (1.2.6.3)$$

Making this change in (1.2.6.2), we obtain

$$r_0 Z(t,s) - r(s) = \int_s^t r(u)\Big[Z_u(u,s) - A(u)Z(u,s)$$
$$- \int_s^u B(u,v)Z(v,s)\,dv\Big]\,du. \qquad (1.2.6.4)$$

1.2. EXISTENCE AND RESOLVENTS

By (1.2.3.7), the integrand is zero. Hence,

$$r(s) = r_0 Z(t, s). \qquad (1.2.6.5)$$

On the other hand,

$$r(s) = r_0 H(t, s)$$

from (1.2.5.13). Therefore, by uniqueness of the solution $r(s)$,

$$r_0 H(t, s) = r_0 Z(t, s). \qquad (1.2.6.6)$$

Now let r_0 be the transpose of the ith basis vector e^i. Then (1.2.6.6) implies that the ith rows of $H(t, s)$ and $Z(t, s)$ are equal for $0 \leq s \leq t$. The theorem follows as t is arbitrary.

Chapter 2

The Remarkable Resolvent

2.1 Perfect Harmony

Most of this section is concerned with basic classical type results for an integral equation $x(t) = a(t) - \int_0^t C(t,s)x(s)ds$ whose kernel satisfies a condition typified by

$$\sup_{t \geq 0} \int_0^t |C(t,s)|ds \leq \alpha < 1.$$

Such conditions generally promote the idea that $a(t)$ and the solution $x(t)$ lie in the same space. Indeed, the variation of parameters formula shows x, a, and $\int_0^t R(t,s)a(s)ds$ all in the same space; that is what we hope the reader will focus on when reading every theorem in this section. While such results are classical and fundamental, they are not astonishing. We call them pedestrian.

Nowhere in mathematics can one find a more mysterious function than the resolvent. The scalar second order equation $x'' + tx = 0$ can be written as a scalar integral equation and Ritt (1966) has shown that its solution is arbitrarily complicated. Thus, one may infer that the resulting resolvent, $R(t,s)$, is also arbitrarily complicated. Yet, such resolvents can have precise control over vast vector spaces of unbounded and badly behaved functions. For a function ϕ from such a space it can happen that $\int_0^t R(t,s)\phi(s)ds$ is virtually an exact copy of $\phi(t)$. On the other hand, for the same $R(t,s)$ there are vector spaces of small and well-behaved functions over which R has virtually no control in that integral. The goal of the investigator is to determine conditions on $C(t,s)$ and vector spaces that go with such C resulting in precise or poor control. Such information is critical in applied

2.1. PERFECT HARMONY

problems; we may identify large perturbations with virtually no effect, while certain small perturbations can lead to disaster.

We are concerned again with the three equations

$$x(t) = a(t) - \int_0^t C(t,s)x(s)ds, \qquad (2.1.1)$$

$$R(t,s) = C(t,s) - \int_s^t R(t,u)C(u,s)du, \qquad (2.1.2)$$

and

$$x(t) = a(t) - \int_0^t R(t,s)a(s)ds \qquad (2.1.3)$$

where $a : [0,\infty) \to \Re^n$ is continuous and $C(t,s)$ is an $n \times n$ matrix of functions continuous for $0 \leq s \leq t < \infty$. Refer back to Theorem 1.2.1 for existence and uniqueness. We will consider many results in which knowledge of R will tell us much about x from (2.1.3). But there are other reasons for studying R. Frequently, embedded in $a(t)$ is a nonlinear function or functional of x and by passing to (2.1.3) we obtain a new integral equation requiring the properties of $R(t,s)$. See, for example, Section 2.8.

Much of this chapter will concern a mapping $P : V \to W$ where V and W are vector spaces of functions $\phi : [0,\infty) \to \Re^n$ with P defined by

$$(P\phi)(t) = \phi(t) - \int_0^t R(t,s)\phi(s)ds. \qquad (2.1.4)$$

The spaces V and W will say much about the character of R.

Notation. Throughout this book, BC will denote the vector space of bounded and continuous functions $\phi : [0,\infty) \to \Re^n$.

Definition 2.1.1. Let P, defined by (2.1.4), map a vector space V into a vector space W.

(i) The resolvent $R(t,s)$ is said to generate an approximate identity on V if $W = BC$.

(ii) Let the resolvent $R(t,s)$ generate an approximate identity on V. Then $R(t,s)$ generates an asymptotic identity on V if $\phi \in V$ implies that for P defined by (2.1.4), then $(P\phi)(t) \to 0$ as $t \to \infty$.

(iii) The resolvent $R(t,s)$ is said to generate an L^p approximate identity on V if for P defined by (2.1.4) there is a positive integer p with $P : V \to L^p$.

There is a major result of Perron (1930) which plays a central role here.

Theorem 2.1.1. *(Perron) Let H be an $n \times n$ matrix of functions continuous on $[0, \infty) \to \Re$. Then*

$$\sup_{t \geq 0} \int_0^t |H(t,s)| ds < \infty$$

if and only if $\int_0^t H(t,s)\phi(s)ds$ is bounded for every $\phi \in \mathcal{BC}$.

Theorem 2.1.2. *(The Fundamental Theorem) Every solution of (2.1.1) is in $\mathcal{BC}$ for every function $a \in \mathcal{BC}$ if and only if*

$$\sup_{t \geq 0} \int_0^t |R(t,s)| ds < \infty. \qquad (2.1.5)$$

Proof. By (2.1.3) it is clear that if (2.1.5) holds then the solution is bounded. But by (2.1.3) again, if x is bounded for every bounded a then by Perron's theorem (2.1.5) holds.

The classical idea is that for $C(t,s)$ well-behaved then the solution of (2.1.1) follows $a(t)$. We will go through the ideas beginning with a repeat of the Adam and Eve idea. The idea for part (iii) below comes from Strauss (1970) and is found in Islam and Neugebauer (2008) with an application to the aforementioned result of Miller in Section 2.6.6.

Theorem 2.1.3. *Let $C(t,s)$ be an $n \times n$ matrix and suppose that there is an $\alpha < 1$ with*

$$\sup_{t \geq 0} \int_0^t |C(t,s)| ds \leq \alpha < 1. \qquad (2.1.6)$$

(i) If $a \in \mathcal{BC}$ so is the solution x of (2.1.1); hence, (2.1.5) holds.
(ii) Suppose, in addition, that for each $T > 0$ then $\int_0^T |C(t,s)| ds \to 0$ as $t \to \infty$. If $a(t) \to 0$ as $t \to \infty$, so does $x(t)$ and $\int_0^t R(t,s)a(s)ds$. Also, $\int_0^T |R(t,s)| ds \to 0$ as $t \to \infty$.
(iii) $\int_0^t |R(t,s)| ds \leq \frac{\alpha}{1-\alpha}$.

Proof. For (i) we define a mapping $Q : \mathcal{BC} \to \mathcal{BC}$ by

$$(Q\phi)(t) = a(t) - \int_0^t C(t,s)\phi(s)ds.$$

By (2.1.6) it is a contraction using the operator norm on C, as in (1.2.15). There is a unique fixed point in $\mathcal{BC}$ which satisfies (2.1.1).

2.1. PERFECT HARMONY

For (ii), we add to the mapping set, say M, the condition that for each $\phi \in M$ then $\phi(t) \to 0$ as $t \to \infty$. Then

$$|(Q\phi)(t)| \leq |a(t)| + \int_0^t |C(t,s)\phi(s)|ds.$$

We will show that the last term tends to zero as $t \to \infty$. For a given $\epsilon > 0$ and for $\phi \in M$, find T such that $|\phi(t)| < \epsilon$ if $t \geq T$ and find J with $|\phi(t)| \leq J$ for all $t \geq 0$. For this fixed T, find $\eta > T$ such that $t \geq \eta$ implies that $\int_0^T |C(t,s)|ds \leq \epsilon/J$. Then $t \geq \eta$ implies that

$$\int_0^t |C(t,s)\phi(s)|ds \leq \int_0^T |C(t,s)\phi(s)|ds + \int_T^t |C(t,s)\phi(s)|ds$$
$$\leq (J\epsilon/J) + \alpha\epsilon < 2\epsilon.$$

Thus, $Q : M \to M$ and the fixed point satisfies $x(t) \to 0$ for every continuous function $a(t)$ which tends to zero. We can also write

$$x(t) = a(t) - \int_0^t R(t,s)a(s)ds$$

and so $\int_0^t R(t,s)a(s)ds \to 0$ for every continuous $a(t)$ which tends to zero. The last part of (ii) is a result of Strauss (1970) and the proof will not be given here.

For (iii), we take the absolute values in (2.1.2) and integrate to obtain

$$\int_0^t |R(t,s)|ds \leq \int_0^t |C(t,s)|ds + \int_0^t \int_s^t |R(t,u)C(u,s)|duds$$
$$\leq \alpha + \int_0^t \int_0^u |C(u,s)|ds |R(t,u)|du$$
$$\leq \alpha + \alpha \int_0^t |R(t,u)|du,$$

from which the result follows.

Part (ii) relates to Theorem 2.7.2.6 and (iii) is related to Theorem 2.6.3.1. See also Theorem 2.2.10 and 2.6.2.1.

Remark 2.1.1. Notice that under these conditions then the mapping P defined in (2.1.4) maps $\mathcal{BC}$ into itself. Thus, $R(t,s)$ generates an approximate identity on $\mathcal{BC}$. For $\phi \in \mathcal{BC}$ then $\int_0^t R(t,s)\phi(s)ds \in \mathcal{BC}$. However complicated $R(t,s)$ may be, it is still true that the integration operation yields a function in $\mathcal{BC}$. This is hardly remarkable, but we ask the reader to watch this process as we progress.

Now, we will see an example in which the solution tends to zero so that $R(t,s)$ will generate an asymptotic identity.

Example 2.1.2. If $r : [0, \infty) \to (0, 1]$ with $r(t) \downarrow 0$, with

$$\sup_{t \geq 0} \int_0^t |C(t,s)|r(s)/r(t)ds \leq \alpha < 1, \qquad (2.1.7)$$

and with

$$|a(t)| \leq kr(t) \qquad (2.1.8)$$

for some $k > 0$, then the unique solution $x(t)$ of (2.1.1) also satisfies $|x(t)| \leq k^* r(t)$ for some $k^* > 0$. Moreover, the resolvent $R(t,s)$ in (2.1.2) generates an asymptotic identity on the space of functions $\phi : [0, \infty) \to \Re^n$ with $\sup_{t \geq 0} \left| \frac{\phi(t)}{r(t)} \right| < \infty$.

Proof. The proof is based on a weighted norm. Let $(M, |\cdot|_r)$ denote the Banach space of continuous functions $\phi : [0, \infty) \to \Re^n$ with the property that

$$|\phi|_r := \sup_{t \geq 0} \frac{|\phi(t)|}{r(t)} < \infty.$$

Define $Q : M \to M$ by $\phi \in M$ implies that

$$(Q\phi)(t) = a(t) - \int_0^t C(t,s)\phi(s)ds.$$

We have

$$|(Q\phi)(t)|/r(t) \leq |a(t)|/r(t) + \int_0^t |C(t,s)|r(s)/r(t)|\phi(s)|/r(s)ds$$

$$\leq k + |\phi|_r \int_0^t |C(t,s)|r(s)/r(t)ds$$

$$\leq k + \alpha|\phi|_r$$

so $Q\phi \in M$. To see that Q is a contraction in that norm we have immediately that

$$|[(Q\phi)(t) - (Q\eta)(t)]|/r(t) \leq \alpha|\phi - \eta|_r$$

for $\phi, \eta \in M$. Hence, there is a fixed point in M and so it has the required properties. As $x(t)$ in (2.1.3) tends to zero for $a(t)$ satisfying (2.1.8), so does $P\phi$ in (2.1.4). This completes the proof.

2.1. PERFECT HARMONY

Equation (2.1.1) can have periodic solutions, but only under exceptionally rare conditions. Those conditions involve certain orthogonality relations which usually will not hold throughout an entire vector space of large dimension. The natural solution is an asymptotically periodic solution. It will help to see this if we first change (2.1.1) so that the existence of a periodic solution can be proved. That will be our next result.

Theorem 2.1.5. *Let*

$$x(t) = a(t) - \int_{-\infty}^{t} C(t,s)x(s)ds$$

in which $a : \Re \to \Re^n$ *is continuous, while* $C(t,s)$ *is continuous on* $\Re \times \Re$, *and there is a positive constant* T *with*

$$a(t+T) = a(t) \text{ and } C(t+T, s+T) = C(t,s).$$

Suppose also that $\int_{-\infty}^{t} |C(t,s)|ds$ *is continuous and there is an* $\alpha < 1$ *with*

$$\sup_{0 \leq t \leq T} \int_{-\infty}^{t} |C(t,s)|ds \leq \alpha < 1.$$

Then the equation has a T-*periodic solution.*

Proof. Let $(X, |\cdot|)$ be the Banach space of continuous T-periodic functions ϕ on $\Re$ into $\Re^n$ with the supremum norm and define $H : X \to X$ by $\phi \in X$ implies

$$(H\phi)(t) = a(t) - \int_{-\infty}^{t} C(t,s)\phi(s)ds.$$

A translation readily establishes that $H\phi$ is T-periodic. Next, if $\phi, \eta \in X$ then for $0 \leq t \leq T$ we have

$$|(H\phi)(t) - (H\eta)(t)| \leq \int_{-\infty}^{t} |C(t,s)||\phi(s) - \eta(s)|ds$$

$$= \int_{-\infty}^{t} |C(t,s)||\phi(s) - \eta(s)|ds$$

$$\leq |\phi - \eta| \int_{-\infty}^{t} |C(t,s)|ds$$

$$\leq \alpha|\phi - \eta|.$$

Hence, H is a contraction and there is a unique fixed point solving the equation and residing in X.

For (2.1.1), once we determine the proper space, then contraction mappings can provide a seemingly one-step solution of the problem. If we write (2.1.1) as

$$x(t) = a(t) - \int_{-\infty}^{t} C(t,s)x(s)ds + \int_{-\infty}^{0} C(t,s)x(s)ds$$

then

$$a(t) - \int_{-\infty}^{t} C(t,s)x(s)ds$$

suggests the periodic function just considered, while

$$\int_{-\infty}^{0} C(t,s)x(s)ds$$

can readily be expected to tend to zero for any bounded function x. It is then natural to expect a solution $x = p + q$ where p is periodic and q tends to zero. Moreover, a space of such functions with the supremum norm is a Banach space, $(Y, \|\cdot\|)$. We note that the natural mapping defined from (2.1.1) will map $Y \to Y$.

Let $\mathcal{P}_T$ be the set of continuous T-periodic functions on $\Re$ into $\Re^n$ and suppose that for $\phi \in \mathcal{P}_T$ then

$$\int_{-\infty}^{0} C(t,s)\phi(s)ds \to 0 \text{ as } t \to \infty \qquad (2.1.9)$$

and is continuous. Let Q be the set of continuous functions $q : [0, \infty) \to \Re^n$ such that $q(t) \to 0$ as $t \to \infty$. For each $q \in Q$ let

$$\int_{0}^{t} C(t,s)q(s)ds \to 0 \text{ as } t \to \infty. \qquad (2.1.10)$$

We will need the following lemma in the proof of the next result.

Lemma 2.1.1. *Let $(Y, \|\cdot\|)$ be the space of continuous functions $\phi : [0, \infty) \to \Re^n$ such that $\phi \in Y$ implies there is a $p \in \mathcal{P}_T$ and $q \in Q$ with $\phi = p + q$. Then $(Y, \|\cdot\|)$ is a Banach space.*

2.1. PERFECT HARMONY

Proof. Let $\{p_n + q_n\}$ be a Cauchy sequence in $(Y, \|\cdot\|)$. Now for each $\epsilon > 0$ and each $q \in Q$ there is an $L > 0$ such that $t \geq L$ implies that $|q(t)| < \epsilon/4$. Given $\epsilon > 0$ there is an N such that for $n, m \geq N$ then

$$|p_n(t) + q_n(t) - p_m(t) - q_m(t)| < \epsilon/2.$$

Fix $n, m \geq N$; for $\epsilon/4$ find L such that $t \geq L$ implies that both $|q_n(t)| < \epsilon/4$ and $|q_m(t)| < \epsilon/4$. Then $t \geq L$ implies that

$$|p_n(t) - p_m(t)| - |q_n(t) - q_m(t)| \leq |p_n(t) + q_n(t) - p_m(t) - q_m(t)| < \epsilon/2$$

so that $t \geq L$ implies that

$$|p_n(t) - p_m(t)| < (\epsilon/2) + |q_n(t)| + |q_m(t)| < \epsilon.$$

But p_n and p_m are periodic so the end terms of the last inequality hold for all t. As this is true for every pair with $n \geq N$ and $m \geq N$, it follows that $\{p_n\}$ is a Cauchy sequence. This, in turn, shows the same for $\{q_n\}$. As both $\mathcal{P}_T$ and Q are complete in the supremum norm, Y is complete.

Theorem 2.1.6. *Let $C(t+T, s+T) = C(t, s)$, $a \in \mathcal{P}_T$, and let (2.1.9) and (2.1.10) hold. Suppose also that there is an $\alpha < 1$ with $\int_0^t |C(t, s)| ds \leq \alpha$. Then (2.1.1) has a solution $x(t) = p(t) + q(t)$ where $p \in \mathcal{P}_T$ and $q \in Q$.*

Proof. Let $(Y, \|\cdot\|)$ be the Banach space of functions $\phi = p + q$ where $p \in \mathcal{P}_T$ and $q \in Q$ with the supremum norm. Define a mapping $H : Y \to Y$ by $\phi = p + q \in Y$ implies that

$$(H\phi)(t) = a(t) - \int_0^t C(t, s)[p(s) + q(s)] ds$$

$$= \left[a(t) - \int_{-\infty}^t C(t, s) p(s) ds \right]$$

$$+ \left[\int_{-\infty}^0 C(t, s) p(s) ds - \int_0^t C(t, s) q(s) ds \right]$$

$$=: B\phi + A\phi.$$

This defines operators A and B on Y. Note that $B : Y \to \mathcal{P}_T \subset Y$ and $A : Y \to Q \subset Y$.

But from the first line of this array we see that H is a contraction with unique fixed point $\phi \in Y$ and that proves the result.

The same result holds when $a \in Y$.

We now turn to problems in which the first coordinate of C is integrated. Moreover, we change from contraction mappings to Liapunov functionals.

2. THE REMARKABLE RESOLVENT

The reader is urged to return to Section 1.1 and review the idea of a Liapunov functional. Our first result in this set is just like Theorem 1.1.3, but now it is for systems. There will be interesting contrasts in the results as we change Liapunov functionals.

Theorem 2.1.7. *Suppose that $a \in L^1[0, \infty)$ and $\int_{t-s}^{\infty} |C(u+s, s)| du$ is continuous for $0 \leq s \leq t < \infty$. If there is an $\alpha < 1$ with $\int_0^{\infty} |C(u+t, t)| du \leq \alpha$ then the solution $x(t)$ of (2.1.1) is in $L^1[0, \infty)$ and the resolvent, $R(t, s)$, of (2.1.2) generates an L^1 approximate identity on L^1.*

Proof. Define a Liapunov functional

$$V(t) = \int_0^t \int_{t-s}^{\infty} |C(u+s, s)| du |x(s)| ds.$$

We will take the derivative of V along the unique solution of (2.1.1) and need to unite the Liapunov functional with the integral equation. The inequality accomplishing this will now be prepared. From (2.1.1) we have

$$|x(t)| \leq |a(t)| + \int_0^t |C(t, s) x(s)| ds.$$

Now

$$V'(t) \leq \int_0^{\infty} |C(u+t, t)| du |x(t)| - \int_0^t |C(t, s) x(s)| ds$$
$$\leq \alpha |x(t)| - |x(t)| + |a(t)| = (\alpha - 1)|x(t)| + |a(t)|.$$

An integration from 0 to t, use of $V(t) \geq 0$, and use of $a \in L^1$ will yield $x \in L^1$. If we look at (2.1.3) and (2.1.4) we see that $P\phi \in L^1$ for each $\phi \in L^1$ and that completes the proof.

REMARK. Theorems 2.1.3(ii) and 2.1.7 are nonconvolution counterparts of a pair of classical results. The first states that the convolution of an L^1 function ($R(t, s)$) with a function tending to zero ($a(t)$) tends to zero. The second states that the convolution of two L^1 functions is an L^1 function. These two results are of vital importance in problems of the type studied in Section 2.6.7.

We will now add to the conditions of Theorem 2.1.7 classical conditions for boundedness of solutions of (2.1.1). Notice how a change in the Liapunov functional so that x^2 is in the integrand yields the solution in L^2.

Theorem 2.1.8. *Let (2.1.1) be a scalar equation with $\int_{t-s}^{\infty} |C(u+s, s)| du$ continuous. Suppose there exist $\alpha < 1$ and $\beta < 1$ with $\int_0^{\infty} |C(u+t, t)| du \leq$*

2.1. PERFECT HARMONY

α and $\sup_{t\geq 0} \int_0^t |C(t,s)|ds \leq \beta$. Then $R(t,s)$ generates an L^1, an L^2, and an L^∞ approximate identity on the spaces L^1, L^2, and L^∞ respectively. If $\phi = \phi_1 + \phi_2 + \phi_3$ where $\phi_1 \in L^1$, $\phi_2 \in L^2$, and $\phi_3 \in L^\infty$, then $P\phi = \psi_1 + \psi_2 + \psi_3$ where $\psi_1 \in L^1$, $\psi_2 \in L^2$, and $\psi_3 \in L^\infty$.

Proof. First we prepare the inequality which will unite the integral equation to the Liapunov functional. Notice that for any $\epsilon > 0$ there is an $M > 0$ so that by squaring both sides of (2.1.1) we can say that

$$x^2(t) \leq Ma^2(t) + (1+\epsilon)\left(\int_0^t C(t,s)x(s)ds\right)^2$$
$$\leq Ma^2(t) + (1+\epsilon)\int_0^t |C(t,s)|ds \int_0^t |C(t,s)|x^2(s)ds$$
$$\leq Ma^2(t) + (1+\epsilon)\beta \int_0^t |C(t,s)|x^2(s)ds$$
$$= Ma^2(t) + \int_0^t |C(t,s)|x^2(s)ds$$

where we choose ϵ so that $(1+\epsilon)\beta = 1$. This means that

$$-\int_0^t |C(t,s)|x^2(s)ds \leq Ma^2(t) - x^2(t)$$

which will be our fundamental uniting inequality.

Next, define a Liapunov functional by

$$V(t) = \int_0^t \int_{t-s}^\infty |C(u+s,s)|du\, x^2(s)ds$$

so that

$$V'(t) = \int_0^\infty |C(u+t,t)|du\, x^2(t) - \int_0^t |C(t,s)|x^2(s)ds$$
$$\leq Ma^2(t) - x^2 + \int_0^\infty |C(u+t,t)|du\, x^2$$
$$\leq Ma^2(t) - (1-\alpha)x^2(t).$$

Hence, an integration yields

$$(1-\alpha)\int_0^t x^2(s)ds \leq M \int_0^t a^2(s)ds$$

and $x \in L^2[0,\infty)$. The L^∞ conclusion is Theorem 2.1.3, while the L^1 conclusion is Theorem 2.1.7. The results on the mappings then follow from the linearity.

2. THE REMARKABLE RESOLVENT

The result can be taken much further.

Lemma 2.1.2. *Let (2.1.1) be a scalar equation. Suppose there is an $\alpha < 1$ with*

$$\sup_{t \geq 0} \int_0^t |C(t,s)| ds \leq \alpha.$$

Consider equation (2.1.1). There is an $M > 0$ and for each integer $n > 0$ we have

$$x^{2^n}(t) \leq M^{2^n-1} a^{2^n}(t) + \int_0^t |C(t,s)| x^{2^n}(s) ds.$$

Proof. In (2.1.1) we square both sides to obtain

$$x^2(t) = a^2(t) - 2a(t) \int_0^t C(t,s) x(s) ds + \left(\int_0^t C(t,s) x(s) ds \right)^2.$$

Find $\epsilon > 0$ with $(1+\epsilon)\alpha = 1$ and then find $M > 1$ with $2|a(t)||y| \leq (M-1)a^2(t) + \epsilon y^2$. Thus,

$$x^2(t) \leq M a^2(t) + (1+\epsilon) \left(\int_0^t C(t,s) x(s) ds \right)^2$$

$$\leq M a^2(t) + (1+\epsilon) \int_0^t |C(t,s)| ds \int_0^t |C(t,s)| x^2(s) ds$$

$$\leq M a^2(t) + \int_0^t |C(t,s)| x^2(s) ds$$

where we have used the Schwarz inequality. Next, suppose there is a positive integer k with

$$x^{2k}(t) \leq M^{2k-1} a^{2k}(t) + \int_0^t |C(t,s)| x^{2k}(s) ds.$$

Squaring yields

$$x^{4k}(t) \leq M^{4k-2}a^{4k}(t)$$
$$+ 2M^{2k-1}a^{2k}(t)\int_0^t |C(t,s)|x^{2k}(s)ds$$
$$+ \left(\int_0^t |C(t,s)|x^{2k}(s)ds\right)^2$$
$$\leq M^{4k-2}a^{4k}(t) + (M-1)\left[M^{2k-1}a^{2k}(t)\right]^2$$
$$+ (1+\epsilon)\left(\int_0^t |C(t,s)|x^{2k}(s)ds\right)^2$$
$$\leq M^{4k-2}a^{4k}(t)(M-1+1)$$
$$+ (1+\epsilon)\int_0^t |C(t,s)|ds \int_0^t |C(t,s)|x^{4k}(s)ds$$
$$\leq M^{4k-1}a^{4k}(t) + \int_0^t |C(t,s)|x^{4k}(s)ds.$$

That is, if $2k = 2^n$, then $4k = (2)2^n = 2^{n+1}$ which establishes the induction.

Theorem 2.1.9. *Let (2.1.1) be a scalar equation with $\int_{t-s}^{\infty} |C(u+s,s)|du$ continuous. Suppose there are constants $\alpha < 1$ and $\beta < 1$ with*

$$\int_0^t |C(t,s)|ds \leq \alpha \quad \text{and} \quad \int_0^\infty |C(u+t,t)|du \leq \beta.$$

If there is an $n > 0$ with $a \in L^{2^n}[0,\infty)$ then the solution of (2.1.1) satisfies $x \in L^{2^n}[0,\infty)$.

Proof. Define a Liapunov functional

$$V(t) = \int_0^t \int_{t-s}^\infty |C(u+s,s)|du\, x^{2^n}(s)ds$$

so that

$$V'(t) = \int_0^\infty |C(u+t,t)|du\, x^{2^n}(t) - \int_0^t |C(t,s)|x^{2^n}(s)ds$$
$$\leq \beta x^{2^n}(t) - x^{2^n}(t) + M^{2^n-1}a^{2^n}(t).$$

Thus, an integration yields

$$0 \leq V(t) \leq V(0) - (1-\beta)\int_0^t x^{2^n}(s)ds + M^{2^n-1}\int_0^\infty a^{2^n}(t)dt,$$

as required.

We are now going to look at problems in which we ask much about the derivatives of C, but little about the magnitude of C. It uses what may be called a perfect Liapunov functional. It has a type of lower wedge, but the casual reader would never see it. Its derivative along the solution is accomplished without any type of inequality; it is a perfect match for the equation. When $a(t) = 0$, it can display all the classical wedges above and below the Liapunov functional and above the derivative of the functional.

Theorem 2.1.10. *Let (2.1.1) be a scalar equation where*

$$C(t,s) \geq 0, \quad C_s(t,s) \geq 0, \quad C_t(t,s) \leq 0, \quad C_{st}(t,s) \leq 0 \quad (2.1.11)$$

and they are all continuous. Then along the solution of (2.1.1) the functional

$$V(t) = \int_0^t C_s(t,s)\left(\int_s^t x(u)du\right)^2 ds + C(t,0)\left(\int_0^t x(s)ds\right)^2 \quad (2.1.12)$$

satisfies

$$V'(t) \leq -x^2(t) + a^2(t). \quad (2.1.13)$$

(i) *If $a \in L^2[0,\infty)$, so are x and $\int_0^t R(t,s)a(s)ds$; moreover, $V(t)$ is bounded.*

(ii) *If there are constants B and K with*

$$\sup_{t \geq 0} \int_0^t C_s(t,s)ds = B < \infty \quad \text{and} \quad \sup_{t \geq 0} C(t,0) = K < \infty \quad (2.1.14)$$

then along the solution of (2.1.1) we have

$$\left(\int_0^t R(t,s)a(s)ds\right)^2 = (a(t) - x(t))^2 \leq 2(B+K)V(t) \quad (2.1.15)$$

where (2.1.15) does not require $a \in L^2$. However, if $a \in L^2$ and bounded then both $V(t)$ and x are bounded.

2.1. PERFECT HARMONY

Proof. We have

$$V(t) = \int_0^t C_s(t,s)\left(\int_s^t x(u)du\right)^2 ds + C(t,0)\left(\int_0^t x(s)ds\right)^2$$

and differentiate to obtain

$$V'(t) = \int_0^t C_{st}(t,s)\left(\int_s^t x(u)du\right)^2 ds + 2x\int_0^t C_s(t,s)\int_s^t x(u)du\,ds$$
$$+ C_t(t,0)\left(\int_0^t x(s)ds\right)^2 + 2xC(t,0)\int_0^t x(s)ds.$$

We now integrate the third-to-last term by parts to obtain

$$2x\left[C(t,s)\int_s^t x(u)du\Big|_0^t + \int_0^t C(t,s)x(s)ds\right]$$
$$= 2x\left[-C(t,0)\int_0^t x(u)du + \int_0^t C(t,s)x(s)ds\right].$$

Cancel terms, use the sign conditions, and use (2.1.1) in the last step of the process to unite the Liapunov functional and the equation obtaining

$$V'(t) = \int_0^t C_{st}(t,s)\left(\int_s^t x(u)du\right)^2 ds + C_t(t,0)\left(\int_0^t x(s)ds\right)^2$$
$$+ 2x[a(t) - x(t)] \leq 2xa(t) - 2x^2(t) \leq a^2(t) - x^2(t).$$

From this we obtain

$$0 \leq V(t) \leq V(0) + \int_0^t a^2(s)ds - \int_0^t x^2(s)ds;$$

when $a \in L^2[0,\infty)$ then $x \in L^2[0,\infty)$ and V is bounded. Moreover, by the Schwarz inequality we have

$$\left(\int_0^t C_s(t,s)\int_s^t x(v)dv\,ds\right)^2$$
$$\leq \int_0^t C_s(t,s)ds \int_0^t C_s(t,s)\left(\int_s^t x(v)dv\right)^2 ds$$
$$\leq B\int_0^t C_s(t,s)\left(\int_s^t x(v)dv\right)^2 ds + BC(t,0)\left(\int_0^t x(s)ds\right)^2$$
$$= BV(t).$$

But
$$\left(\int_0^t C_s(t,s)\int_s^t x(v)dv\,ds\right)^2$$
$$= \left(C(t,s)\int_s^t x(v)dv\Big|_0^t + \int_0^t C(t,s)x(s)ds\right)^2$$
$$= \left(-C(t,0)\int_0^t x(v)dv + \int_0^t C(t,s)x(s)ds\right)^2$$
$$= \left(a(t) - x(t) - C(t,0)\int_0^t x(v)dv\right)^2$$
$$\geq (1/2)(a(t) - x(t))^2 - \left(C(t,0)\int_0^t x(v)dv\right)^2.$$

This yields $(1/2)(x(t) - a(t))^2 \leq (B+K)V(t)$. The left side of (2.1.15) is the variation of parameters formula.

Investigators have always relied on variations of (2.1.13) to bound the Liapunov functional and, hence, the solution. We will obtain a relation from (2.1.13) and substitute that into (2.1.12). In order to let $a(t)$ become large, we will discover three new things. First, we show how to replace $x(t)$ by $a(t)$ in the Liapunov functional. In the coming sections we show how to replace $x(t)$ by $a'(t)$ and allow $a'(t)$ to be bounded and continuous. Finally, we show how to replace $\int_0^t x(s)ds$ by $a(t)$.

We begin by replacing x in the Liapunov functional by $a(t)$. This will give us a condition to ensure that $V(t)$ is bounded. Then notice that when V is bounded and when (2.1.15) holds then x is bounded if and only if a is bounded. We will, thereby, obtain a condition showing that $x(t)$ follows $a(t)$ regardless of the behavior of $a(t)$.

Notice that if $\int_0^t C_s(t,s)(t-s)^2 ds + C(t,0)t^2$ is bounded then there is a vector space of functions $a(t)$ satisfying (2.1.16), below. That space includes continuous functions $a = \phi + \psi$ where ϕ is bounded and $\psi \in L^2[0,\infty)$.

Theorem 2.1.11. *Let (2.1.1) be a scalar equation and let (2.1.11) and (2.1.14) hold. If, in addition, there is a constant M with*
$$\int_0^t C_s(t,s)(t-s)\int_s^t a^2(u)du\,ds + C(t,0)t\int_0^t a^2(s)ds \leq M \quad (2.1.16)$$
then $V(t)$ is bounded along the solution of (2.1.1), where V is defined in (2.1.12). Noting (2.1.15), we have that $\left(\int_0^t R(t,s)a(s)ds\right)^2 = (a(t)-x(t))^2$

2.1. PERFECT HARMONY

is bounded so x is bounded if and only if a is bounded. Finally, when (2.1.16) holds for every $a \in \mathcal{BC}$ then $\int_0^t R(t,s)a(s)ds$ is bounded for every $a \in \mathcal{BC}$ so $\sup_{t \geq 0} \int_0^t |R(t,s)|ds < \infty$.

Proof. Focus on (2.1.13). Suppose that $V(t)$ is not bounded. Then there is a monotone increasing sequence $\{t_n\} \to \infty$ with $V(s) \leq V(t_n)$ for $0 \leq s \leq t_n$. Let t denote any such t_n and let $0 \leq s \leq t$. From (2.1.13)

$$0 \leq V(t) - V(s) \leq \int_s^t a^2(u)du - \int_s^t x^2(u)du$$

so that

$$\int_s^t x^2(u)du \leq \int_s^t a^2(u)du.$$

If we use the Schwarz inequality on both integrals of x in (2.1.12) we obtain

$$V(t) \leq \int_0^t C_s(t,s)(t-s)\int_s^t a^2(u)duds + C(t,0)t\int_0^t a^2(u)du \leq M$$

by (2.1.16). Hence, V is bounded and we apply (2.1.15). As $\int_0^t R(t,s)a(s)ds$ is bounded for every $a \in \mathcal{BC}$, it follows from Perron's theorem that $\sup_{t \geq 0} \int_0^t |R(t,s)|ds < \infty$.

In this example we can take $a^2(t)$ to be the sum of a bounded function and an L^1-function which could have a sequence of spikes of magnitude going to infinity. By (2.1.15) the solution will follow those spikes in a very faithful way, differing only by a fixed bounded function.

Example 2.1.1. Let $C(t,s) = e^{-(t-s)}$ and $a^2(t) = \gamma + \mu(t)$ where γ is a fixed positive constant and $\mu \in L^1[0,\infty)$. Condition (2.1.14) becomes

$$\int_0^t e^{-(t-s)}ds \leq 1 =: B,$$

while $C(t,0) = e^{-t} \leq 1 := K$. Then (2.1.16) is

$$V(t) \leq \int_0^t e^{-(t-s)}(t-s)\int_s^t [\gamma + \mu(u)]duds + e^{-t}t\int_0^t [\gamma + \mu(s)]ds$$

which is bounded. Using this in (2.1.15) yields $(x(t) - a(t))^2$ bounded so $x(t)$ follows $a(t)$ on those spikes going off to infinity. Note that $C(t,s) = ke^{-(t-s)}$ works for any $k > 0$.

In the following exercise we see that very different kernels can be put together in a smooth way and the Liapunov functionals simply added in the study of integral equations. It works in the same way with nonlinearities, but this method often fails for nonlinear integrodifferential equations.

Exercise 2.1.1. Consider the equation

$$x(t) = a(t) - \int_0^t C(t,s)x(s)ds - \int_0^t D(t,s)x(s)ds.$$

Let the conditions on C from Theorem 2.1.10 hold, $\int_0^t |C(t,s)|ds \leq \alpha < 1$, and let $\int_0^\infty |D(u+t,t)|du \leq \beta < 1$. If $a \in L^2$ then prove that the solution x is also. To do this, define

$$V(t) = \int_0^t C_s(t,s)\left(\int_s^t x(u)du\right)^2 ds + C(t,0)\left(\int_0^t x(s)ds\right)^2$$
$$+ \int_0^t \int_{t-s}^\infty |D(u+s,s)|du|x(s)|^2 ds$$

and show that the derivative satisfies

$$V'(t) \leq Ma^2(t) - \beta x^2(t)$$

for a certain positive constant M.

We are now going to look at some problems in which the kernel is related to a convolution kernel. The proofs are no longer particularly simple, but they do show us that with more work we can obtain more information.

In (2.1.1) suppose that

$$\int_{t-s}^\infty |C(u+s,s)|du \qquad (2.1.17)$$

is continuous for $0 \leq s \leq t < \infty$ and that there is a number $\alpha < 1$ with

$$\int_0^\infty |C(u+t,t)|du \leq \alpha \qquad (2.1.18)$$

for $0 \leq s \leq t < \infty$.

2.1. PERFECT HARMONY

Theorem 2.1.12. *Let (2.1.17) and (2.1.18) hold, let $a(t)$ be bounded and continuous, and suppose there is a differentiable and decreasing function $\Phi : [0, \infty) \to (0, \infty)$ with $\Phi \in L^1[0, \infty)$, and*

$$\Phi(t-s) \geq \int_{t-s}^{\infty} |C(u+s,s)|\,du. \qquad (2.1.19)$$

If $x(t)$ is the unique solution of (2.1.1) and if

$$V(t) := \int_0^t \int_{t-s}^{\infty} |C(u+s,s)|\,du\,|x(s)|\,ds, \qquad (2.1.20)$$

then $V(t)$ is bounded. If, in addition, there is a $K > 0$ with

$$\int_{t-s}^{\infty} |C(u+s,s)|\,du \geq K|C(t,s)| \qquad (2.1.21)$$

then $x(t)$ is bounded and $\sup_{t\geq 0} \int_0^t |R(t,s)|\,ds < \infty$.

Proof. Note that in (2.1.20) we have

$$V'(t) = \int_0^{\infty} |C(u+t,t)|\,du\,|x(t)| - \int_0^t |C(t,s)x(s)|\,ds$$

and from (2.1.1) that

$$|x(t)| - |a(t)| \leq \int_0^t |C(t,s)x(s)|\,ds$$

or by (2.1.18)

$$V'(t) \leq \alpha|x(t)| + |a(t)| - |x(t)| =: -\delta|x(t)| + |a(t)| \qquad (2.1.22)$$

for $\delta > 0$. Thus, for $0 \leq s \leq t < \infty$ we can write

$$\frac{dV(s)}{ds}\Phi(t-s) \leq -\delta|x(s)|\Phi(t-s) + |a(s)|\Phi(t-s).$$

Suppose there is a $t > 0$ satisfying

$$V(t) = \max_{0\leq s\leq t} V(s).$$

Then
$$\int_0^t \frac{dV(s)}{ds}\Phi(t-s)ds = V(s)\Phi(t-s)\Big|_0^t - \int_0^t V(s)\frac{d}{ds}\Phi(t-s)ds$$
$$= V(t)\Phi(0) - \int_0^t V(s)\frac{d}{ds}\Phi(t-s)ds$$
$$\geq V(t)\Phi(0) - V(t)\int_0^t \frac{d}{ds}\Phi(t-s)ds$$
$$= V(t)\Phi(0) - V(t)\Phi(0) + V(t)\Phi(t)$$
$$= V(t)\Phi(t).$$

Hence,
$$V(t)\Phi(t) \leq -\delta \int_0^t \Phi(t-s)|x(s)|ds + \int_0^t |a(s)|\Phi(t-s)ds$$
(and by (2.1.19))
$$\leq -\delta V(t) + \|a\|k$$

for some $k > 0$ and $\|a\|$ the supremum of a. Thus,
$$V(t)[\Phi(t) + \delta] \leq \|a\|k$$

and $V(t)$ is bounded.

If (2.1.21) holds, then $V(t) \geq K[|x(t)| - |a(t)|]$. As V is bounded, so is $x(t)$. But $x(t) = a(t) - \int_0^t R(t,s)a(s)ds$ is bounded for every bounded continuous $a(t)$ and so $\int_0^t R(t,s)a(s)ds$ is bounded for every bounded and continuous $a(t)$. By Perron's theorem $\sup_{t\geq 0} \int_0^t |R(t,s)|ds < \infty$.

2.2 A Strong Resolvent: the scalar case

In this section we will assume that C has at least one partial derivative and often we will differentiate a as well. Frequently we will ask that a' be bounded or L^p and we will conclude that the solution of (2.1.1) is at least bounded. The applied mathematician will correctly object that uncertainties and even stochastic elements make such behavior of a difficult or impossible to detect. But there is a practical way around this objection. For a fixed $C(t,s)$ there are many simple ways to establish

$$\sup_{t\geq 0} \int_0^t |R(t,s)|ds < \infty. \tag{2.1.5}$$

The tactic then is as follows. Fix $C(t,s)$ and obtain (2.1.5) for a problem of interest; significant details are found in the next section. Now our real

2.2. A STRONG RESOLVENT: THE SCALAR CASE

problem of interest is $x(t) = b(t) - \int_0^t C(t,s)x(s)ds$, where $b(t)$ may be a large and badly behaved function. Select a nice function, $a(t)$, which is close to $b(t)$ and satisfies one of our subsequent results; that is, the solution of $y(t) = a(t) - \int_0^t C(t,s)y(s)ds$ is at least bounded. We suppose that there is a $K > 0$ with $|a(t) - b(t)| \leq K$ for all $t \geq 0$. Then using the same $R(t,s)$ which depends only on $C(t,s)$ we have

$$x(t) = b(t) - \int_0^t R(t,s)b(s)ds$$

and

$$y(t) = a(t) - \int_0^t R(t,s)a(s)ds.$$

Now,

$$|x(t) - y(t)| \leq |a(t) - b(t)| + \int_0^t |R(t,s)||a(s) - b(s)|ds$$

$$\leq K\left[1 + \sup_{t \geq 0} \int_0^t |R(t,s)|ds\right].$$

Thus, our results here may seem to demand much from $a(t)$, but when (2.1.5) holds they apply to a much larger class of functions.

Our basic equations are the same as in the last section but we restrict our attention to the scalar case. We are concerned with the scalar equations

$$x(t) = a(t) - \int_0^t C(t,s)x(s)ds, \tag{2.2.1}$$

$$R(t,s) = C(t,s) - \int_s^t R(t,u)C(u,s)du, \tag{2.2.2}$$

and

$$x(t) = a(t) - \int_0^t R(t,s)a(s)ds \tag{2.2.3}$$

under the same continuity conditions as with (2.1.1).

In the last section a variant of (2.1.6)

$$\sup_{t \geq 0} \int_0^t |C(t,s)|ds \leq \alpha < 1$$

was almost always used. If there is an additive constant or an additive function of t that condition is almost certain to fail. But that kernel can often be cleansed by differentiation. We begin with differentiation with respect to t, denoted by C_1 or C_t, and in the next section use differentiation with respect to s. In the first case we will also need to differentiate a, but as we are often seeking properties of R we can treat a as a test function and bestow upon it any properties which are convenient. In other cases, the exact properties of a are crucial. Thus, we differentiate (2.2.1) and obtain

$$x'(t) = a'(t) - C(t,t)x(t) - \int_0^t C_1(t,s)x(s)ds. \qquad (2.2.4)$$

When (2.2.4) is related to the solution of (2.2.1) then $x(0) = a(0)$. Moreover, unless otherwise stated solutions of (2.2.4) are always on $[0, \infty)$ with the initial function being only $x(0)$.

We will use the classical resolvent equation for (2.2.4) because we know it is $Z(t,s)$ and the equation itself will announce that Z_s exists which we will need for integration by parts. The resolvent equation for (2.2.4) as seen in Section 1.2.1 is

$$Z_s(t,s) = Z(t,s)C(s,s) + \int_s^t Z(t,u)C_1(u,s)du, \quad Z(t,t) = 1, \qquad (2.2.5)$$

and the variation of parameters formulae are

$$x(t) = Z(t,0)x(0) + \int_0^t Z(t,s)a'(s)ds$$

$$= Z(t,0)[x(0) - a(0)] + a(t) - \int_0^t Z_s(t,s)a(s)ds. \qquad (2.2.6)$$

In conjunction with this, keep in mind that the solution of (2.2.1) is also given by (2.2.3).

Suppose that

$$\int_0^t C(s,s)ds \to \infty$$

as $t \to \infty$ and use the variation of parameters formula to write (2.2.4) as

$$x(t) = x(0)e^{\int_0^t -C(s,s)ds} + \int_0^t e^{\int_u^t -C(s,s)ds}a'(u)du$$

$$- \int_0^t e^{\int_u^t -C(s,s)ds} \int_0^u C_1(u,s)x(s)dsdu. \qquad (2.2.7)$$

Remark 2.2.1. Thus, our equation is again an integral equation and it will require the integral of the second coordinate of $C_1(t,s)$ to be small;

2.2. A STRONG RESOLVENT: THE SCALAR CASE

C_1 is cleansed of any additive constants or additive functions of t which might have conflicted with (2.1.6). But, perhaps more to the point, any such constants are now transferred to the exponential which can help the subsequent contraction condition.

In order to make (2.2.6) and (2.1.3) more symmetric we begin with a proposition showing $Z(t,0)$ bounded.

Theorem 2.2.1. *Suppose that $\int_0^t C(s,s)ds$ is bounded below and that there is an $\alpha < 1$ with*

$$\sup_{t\geq 0} \int_0^t e^{\int_u^t -C(s,s)ds} \int_0^u |C_1(u,s)|ds\,du \leq \alpha. \tag{2.2.8}$$

Then $Z(t,0)$ in (2.2.6) is bounded. Moreover:

(a) Every solution of (2.2.1) is bounded for every $a(t)$ with $a'(t)$ bounded and continuous if and only if

$$\sup_{t\geq 0} \int_0^t |Z(t,s)|ds < \infty.$$

(b) Every solution of (2.2.1) is bounded for every bounded and continuous $a(t)$ if and only if

$$\sup_{t\geq 0} \int_0^t |Z_s(t,s)|ds < \infty.$$

Proof. In (2.2.7) to deal with $Z(t,0)$ we let $a(t) = 0$ and use (2.2.7) to define a mapping $Q: \mathcal{BC} \to \mathcal{BC}$. It is a contraction by (2.2.8) with fixed point $x(t) = Z(t,0)x(0)$ which is bounded for each $x(0)$. Parts (a) and (b) now follow exactly as in the proof of Theorem 2.1.2 using (2.2.6).

We will now use this to obtain a remarkable property of the resolvent.

Theorem 2.2.2. *Suppose that (2.2.8) holds, that $\int_0^t C(s,s)ds$ is bounded below, and that*

$$\int_0^t e^{\int_u^t -C(s,s)ds} du \text{ is bounded for } t \geq 0. \tag{2.2.9}$$

Then the unique solution of (2.2.1) is bounded for each continuous function $a(t)$ with $a'(t)$ bounded and continuous; thus, from (2.2.6) we see that $Z_s(t,s)$ and $R(t,s)$ generate an approximate identity on the space of functions $\phi: [0,\infty) \to \Re$ for which $\phi'(t)$ is bounded.

Proof. Use (2.2.7) to define a mapping $Q : \mathcal{BC} \to \mathcal{BC}$ to prove that the solution of (2.2.4) (and, hence, of (2.2.1)) is bounded for every bounded and continuous $a'(t)$. But remember that $x(t) = a(t) - \int_0^t R(t,s)a(s)ds$ which is bounded for every a with $a' \in \mathcal{BC}$; thus, refer to (2.1.4) and see that $R(t,s)$ generates an approximate identity on the space of functions a with $a' \in \mathcal{BC}$. In the second line of (2.2.6) we use the fact that $Z(t,0)$ is bounded and that x is bounded for every a with $a \in \mathcal{BC}$ to see that $a(t) - \int_0^t Z_s(t,s)a(s)ds$ is bounded for all such a. This shows that Z_s also generates an approximate identity on that space.

Example 2.2.1. Let $a(t) = \ln(t+1)$ so that $a'(t) = 1/(t+1)$ and obtain from (2.2.6) with $x(0) = a(0)$ that

$$x(t) = a(t) - \int_0^t Z_s(t,s)a(s)ds = a(t) - \int_0^t R(t,s)a(s)ds$$

is bounded. Thus, $\int_0^t R(t,s)\ln(s+1)ds$ is a fair approximation to $\ln(t+1)$; and, it will get better.

Remark 2.2.2. This theorem tells us that x is bounded when a' is bounded and that is very powerful result. But to implement the opening statement of this section we must have (2.1.5) and, hence, we must know that x is bounded for every bounded a. There is a companion to Theorem 2.2.2 which does exactly that and we consider it in the next section. It turns out that by integrating $\int_0^t C(t,s)x(s)ds$ by parts we obtain an integrodifferential equation parallel to (2.2.4), but containing $a(t)$ instead of $a'(t)$ and the counterpart of Theorem 2.2.2 will yield $\int_0^t x(s)ds$ bounded for each bounded $a(t)$; this, in turn, can be parlayed into $x(t)$ bounded for every bounded $a(t)$.

Remark 2.2.3. If the same resolvent $R(t,s)$ generates an approximate identity on vector spaces V_1 and V_2, then it generates an approximate identity on $V_1 \cup V_2 =: V_3$ in the sense that if $\phi \in V_3$ and if we find $\phi_1 \in V_1$ and $\phi_2 \in V_2$ with $\phi = \phi_1 + \phi_2$, then $P\phi \in \mathcal{BC}$. For example, $\phi(t) = t^{1/3}$ is neither bounded, as in Theorem 2.1.3, nor is ϕ' bounded and continuous, as in Theorem 2.2.2. However, we can write

$$\phi_1(t) = \begin{cases} t^{1/3} - (1/3)t, & \text{for } 0 \leq t \leq 1 \\ (2/3) & \text{for } t \geq 1, \end{cases}$$

and

$$\phi_2(t) = \begin{cases} (1/3)t & \text{for } 0 \leq t \leq 1 \\ t^{1/3} - (2/3) & \text{for } t \geq 1. \end{cases}$$

2.2. A STRONG RESOLVENT: THE SCALAR CASE

Theorem 2.2.3. Let the conditions of Theorem 2.2.2 hold, $a'(t) \to 0$ as $t \to \infty$, and suppose there is a constant $\lambda > 0$ with $-C(t,t) \leq -\lambda$. Suppose also that there is a continuous function $\Phi : [0,\infty) \to [0,\infty)$ with $\Phi \in L^1[0,\infty)$ and $\Phi(u-s) \geq |C_1(u,s)|$. Then the solution $x(t)$ of (2.2.1) tends to zero as $t \to \infty$ and so does $Z(t,0)$. Finally, under these additional conditions the conclusion of Theorem 2.2.2 changes to asymptotic identity.

Proof. In our mapping we add to the mapping set $\mathcal{BC}$ the condition that $\phi(t) \to 0$ as $t \to \infty$. Then notice that

$$\int_0^u |C_1(u,s)\phi(s)|ds \leq \int_0^u |\Phi(u-s)||\phi(s)|ds,$$

is the convolution of an L^1-function with a function tending to zero so it tends to zero. Then

$$\int_0^t e^{\int_u^t -C(s,s)ds} \int_0^u |C_1(u,s)\phi(s)|ds \leq \int_0^t e^{-\lambda(t-u)} \int_0^u |C_1(u,s)\phi(s)|ds$$

which tends to zero for the same reason. Finally,

$$\int_0^t e^{\int_u^t -C(s,s)ds}|a'(u)|du \leq \int_0^t e^{-\lambda(t-u)}|a'(u)|du$$

which tends to zero. This will then show that the modified $\mathcal{BC}$ set will be mapped into itself in (2.2.7).

Theorem 2.2.4. Let (2.2.8) hold and suppose that

$$\int_0^t e^{\int_u^t -C(s,s)ds} du$$

is bounded for $t \geq 0$. In addition, let $a'(t) \to 0$ as $t \to \infty$, suppose that for each $T > 0$

$$\int_0^T e^{\int_u^t -C(s,s)ds} \int_0^u |C_1(u,s)|ds\,du \to 0$$

as $t \to \infty$, and that

$$\int_0^T e^{\int_u^t -C(s,s)ds} du \to 0$$

as $t \to \infty$. Then every solution of (2.2.1) tends to zero as $t \to \infty$ and so does $\int_0^t Z(t,s)\phi(s)ds$ for every continuous ϕ which tends to 0 as $t \to \infty$.

Proof. The mapping is defined from (2.2.7) and, as in the proof of Theorem 2.2.3, we add to the mapping set M the condition that $\phi \in M$ implies that $\phi(t) \to 0$. Now from (2.2.7) we write for $\phi \in M$ the equation

$$(P\phi)(t) = x(0)e^{\int_0^t -C(s,s)ds} + \int_0^t e^{\int_u^t -C(s,s)ds} a'(u)du$$
$$- \int_0^t e^{\int_u^t -C(s,s)ds} \int_0^u C_1(u,s)\phi(s)dsdu.$$

We will show that the last term tends to zero as $t \to \infty$.

Let $\phi \in M$ be fixed, $\epsilon > 0$ be given, and find $T > 0$ with $|\phi(t)| < \epsilon/\alpha$ for $t \geq T$. Thus,

$$\int_T^t e^{\int_u^t -C(s,s)ds} \int_0^u |C_1(u,s)\phi(s)|dsdu \leq \epsilon$$

for $t \geq T$. For that $\phi \in M$, there is a $J > 0$ with $|\phi(t)| \leq J$. For $t > T$ we have

$$\left| \int_0^t e^{\int_u^t -C(s,s)ds} \int_0^u C_1(u,s)\phi(s)dsdu \right|$$
$$\leq \int_0^T e^{\int_u^t -C(s,s)ds} \int_0^u |C_1(u,s)|dsduJ$$
$$+ \int_T^t e^{\int_u^t -C(s,s)ds} \int_0^u |C_1(u,s)\phi(s)|dsdu.$$

The first term on the right is bounded by ϵ for large t and so is the second.

We examine the second term of $P\phi$ and for $|a'(t)| \leq J$ write

$$\left| \int_0^t e^{\int_u^t -C(s,s)ds} a'(u)du \right| \leq \int_0^T e^{\int_u^t -C(s,s)ds} duJ + \int_T^t e^{\int_u^t -C(s,s)ds} |a'(u)|du.$$

The first term on the right tends to zero and the second term can be made small by taking T large since $|a'(t)| \to 0$ and $\int_0^t e^{\int_u^t -C(s,s)ds} du$ is bounded. This means that $P : M \to M$ so the fixed point tends to zero. As

$$x(t) = Z(t,0)x(0) + \int_0^t Z(t,s)a'(s)ds,$$

we argue that for every continuous $a'(t)$ which tends to zero then $x(t) \to 0$, while $Z(t,0) \to 0$ because that corresponds to the case $a'(t) = 0$. Thus, $\int_0^t Z(t,s)\phi(s)ds \to 0$ for every continuous ϕ which tends to zero. This completes the proof.

2.2. A STRONG RESOLVENT: THE SCALAR CASE

Example 2.2.2. Let $g : (-\infty, \infty) \to (0, \infty)$ be bounded and locally Lipschitz. Consider the integral equation

$$x(t) = t + \int_0^t g(x(s))ds - \int_0^t C(t,s)x(s)ds$$

where C satisfies (2.2.8) and (2.2.9). Standard existence theory (given in Chapter 3) will yield a unique solution on $[0, \infty)$ so it is possible to define a unique continuous function

$$a(t) = t + \int_0^t g(x(s))ds$$

with $a'(t)$ being bounded and $a(t) \geq t$. The conditions of Theorem 2.2.2 are satisfied and we have then a list of properties.

(a) The solution $x(t)$ is bounded. This means that $a(t)$ and $\int_0^t C(t,s)x(s)ds$ differ by at most a bounded function. Recall that $a(t) \geq t$.

(b) The variation of parameters formula for the solution is

$$x(t) = a(t) - \int_0^t R(t,s)a(s)ds$$

where R is the resolvent from (2.2.2). That integral differs from $a(t)$ by at most a bounded function and, again, $a(t) \geq t$.

(c) From the second equation in (2.2.6) we have

$$x(t) = t + \int_0^t g(x(s))ds - \int_0^t Z_s(t,s)\left[s + \int_0^s g(x(u))du\right]ds$$

and that quantity is bounded since Z_s generates an approximate identity on functions with bounded derivative. Again, $a(t) \geq t$ and $a(t) - \int_0^t Z_s(t,s)a(s)ds$ is bounded.

(d) From the first equation in (2.2.6) we have

$$x(t) = Z(t,0)x(0) + \int_0^t Z(t,s)[1 + g(x(s))]ds$$

where the last term is bounded. That integral differs from $1 + g(x(t))$ by at most a bounded function, while $Z(t,0)$ is bounded.

Remark 2.2.4. Here is the typical objective. If we can show that $x \in L^2$ for every $a' \in L^2$ then $R(t,s)$ (or $Z_s(t,s)$) generates an L^2 approximate identity on the vector space W of function $a(t)$ for which $a'(t)$ is continuous and L^2. Thus, for example, $a(t) = \ln(t+1)$ qualifies and the solution of

(2.2.1) is L^2, but it does not follow $a(t)$ in any sense at all. The classical idea of investigators is that for well-behaved kernels the solution of (2.2.1) follows $a(t)$. Our kernel can be arbitrarily well-behaved and the solution simply does not follow $a(t)$. There is an entirely different principle at work in all such problems. The principle is that, however complicated $R(t,s)$ may be, the integral strips away the complication and $\int_0^t R(t,s)a(s)ds$ may approximate $a(t)$.

In the result below, notice that the solution $x(t)$ of (2.2.1) is in L^1, but $a(t)$ is merely bounded. Thus, $x(t)$ is not following $a(t)$ and the kernel can be very well-behaved in almost any sense.

Theorem 2.2.5. *Let $\int_{t-s}^{\infty} |C_1(u+s,s)|du$ be continuous for $0 \leq s \leq t$. If there is an $\alpha > 0$ such that*

$$-C(t,t) + \int_0^{\infty} |C_1(u+t,t)|du \leq -\alpha$$

then $a' \in L^1$ implies that the solution $x(t)$ of (2.2.1) is in L^1 and is bounded. Hence, $x \in L^p$ for $p \in [1,\infty]$ and $R(t,s)$ generates an L^p approximate identity on the vector space W of function ϕ with $\phi' \in L^1$. If $\alpha = 0$ then $x(t)$ is bounded.

Proof. Define

$$V(t) = |x(t)| + \int_0^t \int_{t-s}^{\infty} |C_1(u+s,s)|du\,|x(s)|ds.$$

Along solutions of (2.2.4) we have

$$V'(t) \leq |a'(t)| - C(t,t)|x(t)| + \int_0^t |C_1(t,s)x(s)|ds$$
$$+ \int_0^{\infty} |C_1(u+t,t)|du\,|x(t)| - \int_0^t |C_1(t,s)x(s)|ds$$
$$= \left[-C(t,t) + \int_0^{\infty} |C_1(u+t,t)|du\right]|x(t)| + |a'(t)|$$
$$\leq -\alpha|x(t)| + |a'(t)|.$$

An integration and use of V yields

$$|x(t)| \leq V(t) \leq V(0) + \int_0^{\infty} |a'(s)|ds - \alpha \int_0^t |x(s)|ds.$$

The result follows from this.

2.2. A STRONG RESOLVENT: THE SCALAR CASE 79

NOTE The case $\alpha = 0$ can allow large $C(t,s)$ in line with that allowed in Theorem 2.2.7.

Theorem 2.2.6. Let $\int_{t-s}^{\infty} |C_1(u+s,s)| du$ be continuous for $0 \leq s \leq t$. If there exists $\alpha > 0$ such that

$$-2C(t,t) + \int_0^{\infty} |C_1(u+t,t)| du + \int_0^t |C_1(t,s)| ds \leq -\alpha$$

and if $a' \in L^2$, then the solution $x(t)$ of (2.2.1) is bounded and satisfies $x \in L^p$ for $p \in [2, \infty]$. Thus, $R(t,s)$ generates an L^p approximate identity on the vector space W of functions ϕ with $\phi' \in L^2$.

Proof. Define

$$V(t) = x^2(t) + \int_0^t \int_{t-s}^{\infty} |C_1(u+s,s)| du\, x^2(s)\, ds.$$

For any $\epsilon > 0$ there is an $M > 0$ with $2a'(t)x \leq Ma'(t)^2 + \epsilon x^2$. Thus, along any solution of (2.2.4) we have

$$V'(t) = 2a'(t)x(t) - 2C(t,t)x^2(t) - 2x(t)\int_0^t C_1(t,s)x(s)\,ds$$

$$+ \int_0^{\infty} |C_1(u+t,t)| du\, x^2(t) - \int_0^t |C_1(t,s)| x^2(s)\, ds$$

$$\leq 2a'(t)x(t) - 2C(t,t)x^2(t) + \int_0^t |C_t(t,s)|(x^2(t) + x^2(s))\,ds$$

$$+ \int_0^{\infty} |C_1(u+t,t)| du\, x^2(t) - \int_0^t |C_1(t,s)| x^2(s)\, ds$$

$$\leq Ma'(t)^2 + \epsilon x^2(t) - 2C(t,t)x^2(t)$$

$$+ \int_0^t |C_t(t,s)| ds\, x^2(t) + \int_0^{\infty} |C_1(u+t,t)| du\, x^2(t)$$

$$\leq -(\alpha/2)x^2(t) + Ma'(t)^2$$

for $\epsilon = \alpha/2$. It readily follows that $x \in L^2$ and is bounded so the conclusion follows.

We now present a result which asks more about the derivatives of C, but no upper bound on the magnitude.

2. THE REMARKABLE RESOLVENT

Theorem 2.2.7. *Suppose that $C(t,t) \geq \alpha > 0$ and for $H(t,s) = C_t(t,s)$ we suppose that*

$$H(t,s) \geq 0, \ H_s(t,s) \geq 0, \ H_{st}(t,s) \leq 0, \ H_t(t,0) \leq 0. \qquad (2.2.10)$$

If, in addition, $a' \in L^2$, then any solution $x(t)$ of (2.2.4) on $[0,\infty)$ is also in L^2. Thus, $R(t,s)$ generates an L^2 approximate identity on the space of functions ϕ with $\phi' \in L^2$.

Proof. We have

$$x'(t) = a'(t) - C(t,t)x(t) - \int_0^t H(t,s)x(s)ds$$

and we define

$$V(t) = x^2(t) + \int_0^t H_s(t,s)\left(\int_s^t x(u)du\right)^2 ds + H(t,0)\left(\int_0^t x(s)ds\right)^2$$

so that

$$V'(t) = \int_0^t H_{st}(t,s)\left(\int_s^t x(u)du\right)^2 ds + 2x\int_0^t H_s(t,s)\int_s^t x(u)du\,ds$$
$$+ H_t(t,0)\left(\int_0^t x(s)ds\right)^2 + 2x(t)H(t,0)\int_0^t x(s)ds$$
$$+ 2x(t)a'(t) - 2C(t,t)x^2(t) - 2x(t)\int_0^t H(t,s)x(s)ds.$$

Integrate the second term on the right-hand-side and obtain

$$2x(t)\left[H(t,s)\int_s^t x(u)du\bigg|_0^t + \int_0^t H(t,s)x(s)ds\right]$$
$$= 2x(t)\left[-H(t,0)\int_0^t x(u)du + \int_0^t H(t,s)x(s)ds\right].$$

This yields

$$V'(t) \leq 2x(t)a'(t) - 2C(t,t)x^2(t) \leq Da'(t)^2 - Ex^2(t)$$

for appropriate positive numbers D and E. Noting that $x^2 \leq V(t)$ and integrating we obtain $x \in L^\infty$, $x \in L^2$. Thus, as $a' \in L^2$ we obtain $a(t) - \int_0^t R(t,s)a(s)ds \in L^2$.

2.2. A STRONG RESOLVENT: THE SCALAR CASE

Exercise 2.2.1. Combine Theorems 2.2.6 and 2.2.7 by considering

$$x' = a'(t) - C(t,t)x - D(t,t)x - \int_0^t C_t(t,s)x(s) - \int_0^t D_t(t,s)x(s)ds$$

and

$$V(t) = x^2 + \int_0^t \int_{t-s}^\infty |D_t(u+s,s)|du x^2(s)ds$$

$$+ \int_0^t H_s(t,s)\left(\int_s^t x(u)du\right)^2 ds + H(t,0)\left(\int_0^t x(s)ds\right)^2.$$

Obtain a result yielding $a' \in L^2$ implies $x \in L^2$.

We will see that for $a' \in L^p$ then $x \in L^p$ so that (2.2.3) will assure us that the integral so faithfully duplicates $a(t)$ that the error in that duplication is an L^p function. If we were to think of averages, as time goes on the duplication becomes so precise on average that we can hardly tell the difference between the integral in (2.2.3) and $a(t)$. From that point of view, the large function, $a(t) = (t+1)^\beta$, has such a small effect on the solution of (2.2.1) it is almost as if it were absent. The same is true for $a(t) = \sin(t+1)^\beta$ when $0 < \beta < 1$.

Now for more in a very different direction as well as a sufficient condition for (2.1.5) we extend the last result as follows.

Theorem 2.2.8. Let $H(t,s) := C_t(t,s)$, and suppose there is an $\alpha > 0$ with $C(t,t) \geq \alpha$, and let (2.2.10) hold.
(i) If a' is bounded and if there is an $M > 0$ with

$$\int_0^t H_s(t,s)(t-s)\int_s^t |a'(u)|^2 duds + H(t,0)t\int_0^t |a'(u)|^2 du \leq M, \quad (2.2.11)$$

then any solution of (2.2.1) or (2.2.4) on $[0,\infty)$ is bounded. Thus, the resolvent for (2.2.4) satisfies $\sup_{t\geq 0} \int_0^t |Z(t,s)|ds < \infty$ and $Z(t,0)$ is bounded.
(ii) If, in addition to the conditions of (i), we have

$$|C(t,t)| + \int_0^t |C_t(t,s)|ds$$

bounded, then $Z(t,0) \to 0$ as $t \to \infty$.

Proof. Define V as in the last proof and obtain the derivative of V along that same equation as

$$V'(t) \leq 2a'(t)x - 2C(t,t)x^2 \leq D|a'(t)|^2 - Ex^2(t)$$

for positive constants D and E, exactly as in the last proof. Note that x is bounded if V is bounded.

Now, assume $a'(t)$ bounded and let (2.2.11) hold; we will bound V and, hence, x. From V' and the boundedness of $|a'|$ we see that there is a $\mu > 0$ such that if $V'(t) > 0$ then $|x(t)| < \mu$. Suppose, by way of contradiction, that V is not bounded. Then there is a sequence $\{t_n\} \uparrow \infty$ with $V'(t_n) \geq 0$ and $V(t_n) \geq V(s)$ for $0 \leq s \leq t_n$; thus, $|x(t_n)| \leq \mu$. If $0 \leq s \leq t_n$ then

$$0 \leq V(t_n) - V(s) \leq - \int_s^{t_n} Ex^2(u)du + D \int_s^{t_n} |a'(u)|^2 du.$$

Using these values in the formula for V, taking $|x(t_n)| \leq \mu$, and applying the Schwarz inequality yields at $t = t_n$ the inequality

$$V(t) \leq \mu^2 + \int_0^t H_s(t,s)(t-s) \int_s^t (D/E)|a'(u)|^2 du\,ds$$

$$+ H(t,0)t(D/E) \int_0^t |a'(u)|^2 du = \mu^2 + (D/E)M.$$

Thus, $V(t)$ and $x(t)$ are bounded.

Now the variation of parameters formula for (2.2.4) is

$$x(t) = Z(t,0)x(0) + \int_0^t Z(t,s)a'(s)ds.$$

If $a'(t) \equiv 0$, then $V'(t) \leq -Ex^2(t)$ so $x^2 \in L^1[0,\infty)$ and $V(t)$ is bounded so $x(t)$ is bounded. This means that $Z(t,0)$ is bounded and, hence, $\int_0^t Z(t,s)a'(s)ds$ is bounded for every bounded and continuous $a'(t)$. By Perron's theorem, $\sup_{t \geq 0} \int_0^t |Z(t,s)|ds < \infty$. If $|C(t,t)| + \int_0^t |C_t(t,s)|ds$ is bounded, then $x'(t)$ is bounded so $Z(t,0) \to 0$.

It is a very useful result. The solution is bounded and (2.2.11) will yield a computable bound in spite of $a(t)$ being unbounded.

Example 2.2.3. Let $C(t,s) = 2 - e^{-(t-s)}$ and $a(t) = (t+1)^{1/2}$. Then $C(t,t) = 1 =: \alpha > 0$, $C_t(t,s) = e^{-(t-s)} =: H(t,s)$ so (2.2.10) holds. Also, (2.2.11) is

$$\int_0^t e^{-(t-s)}(t-s) \int_s^t \frac{1}{4(u+1)} du\,ds + e^{-t} t \int_0^t \frac{1}{4(u+1)} du$$

which is bounded. By part (i), $x(t)$ is bounded. Hence, $\int_0^t R(t,s)a(s)ds$ closely follows $a(t)$, but $a(t)$ diverges far from $x(t)$. Consider (i) and note that $a(t) = 3t$ qualifies and the solution is bounded. Moreover, if $b(t)$ is any continuous function so that $|a(t) - b(t)|$ is bounded and if (2.1.5) holds

2.2. A STRONG RESOLVENT: THE SCALAR CASE

then the solution of $y(t) = b(t) - \int_0^t C(t,s)y(s)ds$ is bounded. Notice that $x(t) = 3t - \int_0^t R(t,s)3s\,ds$ is bounded. That resolvent has extremely strong properties enabling the integral to closely approximate $3t$.

In preparation for the next result, we note that Hölder's inequality states that if p and q are numbers with $p > 1$, $q > 1$, and $(1/p) + (1/q) = 1$, then

$$|ab| \leq \frac{|a|^p}{p} + \frac{|b|^q}{q}.$$

For our repeated application below we will have n a positive integer, $p = \frac{2n}{2n-1}$, and $q = 2n$.

Theorem 2.2.9. Suppose there is a positive integer n with $a'(t) \in L^{2n}[0,\infty)$, a constant $\alpha > 0$, and a constant $N > 0$ with

$$\frac{2n-1}{2nN^{\frac{2n}{2n-1}}} - C(t,t) + \frac{2n-1}{2n}\int_0^t |C_t(t,s)|ds$$
$$+ \frac{1}{2n}\int_0^\infty |C_t(u+t,t)|du \leq -\alpha.$$

Then the unique solution x of (2.2.1) is bounded and $x \in L^{2n}[0,\infty)$. From (2.2.3) we then have $x(t) = a(t) - \int_0^t R(t,s)a(s)ds \in L^{2n}$.

Proof. For a fixed solution of (2.2.4) on $[0,\infty)$ we define the function

$$V(t) = \frac{x^{2n}(t)}{2n} + \frac{1}{2n}\int_0^t \int_{t-s}^\infty |C_1(u+s,s)|du\,x^{2n}(s)ds.$$

Compute the derivative along a solution of (2.2.4) on $[0, \infty)$ by the chain rule as

$$V'(t) = -C(t,t)x^{2n} - \int_0^t C_t(t,s)x(s)x^{2n-1}(t)ds + x^{2n-1}(t)a'(t)$$

$$+ \frac{1}{2n}\int_0^\infty |C_1(u+t,t)|du x^{2n}(t) - \frac{1}{2n}\int_0^t |C_1(t,s)|x^{2n}(s)ds$$

(Use Hölder's inequality on the 2nd & 3rd terms.)

$$\leq \frac{(2n-1)x^{2n}(t)}{2nN^k} + \frac{(Na'(t))^{2n}}{2n} - C(t,t)x^{2n}$$

$$+ \int_0^t |C_t(t,s)|\left[\frac{(2n-1)x^{2n}(t)}{2n} + \frac{x^{2n}(s)}{2n}\right]ds$$

$$+ \frac{1}{2n}\int_0^\infty |C_1(u+t,t)|du x^{2n}(t) - \frac{1}{2n}\int_0^t |C_1(t,s)|x^{2n}(s)ds$$

$$= \frac{(Na'(t))^{2n}}{2n} + x^{2n}(t)\left[\frac{(2n-1)}{2nN^k} - C(t,t)\right.$$

$$+ \frac{2n-1}{2n}\int_0^t |C_t(t,s)|ds + \frac{1}{2n}\int_0^\infty |C_1(u+t,t)|du\right]$$

$$\leq -\alpha x^{2n}(t) + \frac{N^{2n}}{2n}|a'(t)|^{2n}$$

for large N and for $k = \frac{2n}{2n-1}$.

It follows that

$$\frac{x^{2n}(t)}{2n} \leq V(t) \leq V(0) - \alpha \int_0^t x^{2n}(s)ds + k^* \int_0^\infty (a'(s))^{2n}ds$$

for some $k^* > 0$. This is true for every solution of (2.2.4) on $[0, \infty)$ and, hence, for (2.2.1).

As we continue to study the behavior of the resolvent, it increasingly seems to be a question of stability.

Definition 2.2.1. A function R mapping $[0, \infty) \times [0, \infty)$ into the reals is said to be L^N-stable with respect to a vector space W of specified continuous functions ϕ mapping $[0, \infty)$ into the reals if for each $\phi \in W$ there is an integer n with

$$\phi(t) - \int_0^t R(t,s)\phi(s)ds \in L^n[0, \infty).$$

2.2. A STRONG RESOLVENT: THE SCALAR CASE

In our last theorem, the vector space W consisted of those functions ϕ such that $\phi' \in L^p$ for some $p \in (0, \infty)$.

Remark 2.2.4. *Of course, this is a classical stability concept.* For suppose that ϕ_1 and ϕ_2 are functions with $\phi_1 - \phi_2 \in W$. Then for

$$x_{\phi_1}(t) = \phi_1(t) - \int_0^t C(t,s)x(s)ds$$

and

$$x_{\phi_2}(t) = \phi_2(t) - \int_0^t C(t,s)x(s)ds$$

we have

$$x_{\phi_1} - x_{\phi_2} = \phi_1 - \phi_2 - \int_0^t R(t,s)[\phi_1(s) - \phi_2(s)]ds \in L^n[0,\infty).$$

We are saying that if $\phi_1 - \phi_2 \in W$ then they are "close" and the solutions generated are "close." In our examples we find that the functions $\sin(t+1)^\beta$, $(t+1)^\beta$, and $(t+1)$ are "close." Notice that if our examples are based on Theorem 2.2.9 with $\phi_1' \in L^p$ and $\phi_2' \in L^q$ where $p < q$, since $x^{2n}/2n \leq V(t)$ we have $x(t) = \phi_1(t) - \int_0^t R(t,s)\phi_1(s)ds \in L^p$ and also $x \in L^q$.

We now introduce a Razumikhin technique which begins with a Liapunov function and then deals with functionals by keeping track of past behaviors.

Theorem 2.2.10. *Suppose that there is a $K > 0$ with*

$$\sup_{t \geq 0}\{[|a'(t)| - C(t,t)K + K\int_0^t |C_1(t,s)|ds]\} < 0.$$

Then every solution of (2.2.4) on $[0, \infty)$ (and hence (2.2.1)) is bounded.
Now suppose there is an $\alpha > 0$ with

$$-C(t,t) + \int_0^t |C_1(t,s)|ds \leq -\alpha.$$

(i) Then every solution of (2.2.4) on $[0,\infty)$ and (2.1.1) is bounded for every $a(t)$ with bounded $a'(t)$; moreover, the resolvent for (2.2.4), $Z(t,s)$, satisfies $Z(t,0)$ is bounded and there is a $J > 0$ with $\int_0^t |Z(t,s)|ds < J$.
(ii) If, in addition, $C(t,t) \leq M$ for some $M > 0$ and if $a'(t)$ is continuous (not necessarily bounded), then (1.2.14) holds ($x(t) = a(t) - \int_0^t Z_s(t,s)a(s)ds$) and there is an $L > 0$ with $\int_0^t |Z_s(t,s)|ds \leq L$ for $t \geq 0$.

Proof. Let $x(t)$ be a solution of (2.2.4) on $[0,\infty)$ and suppose there is a $K^* \geq K$ and $t^* > 0$ with $|x(s)| < K^*$ for $0 \leq s < t^*$, but $|x(t^*)| = K^*$. Clearly, $C(t,t) \geq 0$. Then for $V(t) = |x(t)|$ and $0 \leq t < t^*$ we have

$$V'(t) \leq |a'(t)| - C(t,t)|x(t)| + \int_0^t |C_1(t,s)||x(s)|ds$$

and at $t = t^*$ we have

$$V'(t^*) \leq |a'(t^*)| - C(t^*, t^*)K^* + K^* \int_0^{t^*} |C_1(t^*, s)|ds$$

$$\leq |a'(t^*)| - C(t^*, t^*)K + K \int_0^{t^*} |C_1(t^*, s)|ds < 0,$$

a contradiction to $V'(t^*) \geq 0$.

To prove (i), for every bounded continuous $a'(t)$ we find K so that $\alpha K > \sup_{t \geq 0} |a'(t)|$ and the first condition is satisfied and so every solution of (2.2.1) and (2.2.4) on $[0, \infty)$ is bounded for every bounded and continuous $a'(t)$. In particular, if $a'(t) = 0$ then every solution of (2.2.4) is bounded so $Z(t, 0)$ is bounded, as seen in (2.2.6). But for any continuous $a'(t)$ any solution of (2.2.4) on $[0, \infty)$ is given by (2.2.6) and, since $x(t)$ and $Z(t, 0)$ are bounded, so is the integral in (2.2.6). By Perron's theorem $\sup_{t \geq 0} \int_0^t |Z(t,s)|ds < \infty$.

To prove (ii), under these conditions (1.2.14) does hold and we consider Miller's resolvent, (1.2.13), and recall that $H = Z$ by Theorem 1.2.6.1. An integration of that resolvent yields

$$\int_0^t |Z_s(t,s)|ds \leq M \int_0^t |Z(t,s)|ds + \int_0^t \int_s^t |Z(t,u)C_t(u,s)|duds$$

$$\leq MJ + \int_0^t \int_0^u |C_t(u,s)|ds|Z(t,u)|du$$

$$\leq MJ + M \int_0^t |Z(t,u)|du \leq 2MJ.$$

Refer back to Theorem 2.2.5. We are now going to let a' be bounded and get $x(t)$ bounded. We started with

$$x(t) = a(t) - \int_0^t C(t,s)x(s)ds$$

and

$$x(t) = a(t) - \int_0^t R(t,s)a(s)ds.$$

2.2. A STRONG RESOLVENT: THE SCALAR CASE

If we prove that $x \in L^p$ for every $a \in L^q$ then

$$(P\phi)(t) = \phi(t) - \int_0^t R(t,s)\phi(s)ds$$

maps L^q into L^p and $R(t,s)$ generates an L^p approximate identity on the vector space L^q. The resolvent $R(t,s)$ is determined from $C(t,s)$ alone.

If $x \in L^p$ for each a with $a^{(q)} \in L^d$, then $R(t,s)$ generates an L^p approximate identity on the vector space W of functions ϕ such that $\phi^{(q)} \in L^d$. Thus, if $\phi^{(q)} \in L^d$, then ϕ can have arbitrarily rapid growth as $q \to \infty$.

We have seen resolvents generate L^p approximate identities on spaces of functions which grow as fast as $t^{1/2}$. Can we continue and obtain L^p approximate identities on spaces of functions with arbitrarily rapid growth? Our ability to prove such behavior is limited only by our ability to prove that $x \in L^p$ when $a^{(q)} \in L^d$. That is simply a technical problem. Here, we contrive such a problem, allowing a with $a'' \in L^1$. At the same time we show how the Levin functional can be used in a variety of ways.

The following result concerns (2.2.1) in which

$$C(t,t) = \alpha > 0, \ C_1(t,t) = \beta > 0,$$

$$C_{11}(t,s) =: f(t-s) < 0, \ \beta + \int_0^\infty f(u)du > 0, \quad (2.2.12)$$

and

$$\int_{t-s}^\infty f(u)du =: F(t-s), \ F(t-s) < 0,$$

$$F_s(t-s) < 0, \ F_{st}(t-s) > 0, \ F(t) < 0, \ F_t(t) > 0. \quad (2.2.13)$$

A function satisfying these conditions is

$$C(t,s) = 2 + 3(t-s) - e^{-(t-s)}.$$

Moreover, taking $a(t) = t$ will show the desired behavior and will avoid much of the work below.

Theorem 2.2.11. *If (2.2.12) and (2.2.13) hold, then the solution x of (2.2.1) is in L^2 whenever $a'' \in L^1$. Thus, $R(t,s)$ generates an L^2 approximate identity on the vector space of functions $\phi : [0,\infty) \to R$ with $\phi''(t) \in L^1[0,\infty)$.*

2. THE REMARKABLE RESOLVENT

Proof. We have
$$x'(t) = a'(t) - C(t,t)x(t) - \int_0^t C_1(t,s)x(s)ds$$
and
$$x''(t)$$
$$= a''(t) - \alpha x'(t) - C_1(t,t)x(t) - \int_0^t C_{11}(t,s)x(s)ds$$
$$= a''(t) - \alpha x'(t) - \beta x(t) - \int_0^t f(t-s)x(s)ds.$$

Write
$$x'(t) = y(t) - \alpha x(t) + \int_0^t \int_{t-s}^\infty f(u)du\, x(s)ds$$
so
$$x''(t)$$
$$= y' - \alpha x'(t) + \int_0^\infty f(u)du\, x(t) - \int_0^t f(t-s)x(s)ds$$
$$= a''(t) - \alpha x'(t) - \beta x(t) - \int_0^t f(t-s)x(s)ds$$
or
$$y' = a''(t) - \beta x(t) - \int_0^\infty f(u)du\, x(t).$$

This yields the system
$$x'(t) = y(t) - \alpha x(t) + \int_0^t F(t-s)x(s)ds$$
$$y'(t) = a''(t) - \left(\beta + \int_0^\infty f(u)du\right)x(t).$$

A suitable Liapunov functional is
$$V(t) = \frac{x^2(t)}{2} + \frac{y^2(t)}{2\left(\beta + \int_0^\infty f(u)du\right)}$$
$$- (1/2)\int_0^t F_s(t-s)\left(\int_s^t x(u)du\right)^2 ds$$
$$- (1/2)F(t)\left(\int_0^t x(u)du\right)^2$$

so that
$$V'(t) = xy - \alpha x^2 + x \int_0^t F(t-s)x(s)ds + \frac{a''(t)y}{\beta + \int_0^\infty f(u)du}$$
$$- xy - (1/2)\int_0^t F_{st}(t-s)\left(\int_s^t x(u)du\right)^2 ds$$
$$- x\int_0^t F_s(t-s)\int_s^t x(u)du\,ds$$
$$- (1/2)F_t(t)\left(\int_0^t x(u)du\right)^2 - xF(t)\int_0^t x(u)du.$$

Integrating the third-to-last term by parts yields
$$-x\left[F(t-s)\int_s^t x(u)du\Big|_0^t + \int_0^t F(t-s)x(s)ds\right]$$
$$= -x\left[-F(t)\int_0^t x(u)du + \int_0^t F(t-s)x(s)ds\right].$$

Hence,
$$V'(t) = -\alpha x^2 + \frac{a''(t)y}{\beta + \int_0^\infty f(u)du}$$
$$- (1/2)\int_0^t F_{st}(t-s)\left(\int_s^t x(u)du\right)^2 ds$$
$$- (1/2)F_t(t)\left(\int_0^t x(u)du\right)^2$$
$$\leq -\alpha x^2 + \frac{a''(t)y}{\beta + \int_0^\infty f(u)du}$$
$$\leq -\alpha x^2 + K|a''(t)|[V(t)+1]$$

for an appropriate constant K.

We integrate that differential inequality and obtain
$$V(t) \leq V(0)e^{\int_0^t K|a''(s)|ds} - \int_0^t \alpha x^2(s)ds + e^{\int_0^t K|a''(s)|ds}.$$

This yields $x \in L^2$ whenever $a'' \in L^1$.

2.3 An Unusual Scalar Equation

The differentiated form (2.2.4) has been studied for more than one hundred years. It enabled us to prove Theorem 2.2.2 which yields every solution of

(2.2.1) bounded for every a with a' bounded. But in order to utilize the stability scheme described in the opening of Section 2.2 we must know that (2.1.5) holds and that requires that every solution of (2.2.1) be bounded for every bounded a. There is a little known companion of (2.2.4) with $a(t)$ instead of $a'(t)$ and we will be able to prove a counterpart of Theorem 2.2.2 for it, thereby providing the machinery necessary for that stability argument. In fact, every theorem proved in the last section can be proved for our central equation here simply by changing a' to a, $C_t(t,s)$ to $-C_s(t,s)$, and the conclusion from properties of x to the same properties of $\int_0^t x(s)ds$. Thus, our work here will be somewhat abbreviated, but enough detail will be given so that the interested reader may make the indicated changes.

Once more we begin with the scalar equations

$$x(t) = a(t) - \int_0^t C(t,s)x(s)ds, \qquad (2.3.1)$$

$$R(t,s) = C(t,s) - \int_s^t R(t,u)C(u,s)du, \qquad (2.3.2)$$

and

$$x(t) = a(t) - \int_0^t R(t,s)a(s)ds \qquad (2.3.3)$$

under the same continuity conditions as with (2.1.1).

If $C(t,s)$ contains an additive function of t, perhaps a constant, then (2.1.6) must often fail. But such problems can be effectively removed if $C_s(t,s)$ is continuous. In that case we write (2.3.1) as

$$x(t) = a(t) - \int_0^t C(t,s)x(s)ds$$

$$= a(t) - C(t,s)\int_0^s x(u)du \Big|_0^t + \int_0^t C_2(t,s)\int_0^s x(u)duds$$

$$= a(t) - C(t,t)\int_0^t x(u)du + \int_0^t C_2(t,s)\int_0^s x(u)duds.$$

If we let $y(t) = \int_0^t x(u)du$ our equation becomes

$$y'(t) = a(t) - C(t,t)y(t) + \int_0^t C_2(t,s)y(s)ds. \qquad (2.3.4)$$

The covering assumption is that a, C, and C_2 are continuous.

2.3. AN UNUSUAL SCALAR EQUATION

There is good independent reason for studying $\int_0^t x(u)du$, as is discussed by Feller (1941) concerning the renewal equation. The resolvent equation for (2.3.4) is

$$Z_s(t,s) = Z(t,s)C(s,s) - \int_s^t Z(t,u)C_2(u,s)du, \quad Z(t,t) = 1 \quad (2.3.5)$$

with resolvent $Z(t,s)$ and with y satisfying the variation of parameters formula

$$y(t) = Z(t,0)y(0) + \int_0^t Z(t,s)a(s)ds \quad (2.3.6)$$

and (remembering that $y(0) = 0$ since $y(t) = \int_0^t x(u)du$) we have, upon integration by parts,

$$y(t) = \int_0^t a(s)ds - \int_0^t Z_s(t,s) \int_0^s a(u)duds. \quad (2.3.7)$$

In this problem we will see $y(t)$ bounded even when $\int_0^t a(s)ds$ is unbounded, meaning that $Z_s(t,s)$ generates an approximate identity on a space of unbounded functions. In the next example we will get an asymptotic identity. But the crucial point is to note from (2.3.4) that if $y(t)$ is bounded and if $C(t,t)$ and $\int_0^t |C_2(t,s)|ds$ are bounded, then $y'(t) = x(t)$ is bounded.

Theorem 2.3.1. Let $Z(t,s)$ be the solution of (2.3.5). Every solution $y(t) = \int_0^t x(u)du$ of (2.3.4) is bounded for every bounded continuous $a(t)$ if and only if

$$\sup_{t \geq 0} \int_0^t |Z(t,s)|ds < \infty. \quad (2.3.8)$$

Moreover, if (2.3.8) holds then $Z_s(t,s)$ generates an approximate identity on the vector space of continuous functions $\phi : [0, \infty) \to \Re$ for which $\phi'(t)$ is bounded. Finally, if in addition, $y(t) \to 0$ for every $a(t)$ which tends to zero, then $Z_s(t,s)$ generates an asymptotic identity on the vector space of continuous functions $\phi : [0, \infty) \to \Re$ for which $\phi'(t) \to 0$.

Proof. The proof of the first part is like that of Theorem 2.2.1 using (2.3.6) with $y(0) = 0$. The next part, $Z_s(t,s)$ generates an approximate identity, follows from (2.3.7) when we recall that $y(t)$ is bounded for bounded $a(t)$. The last conclusion follows in the same way.

Next, recall that $y(0) = 0$ and write

$$y(t) = \int_0^t e^{\int_u^t -C(s,s)ds} a(u)du$$
$$+ \int_0^t e^{\int_u^t -C(s,s)ds} \int_0^u C_2(u,s)y(s)dsdu. \quad (2.3.9)$$

Theorem 2.3.2. *Suppose that $a(t)$ is bounded, that $\int_0^t e^{\int_u^t -C(s,s)ds} du$ is bounded, and that there exists $\alpha < 1$ with*

$$\sup_{t \geq 0} \int_0^t e^{\int_u^t -C(s,s)ds} \int_0^u |C_2(u,s)|dsdu \leq \alpha. \quad (2.3.10)$$

Then for $x(t)$ the solution of (2.3.1) we have $\int_0^t x(s)ds$ bounded. Thus, $Z_s(t,s)$ of (2.3.5) generates an approximate identity on the space of functions ϕ such that ϕ' is bounded. If $|C(t,t)| + \int_0^t |C_2(t,s)|ds$ is bounded, then $x(t)$ is also.

Proof. Use (2.3.9) and the supremum norm to define a mapping $Q : \mathcal{BC} \to \mathcal{BC}$ by $\phi \in \mathcal{BC}$ implies that

$$(Q\phi)(t) = \int_0^t e^{\int_u^t -C(s,s)ds} a(u)du + \int_0^t e^{\int_u^t -C(s,s)ds} \int_0^u C_2(u,s)\phi(s)dsdu.$$

If $\phi \in \mathcal{BC}$, so is $Q\phi$ by assumption and (2.3.10). Also, Q is a contraction by (2.3.10). Hence, $y(t) = \int_0^t x(s)ds$ is bounded. The final conclusion follows from (2.3.4).

Theorem 2.3.3. *Let the conditions of Theorem 2.3.2 hold. Suppose that for each $T > 0$*

$$\int_0^T e^{\int_u^t -C(s,s)ds} \int_0^u |C_2(u,s)|dsdu \to 0$$

and

$$\int_0^T e^{\int_u^t -C(s,s)ds} du \to 0$$

as $t \to \infty$. If, in addition, $a(t) \to 0$ as $t \to \infty$ then the solution $x(t)$ of (2.3.1) satisfies $y(t) = \int_0^t x(u)du \to 0$ as $t \to \infty$. Also, for the $Z(t,s)$ of (2.3.5) we have $\int_0^t Z(t,s)\phi(s)ds \to 0$ as $t \to \infty$ for every continuous function ϕ which tends to zero as $t \to \infty$.

2.3. AN UNUSUAL SCALAR EQUATION

Proof. Use the mapping from the proof of Theorem 2.3.2, but replace $\mathcal{BC}$ by the complete metric space of continuous $\phi : [0, \infty) \to \Re$ such that $\phi(t) \to 0$ as $t \to \infty$. Use the assumptions and essentially the classical proof that the convolution of an L^1-function with a function tending to zero does, itself, tend to zero. (See the proof of Theorem 2.2.4.) This will show that $(Q\phi)(t) \to 0$ when $\phi(t) \to 0$. The mapping is a contraction as before with unique solution $y(t) = \int_0^t x(u)du \to 0$ as $t \to \infty$. The last conclusion is immediate.

Theorem 2.3.4. *Let $\int_{t-s}^{\infty} |C_2(u+s,s)|du$ be continuous. If there is a positive number α such that*

$$\int_0^{\infty} |C_2(u+s,s)|du - C(t,t) \leq -\alpha$$

and if $a \in L^1$, then $y \in L^1$ and bounded for any solution of (2.3.4). Hence

$$y(t) = \int_0^t a(u)du - \int_0^t Z_s(t,s) \int_0^s a(u)duds \in L^1.$$

Therefore,

$$(P\phi)(t) = \int_0^t \phi(s)ds - \int_0^t Z_s(t,s) \int_0^s \phi(u)duds$$

maps $W \to L^1$ where $\phi \in W$ if ϕ is continuous and $\phi \in L^1$.

Proof. Let

$$V(t) = |y(t)| + \int_0^t \int_{t-s}^{\infty} |C_2(u+s,s)|du|y(s)|ds$$

so that

$$V'(t) \leq |a(t)| - C(t,t)|y(t)| + \int_0^t |C_s(t,s)||y(s))|ds$$

$$+ \int_0^{\infty} |C_2(u+t,t)|du|y(t)| - \int_0^t |C_2(t,s)y(s)|ds$$

$$\leq |a(t)| + \left[-C(t,t) + \int_0^{\infty} |C_2(u+t,t)|du\right]|y(t)|$$

$$\leq |a(t)| - \alpha|y(t)|.$$

Hence, $a \in L^1$ implies $y \in L^1$. Also, V bounded yields y bounded.

Keep in mind that these results are not to be confused with those in Section 2.2. Here, $y = \int_0^t x(s)ds$ and additional work is needed to obtain properties of x.

Theorem 2.3.5. Let $\int_{t-s}^{\infty} |C_2(u+s,s)|du$ be continuous. If there exists $\alpha > 0$ such that

$$\int_0^{\infty} |C_2(u+s,s)|du + \int_0^t |C_2(t,s)|ds - 2C(t,t) \leq -\alpha$$

and if $a \in L^2$ then any solution $y(t)$ of (2.3.4) on $[0, \infty)$ is also in L^2 and bounded. Notice that for $a(t) = 1/(t+1)$, then $\int_0^t a(u)du = \ln(t+1)$ so the terms in the variation of parameter formula (2.3.7) tend to ∞ and, yet, the difference in the terms is in L^2.

Proof. Let

$$V(t) = y^2(t) + \int_0^t \int_{t-s}^{\infty} |C_2(u+s,s)|duy^2(s)ds$$

so that for $\alpha/2 > 0$ there is a positive number M with $2y(t)a(t) \leq (\alpha/2)y^2(t) + Ma^2(t)$ and

$$V'(t) = 2y(t)a(t) - 2C(t,t)y^2(t) + 2y(t)\int_0^t C_s(t,s)y(s)ds$$

$$+ \int_0^{\infty} |C_2(u+t,t)|duy^2(t) - \int_0^t |C_2(t,s)|y^2(s)ds$$

$$\leq Ma^2(t) + (\alpha/2)y^2(t) - 2C(t,t)y^2(t)$$

$$+ \int_0^t |C_s(t,s)|(y^2(t) + y^2(s))ds$$

$$+ \int_0^{\infty} |C_2(u+t,t)|duy^2(t) - \int_0^t |C_2(t,s)|y^2(s)ds$$

$$= Ma^2(t) + \left[(\alpha/2) - 2C(t,t)\right.$$

$$+ \int_0^{\infty} |C_2(u+t,t)|du + \int_0^t |C_s(t,s)|ds\right]y^2(t)$$

$$\leq -(\alpha/2)y^2(t) + Ma^2(t).$$

This yields $y \in L^2$. As V is bounded, so is y.

The following result is a companion to Theorems 2.2.7 and 2.2.8 and is proved in exactly the same way. Notice that we obtain (2.1.5) so that the stability scheme of Section 2.2 can be realized.

Theorem 2.3.6. Let $C(t,t) \geq \alpha > 0$, $H(t,s) = -C_s(t,s)$, and let

$$H(t,s) \geq 0, \ H_t(t,s) \leq 0, \ H_s(t,s) \geq 0, \ H_{st}(t,s) \leq 0; \qquad (2.3.11)$$

if $a \in L^2[0,\infty)$, then any solution y of (2.3.4) is bounded; also, $Z(t,0) \in L^2[0,\infty)$ and bounded so by (2.3.6) for (2.3.4), $\int_0^t Z(t,s)a(s)ds \in L^2[0,\infty)$ and bounded. Define $\lambda(t)$ by

$$\lambda(t) := \int_0^t H_s(t,s)(t-s) \int_s^t a^2(u)du\,ds + H(t,0)t \int_0^t a^2(u)du. \quad (2.3.12)$$

If there is an M with $\lambda(t) < M$ and if $a(t)$ is bounded so is every solution of (2.3.4). If $C(t,t)$ and $\int_0^t C_s(t,s)ds$ are bounded, so is the solution of (2.3.1); in particular, then, $\sup_{t\geq 0} \int_0^t |R(t,s)|ds < \infty$.

Proof. Define V as in the proof of Theorem 2.2.8 with x replaced by y and get

$$V'(t) \leq -Ey^2 + Da^2(t)$$

for positive constants D and E. The constant M in (2.2.11) is defined with a' replaced by a. The theorem is repeated with y bounded. That means that $\int_0^t x(s)ds$ is bounded. The bound on x, the solution of (2.3.1), follows as stated in the theorem. Finally, in that last case x is bounded for every bounded and continuous $a(t)$ so by Perron's theorem (2.1.5) holds.

2.4 A Mathematical David & Goliath: Periodicity

We begin once more with

$$x(t) = a(t) - \int_0^t C(t,s)x(s)ds, \qquad (2.4.1)$$

$$R(t,s) = C(t,s) - \int_s^t R(t,u)C(u,s)du, \qquad (2.4.2)$$

and

$$x(t) = a(t) - \int_0^t R(t,s)a(s)ds \qquad (2.4.3)$$

with the same continuity properties as in (2.1.1) and C_t continuous.

This continues with the notation from Theorem 2.1.6 and Lemma 2.1.1 and we also differentiate (2.4.1) to obtain

$$x' = a'(t) - C(t,t)x(t) - \int_0^t C_t(t,s)x(s)ds. \tag{2.4.4}$$

From (2.4.4) and the variation of parameters formula we have a new integral equation with $x(0) = a(0)$ in the form

$$x(t) = x(0)e^{-\int_0^t C(s,s)ds} + \int_0^t e^{-\int_u^t C(s,s)ds}\left[a'(u) - \int_0^u C_1(u,s)x(s)ds\right]du. \tag{2.4.5}$$

We assume that there is a $T > 0$ with

$$a(t+T) = a(t) \text{ and } C(t+T, s+T) = C(t,s). \tag{2.4.6}$$

Notice that (2.4.6) will bestow many properties on (2.4.5). For example, $C(t,t)$ is periodic since $C(t+T, t+T) = C(t,t)$. Thus,

$$\int_{u+T}^{t+T} C(s,s)ds = \int_u^t C(s,s)ds,$$

as will be needed to show periodicity later. We suppose that there is a number $c^* > 0$ with

$$C(t,t) \geq c^* \tag{2.4.7}$$

and an $\alpha < 1$ with

$$\int_0^t e^{-\int_u^t C(s,s)ds} \int_0^u |C_1(u,s)|dsdu \leq \alpha \tag{2.4.8}$$

which will make the mapping defined from (2.4.5) be a contraction on any space with the supremum norm. Notice again that (2.4.8) is almost like inequality (2.2.8) in Theorem 2.2.1. We are assuming that when we integrate $C_1(t,s)$ we get less than $C(t,s)$; something is lost in the differentiation, possibly a constant.

In order to prove that (2.4.5) has an asymptotically periodic solution in the same way we proved Theorem 2.1.6, (2.4.5) must be decomposed into the mappings A and B as in the proof of Theorem 2.1.6. Recall that in the notation we are using p is periodic and q tends to zero. See Lemma 2.1.1.

2.4. PERIODICITY

Thus, we begin by writing $a'(t) = p^*(t) + q^*(t) \in Y$ and define a mapping from (2.4.5) by $\phi = p + q \in Y$ implies that

$$(P\phi)(t) = a(0)e^{-\int_0^t C(s,s)ds}$$
$$+ \int_0^t e^{-\int_u^t C(s,s)ds}\left[p^*(u) + q^*(u) - \int_0^u C_1(u,s)[p(s) + q(s)]ds\right]du. \tag{2.4.9}$$

The decomposition will be done in the proof of Theorem 2.4.1.

Suppose that $\int_{-\infty}^0 |C_1(t,s)|ds$ is continuous, that

$$\int_{-\infty}^0 |C_1(t,s)|ds \to 0 \text{ as } t \to \infty, \tag{2.4.10}$$

and for $q \in Q$ then

$$\int_0^t C_1(t,s)q(s)ds \to 0 \text{ as } t \to \infty. \tag{2.4.11}$$

It may help to understand these by noting that if C were of convolution type then (2.4.10) would say that $C_1 \in L^1[0,\infty)$, while (2.4.11) would then be the classical theorem that the convolution of an L^1 function with a function tending to zero does, itself, tend to zero as $t \to \infty$.

Lemma 2.4.1. *If (2.4.7) holds then*

$$\int_{-\infty}^0 e^{-\int_u^t C(s,s)ds}du \to 0 \text{ as } t \to \infty \tag{2.4.12}$$

and for $q \in Q$ then

$$\int_0^t e^{-\int_u^t C(s,s)ds}q(u)du \to 0 \text{ as } t \to \infty. \tag{2.4.13}$$

Proof. We have $c^* = \min C(t,t)$ and $C(t,t) \in \mathcal{P}_T$. Thus,

$$\int_{-\infty}^0 e^{-\int_u^t C(s,s)ds}du \leq (1/c^*)\int_{-\infty}^0 C(u,u)e^{-\int_u^t C(s,s)ds}du$$
$$= (1/c^*)e^{-\int_u^t C(s,s)ds}\Big|_{-\infty}^0$$
$$= (1/c^*)e^{\int_0^t -C(s,s)ds}$$

which tends to zero as $t \to \infty$.

Next,
$$\int_0^t e^{-\int_u^t C(s,s)ds}|q(u)|du \le \int_0^t e^{-c^*(t-u)}|q(u)|du$$
which is the convolution of an L^1-function with a function tending to zero so it tends to zero.

Theorem 2.4.1. *In (2.4.1) let a' and $C_1(t,s)$ be continuous. Let (2.4.7)-(2.4.8), and (2.4.10)-(2.4.11) hold. Suppose, in addition, that*
$$\int_{-\infty}^t |C_1(t,s)|ds \tag{2.4.14}$$
is bounded and continuous, while $C(t+T, s+T) = C(t,s)$. If $a' \in Y$ so is x, the unique solution of (2.4.1).

Proof. Using (2.4.5) we define a mapping $P : Y \to Y$ by $\phi = p + q \in Y$ implies that
$$(P\phi)(t) = a(0)e^{-\int_0^t C(s,s)ds} + \int_0^t e^{-\int_u^t C(s,s)ds}\left[a'(u) - \int_0^u C_1(u,s)\phi(s)ds\right]du.$$

By (2.4.8) it is clearly a contraction, but we must show that $P : Y \to Y$. Write $a' = p^* + q^*$ and then
$$(P\phi)(t) = \int_{-\infty}^t e^{-\int_u^t C(s,s)ds}\left[p^*(u) - \int_{-\infty}^u C_1(u,s)p(s)ds\right]du$$
$$- \int_0^t e^{-\int_u^t C(s,s)ds}\int_0^u C_1(u,s)q(s)dsdu + a(0)e^{-\int_0^t C(s,s)ds}$$
$$- \int_{-\infty}^0 e^{-\int_u^t C(s,s)ds}\left[p^*(u) - \int_{-\infty}^u C_1(u,s)p(s)ds\right]du$$
$$+ \int_0^t e^{-\int_u^t C(s,s)ds}\int_{-\infty}^0 C_1(u,s)p(s)dsdu$$
$$+ \int_0^t e^{-\int_u^t C(s,s)ds}q^*(u)du.$$

The first term on the right-hand-side is clearly in $\mathcal{P}_T$. In the second term, $\int_0^u C_1(u,s)q(s)ds \in Q$ by (2.4.11). Hence the second term is in Q by (2.4.13). The third term is in Q by (2.4.7). The fourth term is in Q by (2.4.12), (2.4.14), and the fact that $p^* \in \mathcal{P}_T$ and, hence, is bounded. The next to last term is in Q because of (2.4.10) followed by (2.4.11). The last term is in Q by (2.4.13). This completes the proof

2.4. PERIODICITY

Remark 2.4.1. *Notice in the last result that a significant instability can occur at $\beta = 1$. Under conditions on $C(t,s)$ of Theorem 2.2.9 the integral of that resolvent has been faithfully following $\sin(t+1)^\beta$ so that the difference is an L^p function. Suddenly, that relationship breaks completely and the integral with the resolvent seems to "struggle along trying to catch up with $\sin(t+1)$" but always is out of step, lagging by a nontrivial periodic function plus a function tending to zero.*

Corollary 1. *If the conditions of Theorem 2.2.9 hold and if $0 < \beta < 1$ then $\sin(t+1)^\beta - \int_0^t R(t,s)\sin(s+1)^\beta ds \in L^s$ for some $s < \infty$. But at $\beta = 1$, under conditions on $C(t,s)$ of Theorem 2.4.1 then $s = \infty$ and that difference approaches a periodic function.*

We can now state the promised result, a corollary of Theorems 2.4.1 and 2.2.8.

Corollary 2. *Let the conditions on $C(t,s)$ of Theorems 2.4.1 and 2.2.9 hold. For fixed $\beta \in (0,1)$ there is a $p \in \mathcal{P}_T$, $q \in Q$, and $u \in L^s[0,\infty)$ for some $s > 0$ so that the solution of*

$$x(t) = \sin t + (t+1)^\beta - \int_0^t C(t,s)x(s)ds$$

may be written as

$$x(t) = p(t) + q(t) + u(t).$$

Proof. The solution is

$$x(t) = \sin t + (t+1)^\beta - \int_0^t R(t,s)[\sin s + (s+1)^\beta]ds.$$

But

$$(t+1)^\beta - \int_0^t R(t,s)(s+1)^\beta ds =: u(t) \in L^s[0,\infty),$$

while

$$\sin t - \int_0^t R(t,s)\sin s\, ds$$

is the solution described in Theorem 2.4.1 and it has the required form of $p + q$.

Here is an example which should awaken the most complacent among us. It is a mathematical David and Goliath.

Example 2.4.1. Let the conditions on C of Theorem 2.4.1 hold and let $a(t) = t+\sin t$. Then $a' \in \mathcal{P}_T \subset Y$. The solution of (2.4.1) is asymptotically periodic. The large function t is simply absorbed, while $\sin t$ makes its presence felt forever.

Example 2.4.2. For a transparent example linking Theorem 2.2.9 and Theorem 2.4.1, let $k > 0$,

$$C(t,s) = k + \sin^2 s + D(t-s), \; D(t) > 0, \; D'(t) \leq 0.$$

We then have

$$\int_0^t -D'(s)ds = D(0) - D(t) < D(0)$$

and $C(t,t) = k + \sin^2 t + D(0)$ and so we readily verify the inequality in Theorem 2.2.9 holds for large N and n. To satisfy (2.4.8) we have

$$\int_0^t e^{-\int_u^t [k+\sin^2 s + D(0)]ds} \int_0^u -D'(u-s)dsdu < \frac{D(0)}{k+D(0)}$$

and conditions of Theorem 2.4.1 are satisfied.

Notice in Theorem 2.4.1 that we work with $C_1(t,s)$ and that conditions (2.4.10), (2.4.11), and (2.4.14) all concern $C_1(t,s)$. We now consider the transformation of Section 2.3 to obtain a completely parallel result by working with $C_2(t,s)$ and to avoid differentiating $a(t)$. This time one might think of $C(t,s) = k + \sin^2 t + D(t-s)$ and note that $C_s(t,s) = -D'(t-s)$; the kernel has been cleansed of $k + \sin^2 t$ so that conditions of Theorem 2.2.9 can possibly be satisfied.

Review the work in Section 2.3 where we obtain (2.3.4) and designate it here as

$$y'(t) = a(t) - C(t,t)y(t) + \int_0^t C_s(t,s)y(s)ds. \tag{2.4.15}$$

By the variation of parameters formula we have

$$y(t) = \int_0^t e^{-\int_u^t C(s,s)ds} \left[a(u) + \int_0^u C_s(u,s)y(s)ds\right] du \tag{2.4.16}$$

and we will need $\alpha < 1$ with

$$\int_0^t e^{-\int_u^t C(s,s)ds} \int_0^u |C_s(u,s)|dsdu \leq \alpha. \tag{2.4.17}$$

Parallel to (2.4.10) and (2.4.11) we ask that

$$\int_{-\infty}^{0} |C_s(t,s)| ds \to 0 \text{ as } t \to \infty \tag{2.4.18}$$

and

$$\int_{0}^{t} C_s(t,s) q(s) ds \to 0 \text{ as } t \to \infty \text{ for } q \in Q. \tag{2.4.19}$$

Conditions (2.4.7), (2.4.12), (2.4.13), and Lemma 2.4.1 will be the same for both (2.4.4) and (2.4.15), while (2.4.14) will be replaced by the condition that

$$\int_{-\infty}^{t} |C_s(t,s)| ds \tag{2.4.20}$$

is bounded and continuous.

Theorem 2.4.2. *In (2.4.1) let $a(t)$ and $C_s(t,s)$ be continuous. Let (2.4.7), (2.4.17)–(2.4.20) hold. Let $C(t+T, s+T) = C(t,s)$ for some $T > 0$. If $a \in Y$ so is the unique solution of (2.4.1) and of (2.4.16).*

Proof. The proof that $y \in Y$ is completely parallel to that of Theorem 2.4.1. Then consider (2.4.15) with $y \in Y$ so that $y = p + q$. We have

$$\int_{0}^{t} C_s(t,s)[p(s) + q(s)] ds = \int_{-\infty}^{t} C_s(t,s) p(s) ds - \int_{-\infty}^{0} C_s(t,s) p(s) ds$$
$$+ \int_{0}^{t} C_s(t,s) q(s) ds \in Y.$$

It follows that $y' \in Y$.

2.5 Floquet Theory

Consider once more (1.2.7), (1.2.8), and (1.2.10) which we designate here as

$$x'(t) = A(t) x(t) + \int_{0}^{t} B(t,u) x(u) du + f(t), \tag{2.5.1}$$

Becker's resolvent

$$\frac{\partial}{\partial t} Z(t,s) = A(t) Z(t,s) + \int_{s}^{t} B(t,u) Z(u,s) du, \quad Z(s,s) = I, \tag{2.5.2}$$

2. THE REMARKABLE RESOLVENT

and variation of parameters formula

$$x(t) = Z(t,0)x(0) + \int_0^t Z(t,s)f(s)ds. \tag{2.5.3}$$

We examine the resolvent equation and notice that in $Z(t,s)$ it is the case that s is the initial time. Thus, when we ask that

$$\sup_{t\geq 0} \int_0^t |Z(t,s)|ds < \infty$$

then we are integrating with respect to the initial time and that is, at best, disturbing. If we think of the convolution case with $A(t)$ constant, we note that $Z(t,s) = Z(t-s)$ so that

$$\int_0^t |Z(t,s)|ds = \int_0^t |Z(t-s)|ds = \int_0^t |Z(s)|ds :$$

we are integrating with respect to the real time, not the initial time! For a case close at hand, look back at our Theorem 2.2.5 with Liapunov functional

$$V(t) = |x(t)| + \int_0^t \int_{t-s}^\infty |B(u+s,s)|du|x(s)|ds.$$

We can take $a \equiv 0$, $x(0) = 1$, and get

$$V'(t) \leq -\alpha|x(t)| = -\alpha|Z(t)|$$

so that instantly we have $\int_0^\infty |Z(t)|dt < \infty$. In other words, if our equation were only of convolution type and had A constant then we could dispense with that bounded function $a(t)$ which has caused so much trouble and we would never need to introduce Φ as we did in Theorem 2.1.12 and will do again later for an integrodifferential equation. Equation (1.2.10) would immediately give us x bounded for a' bounded.

Floquet theory shows us that whatever can be proved for a constant coefficient system can be proved for a periodic system. Indeed, there is a Liapunov transformation mapping $x' = A(t)x$ into a constant coefficient system, $y' = Ry$. Specifically, for the matrix system

$$x' = A(t)x, \quad A(t+T) = A(t), \quad \exists T > 0,$$

2.5. FLOQUET THEORY

the principal matrix solution is $Z(t) = P(t)e^{Rt}$ where R is a constant matrix and P is periodic and nonsingular; hence, P and P^{-1} are bounded. The solution of $y' = A(t)y + F(t)$ with bounded F is

$$y(t) = Z(t)y(0) + \int_0^t Z(t)Z^{-1}(s)F(s)ds$$

$$= Z(t)y(0) + \int_0^t P(t)e^{R(t-s)}P^{-1}(s)F(s)ds$$

with the integral bounded by $K \int_0^t |e^{Rs}|ds$ for some K. Again, we are able to get by simply by integrating the real time t instead of the initial time s.

Becker, Burton, and Krisztin (1988) set out to do the same for a Volterra equation

$$x' = A(t)x + \int_0^t B(t,s)x(s)ds \tag{2.5.4}$$

under the assumption that A and B are continuous and there is a $T > 0$ with

$$A(t+T) = A(t), \quad B(t+T, s+T) = B(t,s). \tag{2.5.5}$$

This was shown in Burton (2005b; p. 105) to imply that

$$Z(t+T, s+T) = Z(t,s). \tag{2.5.6}$$

The goal is a Floquet theory for Volterra equations in the sense that we can integrate the real time instead of the initial time. That is, we want to show that

$$\int_s^\infty |Z(t,s)|dt$$

bounded implies that

$$\sup_{t \geq 0} \int_0^t |Z(t,s)|ds < \infty.$$

In particular, if

$$\int_s^\infty |Z(t,s)|dt$$

is bounded for $0 \leq s \leq T$, then certainly

$$\int_0^T \int_s^\infty |Z(t,s)|dt\,ds < \infty$$

and it turns out that this, and mild side conditions on B, does imply that $\sup_{t \geq 0} \int_0^t |Z(t,s)|ds < \infty$. This means that in our struggles in Section 2.2 we can set $a'(t) = 0$, resulting in simple arguments.

2. THE REMARKABLE RESOLVENT

We will see this struggle continue throughout the book. There are solutions to this difficulty in place of the periodicity assumptions. The reader may see several examples of this in Section 2.6. Frequently, an integration of the resolvent equation, followed by change of the order of integration, can bring about exactly the situation we want. New results in this direction would be most welcome and we would expect such efforts to be very rewarding.

In preparation for the main result we first prove a special form of Sobolev's inequality.

If $g : [a, b] \to R^n$ has a continuous derivative, then

$$\int_a^t \left(|g(u)| + (b-a)|g'(u)|\right) du \geq (b-a) \max_{a \leq u \leq b} |g(u)|. \qquad (2.5.7)$$

A simple proof proceeds as follows. Let $u_0, u_1 \in [a, b]$ and $m, M \in R$ be defined by

$$m = \min_{a \leq u \leq b} |g(u)| = |g(u_0)|, \quad M = \max_{a \leq u \leq b} |g(u)| = |g(u_1)|.$$

Then

$$\int_a^b \left(|g(u)| + (b-a)|g'(u)|\right) du$$
$$\geq (b-a)m + (b-a)\left|\int_{u_0}^{u_1} g'(u)\, du\right|$$
$$\geq (b-a)m + (b-a)|g(u_1) - g(u_0)|$$
$$\geq (b-a)m + (b-a)(|g(u_1)| - |g(u_0)|) = (b-a)M.$$

One may verify that (2.5.7) holds for $n \times n$ matrices using the induced matrix norm.

2.5. FLOQUET THEORY

Theorem 2.5.1. Let A and B be continuous, $Z(t,s)$ satisfy (2.5.2), and let (2.5.5) hold.

(i) If there is a $J > 0$ such that

$$\int_s^t |B(u,s)|\, du \leq J \quad \text{for} \quad 0 \leq s \leq t < \infty \tag{2.5.8}$$

and

$$\int_0^T \int_s^\infty |Z(t,s)|\, dt\, ds < \infty, \tag{2.5.9}$$

then

$$\sup_{t \geq 0} \int_0^t |Z(t,s)|\, ds = M \quad \text{for some} \quad M > 0. \tag{2.5.10}$$

(ii) If there is a $K > 0$ such that

$$\int_0^t |B(t,s)|\, ds \leq K \quad \text{for} \quad t \geq 0, \tag{2.5.11}$$

then (2.5.10) implies (2.5.9).

Proof. If we integrate (2.5.2) from s to t, $s \leq t$, we obtain

$$\int_s^t |\partial Z(u,s)/\partial u|\, du$$
$$\leq \|A\| \int_s^t |Z(u,s)|\, du + \int_s^t \int_s^u |B(u,v)|\, |Z(v,s)|\, dv\, du$$
$$\leq (\|A\| + J) \int_s^t |Z(u,s)|\, du,$$

where $\|A\| = \max_{0 \leq t \leq T} |A(t)|$, as may be seen by interchanging the order of integration.

2. THE REMARKABLE RESOLVENT

Let $t > T$ be fixed. There exists an integer $k \geq 1$ and an $\eta \in [0, T)$ with $t = kT + \eta$. Then,

$$\int_0^t |Z(t,s)|\,ds = \int_0^{kT} |Z(kT+\eta, s)|\,ds + \int_{kT}^{kT+\eta} |Z(kT+\eta, s)|\,ds$$

$$= \int_0^{kT} |Z(kT+\eta, s)|\,ds + \int_0^{\eta} |Z(\eta, u)|\,du$$

(using $Z(t+T, s+T) = Z(t,s)$
and a variable change)

$$= \int_0^T \sum_{i=1}^k |Z(iT+\eta, s)|\,ds + \int_0^{\eta} |Z(\eta, u)|\,du$$

(by induction)

$$\leq \alpha + \int_0^T \sum_{i=1}^k \max_{iT \leq u \leq (i+1)T} |Z(u,s)|\,ds,$$

where $\alpha = \sup_{0 \leq u \leq T} \int_0^u |Z(u,s)|\,ds$. Applying (2.5.7), we have

$$\int_0^t |Z(t,s)|\,ds$$

$$\leq \alpha + \int_0^T \sum_{i=1}^k \int_{iT}^{(i+1)T} \left[\frac{1}{T}|Z(u,s)| + \left|\frac{\partial Z(u,s)}{\partial u}\right|\right] du\,ds$$

$$\leq \alpha + \int_0^T \int_s^{\infty} \left[\frac{1}{T}|Z(u,s)| + \left|\frac{\partial Z(u,s)}{\partial u}\right|\right] du\,ds$$

$$\leq \alpha + \left(\|A\| + J + \frac{1}{T}\right) \int_0^T \int_s^{\infty} |Z(u,s)|\,du\,ds.$$

Since t is arbitrary, (2.5.9) implies (2.5.10).

Now assume that (2.5.10) and (2.5.11) hold. In order to prove (2.5.9), by Fubini's theorem and the continuity of $Z(t,s)$, it suffices to show that

$$\int_T^{\infty} \int_0^T |Z(t,s)|\,ds\,dt < \infty.$$

Let

$$r(t) = \int_0^T |Z(t,s)|\,ds \quad \text{for} \quad t \geq T.$$

2.5. FLOQUET THEORY

Then for $t_2 \geq t_1 \geq T$ we have

$$|r(t_2) - r(t_1)| = \left| \int_0^T \left(|Z(t_2, s)| - |Z(t_1, s)| \right) ds \right|$$

$$\leq \int_0^T |Z(t_2, s) - Z(t_1, s)| \, ds$$

$$= \int_0^T \left| \int_{t_1}^{t_2} (\partial Z(t, s)/\partial t) \, dt \right| ds$$

$$\leq \int_0^T \int_{t_1}^{t_2} |\partial Z(t, s)/\partial t| \, dt \, ds.$$

Changing the order of integration yields

$$|r(t_2) - r(t_1)| \leq \int_{t_1}^{t_2} \int_0^T |\partial Z(t, s)/\partial t| \, ds \, dt$$

$$= \int_{t_1}^{t_2} \int_0^T \left| A(t) Z(t, s) + \int_s^t B(t, v) Z(v, s) \, dv \right| ds \, dt$$

$$\leq \int_{t_1}^{t_2} \|A\| \int_0^T |Z(t, s)| \, ds \, dt + \int_{t_1}^{t_2} \int_0^t \int_s^t |B(t, v)| \, |Z(v, s)| \, dv \, ds \, dt$$

$$\leq \int_{t_1}^{t_2} \|A\| \int_0^t |Z(t, s)| \, ds \, dt + \int_{t_1}^{t_2} \int_0^t \int_0^v |Z(v, s)| \, ds |B(t, v)| \, dv \, dt$$

$$\leq \int_{t_1}^{t_2} (\|A\|M + KM) \, dt \leq (\|A\|M + KM)(t_2 - t_1).$$

This shows that $r(t)$ is Lipschitz continuous with Lipschitz constant $L = (\|A\| + K)M$.

Now (2.5.10) and $Z(t + T, s + T) = Z(t, s)$ imply that

$$\sum_{i=1}^{\infty} r(iT + \eta) \leq M$$

for all $\eta \in [0, T)$.

Let $k > 0$ be an integer. It follows from the Lipschitz condition on r that

$$r(iT + \eta + u) \leq r(iT + \eta) + L(T/k)$$

for $i = 1, 2, \ldots, \eta \in [0, T)$, $u \in [0, T/k]$. Thus,

$$\int_T^{kT} r(t)dt = \sum_{i=1}^{k-1}\sum_{j=0}^{k-1} \int_{iT+j(T/k)}^{iT+(j+1)(T/k)} r(t)\,dt$$

$$\leq \sum_{i=1}^{k-1}\sum_{j=0}^{k-1} (T/k)\{r(iT+j[T/k]) + L(T/k)\}$$

$$\leq \left\{(T/k)\sum_{j=0}^{k-1}\sum_{i=1}^{k-1} r(iT+j[T/k])\right\} + (k-1)k(T/k)L(T/k)$$

$$\leq TM + LT^2 < \infty.$$

Since k is arbitrary, $r \in L^1[T, \infty)$ and the proof is complete.

Corollary. Let A and B be continuous, $Z(t, s)$ satisfy (2.5.2), and let (2.5.5) and (2.5.8) hold. If there is an $E > 0$ such that

$$\int_s^\infty |Z(t,s)|\,dt \leq E \quad \text{for all} \quad s \in [0, T], \tag{2.5.12}$$

then

$$\sup_{t \geq 0} \int_0^t |Z(t,s)|\,ds = M \quad \text{for some} \quad M > 0 \tag{2.5.10}$$

is satisfied.

Proof. If we integrate (2.5.12) from 0 to T, the value is bounded by ET.

Theorem 2.5.2. *Suppose that* (2.5.5) *and* (2.5.8) *hold with*

$$\int_0^\infty |Z(t,0)|\,dt < \infty.$$

Then $Z(t, 0) \to 0$ *as* $t \to \infty$.

2.5. FLOQUET THEORY

Proof. We showed in the proof of Theorem 2.5.1 that

$$\int_0^\infty |\partial Z(u,0)/\partial u|\, du \leq (\|A\| + J) \int_0^\infty |Z(u,0)|\, du\,.$$

A similar result holds for each jth column of $Z(t,0)$, say $z(t,0,e_j)$. If the theorem is false, there is a j, an $\epsilon > 0$, and a sequence $\{t_n\} \to \infty$ with $|z(t_n,0,e_j)| \geq \epsilon$. Also,

$$z(t,0,e_j) = e_j + \int_0^t z'(u,0,e_j)\, du$$

so that $t_n \leq t \leq t_n + 1$ implies that

$$|z(t,0,e_j) - z(t_n,0,e_j)| \leq \int_{t_n}^t |z'(u,0,e_j)|\, du < \epsilon/2$$

for large n. Hence $|z(t,0,e_j)| \geq \epsilon/2$ for $t_n \leq t \leq t_n + 1$, contradicting $z(t,0,e_j) \in L^1$. This completes the proof.

Example 2.5.1. *Consider (2.5.1) and suppose that $f(t) \equiv 0$. If there is an $\alpha > 0$ with*

$$A(t) + \int_0^\infty |B(u+t,t)|\, du \leq -\alpha$$

then for $0 \leq s \leq t < \infty$ we have

$$\int_s^t |Z(u,s)|\, du \leq 1/\alpha$$

and the conditions of the corollary are satisfied.

Proof. Becker's resolvent may be written as

$$z'(t) = A(t)z(t) + \int_s^t B(t,u)z(u)\, du, \quad z(s) = 1.$$

Thus, $z(t) = Z(t,s)$. Define the Liapunov functional

$$V(t) = |z(t)| + \int_s^t \int_{t-v}^\infty |B(u+v,v)|\, du\, |z(v)|\, dv$$

2. THE REMARKABLE RESOLVENT

so that

$$V'(t) \leq A(t)|z(t)| + \int_s^t |B(t,u)||z(u)|du$$
$$+ \int_0^\infty |B(u+t,t)|du|z(t)| - \int_s^t |B(t,v)||z(v)|dv$$
$$= \left[A(t) + \int_0^\infty |B(u+t,t)|du\right]|z(t)|$$
$$\leq -\alpha|z(t)|.$$

An integration yields the result.

Here is one of the main applications of Theorem 2.5.1. It is known that there are examples of (2.5.1) which do have periodic solutions; indeed, when $B = 0$ they are very common. But for a general B they are rare. The reason for that is that the right-hand-side of (2.5.1) is generally not periodic even when f is periodic. But (2.5.1) can have solutions which are asymptotically periodic in the sense described below.

Under the conditions of Theorem 2.5.1, the following is shown in Burton (1985; p. 102) or (2005b; p. 105). Suppose that

$$\lim_{n \to \infty} \int_{-nT}^t |B(t,s)|\,ds = \int_{-\infty}^t |B(t,s)|\,ds \qquad (2.5.13)$$

is bounded and continuous in t, that

$$\int_0^t |Z(t,s)|\,ds \leq M \quad \text{for} \quad t \geq 0, \qquad (2.5.14)$$

that $Z(t,0) \to 0$ as $t \to \infty$, and that $f(t+T) = f(t)$. Then there exists a sequence of positive integers $\{n_j\}$ such that the function x defined in (2.5.3) satisfies

$$x(t + n_j T, 0, x_0) \to \int_{-\infty}^t Z(t,s)f(s)\,ds := y(t), \quad j \to \infty, \qquad (2.5.15)$$

where $y(t)$ is a T-periodic solution of

$$\frac{dy(t)}{dt} = A(t)y + \int_{-\infty}^t B(t,s)y(s)\,ds + f(t) \qquad (2.5.16)$$

on $(-\infty, \infty)$.

Example 2.5.1 will lead us to two more sections in which we focus on Liapunov functionals for the resolvent equations themselves instead of on the differential equation.

2.6 Liapunov Functionals for Resolvents

We can watch how Liapunov functions have evolved. With small changes at each step the Liapunov function for

$$x'(t) = -x(t)$$

is applied to

$$x'(t) = -x(t) + x(t-1),$$

then to

$$x'(t) = -x(t) + \int_{t-1}^{t} a(s)x(s)ds,$$

and then to

$$x'(t) = -x(t) + \int_{0}^{t} a(t,s)x(s)ds,$$

with many nonlinearities added along the way. In just the last section we saw the Liapunov functional for

$$x'(t) = A(t)x(t) + \int_{0}^{t} B(t,u)x(u)du$$

being changed minutely to yield properties of the resolvent

$$\frac{\partial}{\partial t}Z(t,s) = A(t)Z(t,s) + \int_{s}^{t} B(t,u)Z(u,s)du, \quad Z(s,s) = I.$$

It was so simple because the main change was only the lower limit on the integral. Turning to integral equations, we have seen these same Liapunov functionals being modified to cover integral equations of the form

$$x(t) = a(t) - \int_{0}^{t} C(t,s)x(s)ds$$

and we have deduced properties of the resolvent, $R(t,s)$, by studying that result together with the variation of parameters formula

$$x(t) = a(t) - \int_{0}^{t} R(t,s)a(s)ds.$$

Thus, the properties of the resolvent were always intimately connected with properties of $a(t)$. We had enormous flexibility. No connection between $a(t)$ and $C(t,s)$ was required.

But the resolvent equation involves only $C(t,s)$, totally independent of any forcing function $a(t)$. Surely, we can get closer to the basic properties of $R(t,s)$ by studying

$$R(t,s) = C(t,s) - \int_s^t R(t,u)C(u,s)du.$$

Here, the forcing function $C(t,s)$ is the same as the kernel. All of our flexibility has vanished. Thus, the resolvent equation seems so different than the resolvent of the integrodifferential equation that we can hardly believe that the old Liapunov functionals can be advanced to it. But they can. And that is the substance of this section. The necessary changes are surprisingly small and the reader will anticipate most of the results as we move along.

In fact, the Liapunov functionals in this case are surprisingly superior to those in the former cases. The ones we obtain here are positive definite with negative definite derivatives (with respect to the quantity $C(t,s) - R(t,s)$). Moreover, we are now in a position to use all the theory developed in Section 8.3 of Burton (2005c) to attack the problem of showing that $C(t,s) - R(t,s)$ tends to zero. While $x(t)$ has so frequently followed $a(t)$ in previous work, here we see that $R(t,s)$ follows $C(t,s)$.

2.6.1 Introduction

We return to the three scalar equations

$$x(t) = a(t) - \int_0^t C(t,s)x(s)ds, \tag{2.6.1}$$

$$R(t,s) = C(t,s) - \int_s^t R(t,u)C(u,s)du, \tag{2.6.2}$$

$$\left(\text{equivalently } R(t,s) = C(t,s) - \int_s^t C(t,u)R(u,s)du\right)$$

and variation of parameters formula

$$x(t) = a(t) - \int_0^t R(t,s)a(s)ds. \tag{2.6.3}$$

We have seen many results supporting the idea that if C is a "nice" kernel, then $x(t)$ follows $a(t)$. Here, we notice that it is also true that $R(t,s)$ follows $C(t,s)$. In fact, for fixed s then $C(t,s)$ is an excellent $L^1[s,\infty)$ approximation to $R(t,s)$ and this should be useful in many problems. The first step in studying this is to note the fundamental property

2.6. LIAPUNOV FUNCTIONALS FOR RESOLVENTS 113

that $\int_s^t |R(u,s)|^p du \leq K \int_s^t |C(u,s)|^p du$ for some positive integer p. The real surprise is that this relation can be obtained either from properties of the derivatives of C with respect to both t and s or from integrals of C with respect to both t and s.

For much of the work here it will more than suffice to ask that a and C be continuous on $[0,\infty)$ and $[0,\infty) \times [0,\infty)$, respectively. However, there will be times when we will indicate certain partial derivatives of C and it will be assumed that these are continuous. The main requirement would be that the Liapunov functional defined in (2.6.10) can be differentiated by Leibnitz's rule.

Our objective is to construct Liapunov functionals for (2.6.2) from which we can deduce properties of $R(t,s)$ so that we will be able to obtain properties of the solution, $x(t)$, from the integral in (2.6.3).

It is clear that we need to know properties of the integral of R with respect to s. And it is difficult because s is the initial time. Thus, it will often be convenient to use our fundamental Theorems 2.1.2 and 2.1.3 which we state here for ready reference.

Theorem 2.6.1.1. *The solution of (2.6.1) is bounded for every bounded continuous function $a(t)$ if and only if*

$$\sup_{t \geq 0} \int_0^t |R(t,s)| ds < \infty. \tag{2.6.4a}$$

Theorem 2.6.1.2. *If there is a number $\alpha < 1$ such that*

$$\sup_{0 \leq t < \infty} \int_0^t |C(t,s)| ds \leq \alpha \tag{2.6.5a}$$

then each solution of (2.6.1) is bounded for every bounded and continuous function a. (See Theorem 2.1.3.)

Thus, (2.6.5a) implies (2.6.4a) and that will be used repeatedly. We will, however, focus on cases in which (2.6.5a) does not hold.

There is a long line of results in integral equation theory in which the objective is to show that the solution is in L^p for some p and there are important problems in which the objective is to study the integral of the solution, $x(t)$, of (2.6.1), as may be seen for example in Feller (1941). This section begins with the idea that if we are willing to work with $\int_0^t x(s) ds$ instead of just $x(t)$, then we can work with the integral of R with respect to t rather than with respect to s.

The reason that is important here is that we can construct Liapunov functionals which will give us direct information about $\int_0^t |R(u,s)| du$, as opposed to the indirect information in Theorem 2.6.1.1. This is possible entirely because Miller (1971a; p. 193 and 200) has shown that

$$\int_s^t R(t,u)C(u,s)du = \int_s^t C(t,u)R(u,s)du$$

so that (2.6.2) can be written as

$$R(t,s) = C(t,s) - \int_s^t C(t,u)R(u,s)du. \tag{2.6.2*}$$

The work will focus on four sets of conditions.

Set #1 asks that there is a positive number α such that

$$C(t,t) - \int_0^t |C_s(t,s)|ds \geq \alpha$$

and

$$C(t,t) + \int_0^t |C_s(t,s)|ds$$

is bounded. This will yield (2.6.4a) and its proof is parallel to the classical proof of Theorem 2.6.1.2. It will also yield $|R(t,s)|$ bounded, a central result in this discussion.

Set #2 asks that there exist an $\alpha < 1$ with

$$\sup_{t \geq 0} \int_0^\infty |C(t+v,t)|dv \leq \alpha \tag{2.6.5b}$$

and yields

$$\sup_{0 \leq s \leq t < \infty} \int_s^t |R(v,s)|dv < \infty, \tag{2.6.4b}$$

so it is a counterpart to (2.6.5a) and (2.6.4a) with corresponding applications, both alone and with (2.6.5a).

Set #3 abandons the integral conditions, asking instead that

$$C(t,s) \geq 0, \quad C_s(t,s) \geq 0,$$
$$C_t(t,s) \leq 0, \quad C_{st}(t,s) \leq 0.$$

Surprisingly, the derivative conditions yield virtually the same as our integral conditions.

Set #4 allows us to "cleanse" the kernel of additive functions of t, frequently constants, asking that $\int_0^\infty |C_t(u+t,t)|du - C(t,t) \leq -\alpha$ for some positive number α, yielding results parallel to those with Set #2.

2.6. LIAPUNOV FUNCTIONALS FOR RESOLVENTS

MOTIVATION

We now introduce some ideas which will be developed in the coming subsections. We will be concerned with cases in which (2.6.4a) holds, or $|R(t,s)|$ is bounded, or $\int_s^t |R(u,s)|^p du$ is bounded. These simple results direct our later investigations. Here are some of the consequences of the properties which we will prove.

Theorem 2.6.1.3. *If there is a number M with $|R(t,s)| \leq M$ for $0 \leq s \leq t < \infty$ and if $a \in L^1[0,\infty)$, then the solution x of (2.6.1) satisfies $|x(t) - a(t)| \leq K$ for some $K > 0$ and all $t \geq 0$.*

This is an immediate consequence of (2.6.3) since

$$|x(t) - a(t)| \leq \int_0^t |R(t,s)||a(s)|ds \leq M \int_0^\infty |a(s)|ds.$$

Thus, we seek conditions under which $|R(t,s)| \leq M$.

We will see that there is a certain duality between a and x, as there is between R and C.

Theorem 2.6.1.4. *If (2.6.4a) holds, if $\int_s^t |R(u,s)|du$ is bounded for $0 \leq s \leq t < \infty$, and if $a \in L^{2^n}[0,\infty)$ for some positive integer n, then $x \in L^{2^n}[0,\infty)$, where x solves (2.6.1).*

Proof. There is an M with $\int_0^t |R(t,s)|ds \leq M$. If $a \in L^2$ then

$$x^2(t) \leq 2\left(a^2(t) + \left(\int_0^t R(t,s)a(s)ds\right)^2\right)$$

$$\leq 2\left(a^2(t) + \int_0^t |R(t,s)|ds \int_0^t |R(t,s)|a^2(s)ds\right)$$

$$\leq 2\left(a^2(t) + M \int_0^t |R(t,s)|a^2(s)ds\right).$$

An integration and change of order of integration yields

$$(1/2)\int_0^t x^2(s)ds \leq \int_0^t a^2(s)ds + M \int_0^t \int_s^t |R(u,s)|du\, a^2(s)ds.$$

This gives the result for $n = 1$. An induction with repeated squaring and inequalities as above yields the first conclusion.

As (2.6.5a) implies (2.6.4a) we seek conditions to ensure that $\int_s^t |R(u,s)|du$ is bounded.

2. THE REMARKABLE RESOLVENT

Theorem 2.6.1.5. *If $a \in L^1[0, \infty)$ and if $\int_s^t R^2(u,s)du$ is bounded for $0 \leq s \leq u \leq t < \infty$, then $\int_0^t (x(s) - a(s))^2 ds$ is bounded.*

Proof. There is an M with $\int_0^\infty |a(u)|du \leq M$ and $\int_s^t R^2(u,s)du \leq M$ so from (2.6.3) we have

$$(x(t) - a(t))^2 = \left(\int_0^t R(t,s)a(s)ds\right)^2$$

$$\leq \int_0^t |a(s)|ds \int_0^t |a(s)|R^2(t,s)ds$$

$$\leq M \int_0^t |a(s)|R^2(t,s)ds.$$

Upon integration and change of order of integration we have

$$\int_0^t (x(s) - a(s))^2 ds \leq M \int_0^t \int_0^u |a(s)|R^2(u,s)ds\,du$$

$$= M \int_0^t \int_s^t R^2(u,s)du\,|a(s)|ds$$

$$\leq M^2 \int_0^t |a(s)|ds \leq M^3.$$

Thus, we seek conditions under which $\int_s^t R^2(u,s)du$ is bounded. See, for example, Theorem 2.6.5.2 for sufficient conditions for this to hold.

This is readily extended. In the same way we can write

$$(x(t) - a(t))^4 \leq M^2 \left(\int_0^t |a(s)|R^2(t,s)ds\right)^2$$

$$\leq M^3 \int_0^t |a(s)|R^4(t,s)ds.$$

Inductively we can obtain the statement that if $a \in L^1[0,\infty)$ and if $\int_s^t R^{2^n}(u,s)du$ is bounded then $\int_0^t (x(s) - a(s))^{2^n} ds$ is bounded.

From the above considerations we see that $a(t)$ is an approximation to $x(t)$. But then we look at (2.6.1) and (2.6.3) with the thought that $C(t,s)$

2.6. LIAPUNOV FUNCTIONALS FOR RESOLVENTS

is surely an approximation to $R(t,s)$. Indeed it is. Two of our results give conditions under which

$$\int_s^t |R(u,s)|du \leq K \int_s^t |C(u,s)|du, \quad K > 0,$$

and

$$\int_s^t R^2(u,s)du \leq \int_s^t C^2(u,s)du.$$

That leads us to the following results.

Theorem 2.6.1.6. *If there are positive constants M_1 and M_2 with*

$$\int_s^t |R(v,s)|dv \leq M_1 \text{ and } \int_s^t |C(u,s)|du \leq M_2$$

for $0 \leq s \leq t < \infty$, then

$$\int_s^t |R(u,s) - C(u,s)|du \leq M_1 M_2.$$

Proof. From (2.6.2) we have

$$\int_s^t |R(u,s) - C(u,s)|du \leq \int_s^t \int_s^v |R(v,u)C(u,s)|dudv$$
$$= \int_s^t \int_u^t |R(v,u)|dv|C(u,s)|du$$
$$\leq M_1 M_2.$$

See, for example, Theorem 2.6.3.1 for sufficient conditions for this result to hold.

That relation can be far better than the obvious one of $M_1 + M_2$. If, for example, $M_2 < 1$ then $M_1 M_2 < M_1$; R and C converge to each other faster than they converge to zero in L^1. There are problems in which we would like to use $C(t,s)$ as an approximation to $R(t,s)$ so we strive to improve the relationship.

Theorem 2.6.1.7. *Suppose there are positive constants M_1, M_2, and M_3 with*

$$\int_s^t R^2(u,s)du \leq M_1, \quad 0 \leq s \leq t < \infty,$$

$$\int_u^t |C(v,u)|dv \leq M_2, \quad 0 \leq u \leq t < \infty,$$

and

$$\int_s^t |C(t,u)|du \leq M_3, \quad 0 \leq s \leq t < \infty.$$

Then $\int_s^t (R(u,s) - C(u,s))^2 du \leq M_1 M_2 M_3$ for $0 \leq s \leq t < \infty$.

Proof. We have

$$\left(R(t,s) - C(t,s)\right)^2 = \left(\int_s^t C(t,u)R(u,s)du\right)^2$$

$$\leq \int_s^t |C(t,u)|du \int_s^t |C(t,u)|R^2(u,s)du$$

so

$$\int_s^t (R(u,s) - C(u,s))^2 du \leq M_3 \int_s^t \int_s^v |C(v,u)|R^2(u,s)dudv$$

$$\leq M_3 \int_s^t \int_u^t |C(v,u)|R^2(u,s)dvdu$$

$$\leq M_3 \int_s^t R^2(u,s) \int_u^t |C(v,u)|dvdu$$

$$\leq M_3 M_2 M_1.$$

This completes the proof.

Thus, we seek conditions under which $\int_s^t R^2(u,s)du$ is bounded.

In Liapunov theory for differential equations we almost always find the derivative of the Liapunov function negative definite so that a solution is integrable in some sense. That integrability is then parlayed into a supremum bound because the Liapunov function is positive definite. The same happens with Liapunov theory for resolvents, but it is so much more

2.6. LIAPUNOV FUNCTIONALS FOR RESOLVENTS

difficult to see. To begin with we will look at a Liapunov functional for (2.6.2) of the form

$$V(t) = \int_s^t \int_{t-w}^\infty |C(u+w,w)|du|R(w,s)|dw$$

and obtain conditions to ensure that

$$V(t) \leq \int_s^t |R(u,s)|du \leq K \int_s^t |C(u,s)|du$$

for some fixed constant K. If that second integral is bounded, so is the first. But how can we parlay that into boundedness of $R(t,s)$? In fact, that Liapunov functional is positive definite with respect to $(R(t,s) - C(t,s))$. The next result is used in Theorem 2.6.3.2.

Theorem 2.6.1.8. *If there is a positive constant β with*

$$\int_{t-w}^\infty |C(u+w,w)|du \geq \beta|C(t,w)|$$

then

$$\beta|R(t,s) - C(t,s)| \leq V(t) := \int_s^t \int_{t-w}^\infty |C(u+w,w)|du|R(w,s)|dw$$

along the solution $R(t,s)$ of (2.6.2).

Proof. We see immediately that

$$V(t) \geq \beta \int_s^t |C(t,w)||R(w,s)|dw$$
$$\geq \beta|R(t,s) - C(t,s)|$$

where the last line follows from (2.6.2).

In the same vein we have a similar result when the Liapunov functional contains a quadratic term.

Theorem 2.6.1.9. *If there are positive constants β and M with*

$$\int_{t-w}^\infty |C(u+w,w)|du \geq \beta|C(t,w)|$$

and

$$\int_s^t |C(t,u)|du \leq M$$

then

$$(\beta/M)(R(t,s) - C(t,s))^2 \leq V(t) := \int_s^t \int_{t-w}^\infty |C(u+w,w)|duR^2(w,s)dw.$$

Proof. We have

$$(R(t,s) - C(t,s))^2 \leq \left(\int_s^t |C(t,w)R(w,s)| dw\right)^2$$

$$\leq \int_s^t |C(t,w)| dw \int_s^t |C(t,w)| R^2(w,s) dw$$

$$\leq M \int_s^t |C(t,w)| R^2(w,s) dw$$

$$= (M/\beta) \int_s^t \beta |C(t,w)| R^2(w,s) dw$$

$$\leq (M/\beta) \int_s^t \int_{t-w}^\infty |C(u+w,w)| du R^2(w,s) dw$$

$$= (M/\beta) V(t).$$

2.6.2 Another Variation of Parameters Formula

We now obtain two related consequences of (2.6.1) and (2.6.3). In Section 2.2 we integrated (2.6.1) by parts obtaining

$$x(t) = a(t) - C(t,t) \int_0^t x(u) du + \int_0^t C_s(t,s) \int_0^s x(u) du\, ds.$$

When $C_s(t,s) := \frac{\partial C(t,s)}{\partial s}$ is continuous we define $y(t) = \int_0^t x(s) ds$ obtaining

$$y'(t) = a(t) - C(t,t) y(t) + \int_0^t C_s(t,s) y(s) ds, \quad y(0) = 0. \quad (2.6.6)$$

Next, integrate the resolvent equation for (2.6.1) which is (2.6.3) and obtain

$$\int_0^t x(s) ds = \int_0^t a(u) du - \int_0^t \int_0^u R(u,s) a(s) ds\, du$$

or

$$\int_0^t x(s) ds = \int_0^t a(u) du - \int_0^t \int_s^t R(u,s) du\, a(s) ds, \quad (2.6.7)$$

which critically integrates R with respect to its first component. In terms of y we have

$$y(t) = \int_0^t a(u) du - \int_0^t \int_s^t R(u,s) du\, a(s) ds. \quad (2.6.8)$$

2.6. LIAPUNOV FUNCTIONALS FOR RESOLVENTS

Notice from (2.6.6) that if

$$|a(t)| + |C(t,t)| + |y(t)| + \int_0^t |C_s(t,s)|ds$$

is bounded, so is $x(t)$.

Notice also that neither $x(t)$ nor $\int_0^t x(s)ds$ depend on being able to differentiate C. Our equation (2.6.6) is merely a help in showing $x(t)$ bounded from $y(t)$ bounded.

Following Equation (2.3.4) we pointed out that an analog of equation (2.6.6) can lead to a very important alternative to (2.6.5a), for it may happen that $C(t,s)$ has an additive constant which is cleansed in (2.6.6). Thus, for example, if $C_s(t,s) \geq 0$ then

$$\int_0^t |C_s(t,s)|ds = \int_0^t C_s(t,s)ds = C(t,s)\Big|_0^t = C(t,t) - C(t,0).$$

It may then happen that $C(t,t) - C(t,0) \geq \alpha > 0$ and that can be most useful.

The next result is a companion to Theorem 2.2.10.

Theorem 2.6.2.1. *Suppose there is an $\alpha > 0$ such that*

$$C(t,t) - \int_0^t |C_s(t,s)|ds \geq \alpha$$

on $[0,\infty)$. Then every solution of (2.6.6) is bounded for every bounded and continuous function a. In particular, then, if in addition we have

$$C(t,t) + \int_0^t |C_s(t,s)|ds$$

bounded then (2.6.4a) holds.

Proof. Let a be bounded and continuous and let y solve (2.6.6) on $[0,\infty)$. Use the Razumikhin function $V(t) = |y(t)|$ and obtain

$$V'_{(2.6.6)}(t) \leq |a(t)| - |C(t,t)||y(t)| + \int_0^t |C_s(t,s)||y(s)|ds.$$

Let $\|a\| = M$ and find $J > 0$ with $\alpha J > M$. If, by way of contradiction, $y(t)$ is not bounded then there is a $t_0 > 0$ with $|y(t)| < J$ on $[0,t_0)$, but $|y(t_0)| = J$ so that $V'(t_0) \geq 0$. But $C(t_0,t_0) > 0$ and

$$V'_{(2.6.6)}(t_0) \leq M - C(t_0,t_0)J + J\int_0^{t_0} |C_s(t_0,s)|ds \leq M - \alpha J < 0,$$

2. THE REMARKABLE RESOLVENT

a contradiction. Now from (2.6.6) we have that $y'(t) = x(t)$ which is bounded for every bounded and continuous $a(t)$. The result now follows from Theorem 2.6.1.2.

Thus, Theorems 2.6.1.2 and 2.6.2.1 give us two simple ways of showing (2.6.4a).

We can follow the idea in the proof of Theorem 2.6.2.1 to show that $R(t,s)$ is bounded. We refer the reader to Theorem 2.6.1.3 for an application.

Theorem 2.6.2.2. *Suppose there is an $\alpha > 0$ and an $M > 0$ with $C(t,t) - \int_s^t |C_u(t,u)|\,du \geq \alpha$ and $|C(t,s)| \leq M$ for $0 \leq s \leq t < \infty$. Choose $J > 0$ so that $\alpha J > M$. Then the solution of (2.6.2) satisfies*

$$|R(t,s)| \leq M + 2C(t,t)J$$

for $0 \leq s \leq t < \infty$. If, in addition, $a \in L^1[0,\infty)$ then in (2.6.3) $x - a$ is bounded.

Proof. Integrate (2.6.2*) by parts obtaining

$$R(t,s) = C(t,s) - C(t,t)\int_s^t R(v,s)\,dv + \int_s^t C_u(t,u)\int_s^u R(v,s)\,dv\,du$$

and write $H(t,s) = \int_s^t R(v,s)\,dv$ so that this last equation can be written as

$$H_t(t,s) = C(t,s) - C(t,t)H(t,s) + \int_s^t C_u(t,u)H(u,s)\,du. \qquad (2.6.9)$$

Fix s and define a Razumikhin function by

$$V(t) = |H(t,s)|$$

so that the derivative of V along a solution of the last equation on $[s,\infty)$ satisfies

$$V'(t) \leq |C(t,s)| - C(t,t)|H(t,s)| + \int_s^t |C_u(t,u)||H(u,s)|\,du.$$

Let $|H(t,s)| < J$ on an interval $[s,t_0)$ and suppose that $|H(t_0,s)| = J$ so that $V'(t_0) \geq 0$. Then

$$V'(t_0) \leq M - |C(t_0,t_0)|J + J\int_s^{t_0}|C_u(t_0,u)|\,du \leq M - \alpha J < 0,$$

2.6. LIAPUNOV FUNCTIONALS FOR RESOLVENTS

a contradiction. Thus, $|H(t,s)| \leq J$ for all $t \geq s$. Notice that $C(t,t) \geq \int_s^t |C_u(t,u)|du$ and so we have

$$|R(t,s)| = |H_t(t,s)| \leq M + 2JC(t,t),$$

as required. Referring now to (2.6.3) the final conclusion follows readily.

2.6.3 A first Liapunov Functional

One of our goals now is to find properties of $\int_s^t R(u,s)du$ for use in (2.6.7). Notice in (i) below that for fixed s, then $C(t,s)$ becomes an $L^1[s,\infty)$ approximation to $R(t,s)$, while (2.6.12) can be used directly in (2.6.7).

Theorem 2.6.3.1. Let $\int_{t-u}^\infty |C(u+v,u)|dv$ be continuous for $0 \leq u \leq t < \infty$. If (2.6.5b) holds then for $V(t)$ defined by

$$V(t) = \int_s^t \int_{t-u}^\infty |C(u+v,u)|dv|R(u,s)|du \qquad (2.6.10)$$

the derivative of V along the unique solution of (2.6.2) satisfies

$$V'(t) \leq -(1-\alpha)|R(t,s)| + |C(t,s)| \qquad (2.6.11)$$

and for $0 \leq s \leq t < \infty$ we have

$$\int_s^t |R(v,s)|dv \leq \frac{1}{1-\alpha}\int_s^t |C(v,s)|dv. \qquad (2.6.12)$$

Thus, if there is a positive constant M with $\int_s^t |C(v,s)|dv \leq M$ for $0 \leq s \leq t < \infty$ then (2.6.4b) holds and also, as in Theorem 2.6.1.6,

(i) $\quad \int_s^t |R(u,s) - C(u,s)|du \leq \frac{M^2}{1-\alpha}.$

If, in addition, $a \in L^1[0,\infty)$ then

(ii) $\quad (x-a) \in L^1[0,\infty), \quad x \in L^1[0,\infty).$

Proof. To prove the result we have from (2.6.10) that

$$V'(t) = \int_0^\infty |C(v+t,t)|dv|R(t,s)| - \int_s^t |C(t,u)||R(u,s)|du$$

$$\leq \alpha|R(t,s)| - \int_s^t |C(t,u)||R(u,s)|du$$

so that

$$V'(t) \leq \alpha|R(t,s)| + |C(t,s)| - |R(t,s)|$$
$$= -(1-\alpha)|R(t,s)| + |C(t,s)|$$

from which (2.6.11) and (2.6.12) follow. Now (i) is just Theorem 2.6.1.6.

Finally, by changing the order of integration we have

$$\int_0^t |x(s) - a(s)| ds \leq \int_0^t \int_s^t |R(u,s)| du |a(s)| ds$$

$$\leq \frac{1}{1-\alpha} \int_0^t \int_s^t |C(u,s)| du |a(s)| ds \leq \frac{1}{1-\alpha} M \int_0^\infty |a(s)| ds.$$

This completes the proof.

Notice that (ii) is a counterpart of the Adam and Eve theorem. It integrates the first coordinate of C and yields $x \in L^1$, whereas the Adam and Eve theorem integrates the second coordinate of C and yields x bounded.

As discussed following (2.6.4a) and (2.6.5a), when we ask that $\int_s^t |C(t,u)| du \leq \alpha < 1$ we will have every solution of (2.6.1) bounded for every bounded and continuous $a(t)$ and we also will have (2.6.4a). But asking this condition does not seem to yield conditions on x for $a \in L^1$. We seem to need to add that $\int_s^t |C(u,s)| du$ is bounded in order to get the needed condition that $\int_s^t |R(u,s)| du$ be bounded.

We saw in Theorem 2.6.1.8 that $\beta |R(t,s) - C(t,s)| \leq V(t)$. We also saw in Theorem 2.6.3.1 that

$$V'(t) \leq -(1-\alpha)|R(t,s)| + |C(t,s)|.$$

If we add the condition that $\int_s^t |C(u,s| du$ is bounded then the next result will put us in a position to use the theory developed in Section 8.3 of Burton (2005c) to show that $R(t,s)$ converges to $C(t,s)$ pointwise for fixed s.

Theorem 2.6.3.2. *Let (2.6.5b) hold and define V by (2.6.10). Suppose that there are $\beta > 0$, $J > 0$ and $K > 0$ with*

$$\int_s^t (1+\beta)|C(u,s)| du \leq J, \ 0 \leq s \leq t < \infty,$$

$$K|R(t,s) - C(t,s)| \leq V(t),$$

and

$$V'(t) \leq -\beta |R(t,s)| + |C(t,s)|.$$

Then for

$$W(t) = [1 + V(t)] e^{-\int_s^t (1+\beta)|C(u,s)| du}$$

we have

$$[1 + K|R(t,s) - C(t,s)|] e^{-J} \leq W(t)$$

and

$$W'(t) \leq -\beta |R(t,s) - C(t,s)| e^{-J}.$$

2.6. LIAPUNOV FUNCTIONALS FOR RESOLVENTS 125

Proof. Now

$$V'(t) \leq -\beta|R(t,s)| + |C(t,s)|$$
$$\leq -\beta|R(t,s) - C(t,s)| + (\beta+1)|C(t,s)|.$$

Clearly the lower bound on W holds. A calculation verifies the stated derivative of W.

Refer to Theorem 2.6.1.3 for consequences of the next result.

Theorem 2.6.3.3. *Let (2.6.5b) hold, let $\int_{t-u}^{\infty} |C(u+v,u)|dv$ be continuous, let $|C(t,s)| \leq M$ for some $M > 0$ and $0 \leq s \leq t < \infty$, and suppose there is a differentiable function $\Phi : [0, \infty) \downarrow (0, \infty)$ with $\Phi \in L^1[0, \infty)$ and*

$$\Phi(t-u) \geq \int_{t-u}^{\infty} |C(u+v,u)|dv$$

for $0 \leq u \leq t < \infty$. If, in addition, there is a $K > 0$ with

$$\int_{t-u}^{\infty} |C(u+v,u)|dv \geq K|C(t,u)|$$

then $|R(t,s)|$ is bounded.

Proof. As (2.6.5b) holds, so does (2.6.11) for (2.6.10). Fix $s \geq 0$. We first show that $V(t)$, as defined in (2.6.10), is bounded. Suppose that t is chosen so that $V(t) = \max_{s \leq u \leq t} V(u)$. Now for such u we have from (2.6.11) that

$$\frac{dV(u)}{du} \leq -(1-\alpha)|R(u,s)| + |C(u,s)|$$

where $\alpha < 1$ and then

$$\frac{dV(u)}{du}\Phi(t-u) \leq -(1-\alpha)|R(u,s)|\Phi(t-u) + M\Phi(t-u).$$

Thus,

$$\int_s^t \frac{dV(u)}{du}\Phi(t-u)du \leq -(1-\alpha)\int_s^t |R(u,s)|\Phi(t-u)du + M\int_s^t \Phi(t-u)du$$

and then since $V(s) = 0$ we have

$$\int_s^t \frac{dV(u)}{du} \Phi(t-u) du = V(u)\Phi(t-u)\Big|_s^t - \int_s^t V(u) \frac{d\Phi(t-u)}{du} du$$

$$= V(t)\Phi(0) - \int_s^t V(u) \frac{d\Phi(t-u)}{du} du$$

$$\geq V(t)\Phi(0) - V(t) \int_s^t \frac{d\Phi(t-u)}{du} du$$

(since $V(t)$ is the maximum and $\frac{d\Phi(t-u)}{du} \geq 0$)

$$= V(t)\Phi(0) - V(t)[\Phi(0) - \Phi(t-s)]$$

$$= V(t)\Phi(t-s) \geq 0.$$

Hence,

$$(1-\alpha)\int_s^t |R(u,s)|\Phi(t-u) du \leq M \int_0^\infty \Phi(u) du.$$

Then

$$(1-\alpha)V(t) = (1-\alpha)\int_s^t |R(u,s)| \int_{t-u}^\infty |C(u+v,u)| dv du$$

$$\leq (1-\alpha)\int_s^t |R(u,s)|\Phi(t-u) du$$

$$\leq M \int_0^\infty \Phi(u) du$$

or $V(t)$ is bounded. But

$$V(t) = \int_s^t |R(u,s)| \int_{t-u}^\infty |C(u+v,u)| dv du$$

$$\geq K \int_s^t |C(t,u)||R(u,s)| du$$

$$\geq K[|R(t,s)| - |C(t,s)|]$$

so that boundedness of $V(t)$ and $|C(t,s)|$ yields the boundedness of $R(t,s)$.

Refer to Theorems 2.6.1.5 and 2.6.1.7 for consequences of the next result. Integral bounds on R^2 such as obtained below can also be used in the alternate form of (2.6.2) with the Schwarz inequality to obtain a pointwise bound on $R(t,s)$ of the type used in Theorem 2.6.1.3.

2.6. LIAPUNOV FUNCTIONALS FOR RESOLVENTS

Theorem 2.6.3.4. *If (2.6.5a) and (2.6.5b) hold and if $\int_{t-u}^{\infty} |C(u+v,u)|dv$ is continuous then there is a constant M with*

$$\int_s^t R^2(u,s)du \leq \frac{M}{1-\alpha} \int_s^t C^2(u,s)du.$$

Proof. Find an $\epsilon > 0$ such that $(1+\epsilon)\alpha = 1$ and then find $M > 0$ such that

$$R^2(t,s) \leq MC^2(t,s) + (1+\epsilon)\left(\int_s^t |C(t,u)R(u,s)|du\right)^2$$

$$\leq MC^2(t,s) + (1+\epsilon)\int_s^t |C(t,u)|du \int_s^t |C(t,u)|R^2(u,s)du$$

$$\leq MC^2(t,s) + \int_s^t |C(t,u)|R^2(u,s)du.$$

Next, define

$$V(t) = \int_s^t \int_{t-u}^{\infty} |C(u+v,u)|dv R^2(u,s)du$$

so that

$$V'(t) = \int_0^{\infty} |C(t+v,t)|dv R^2(t,s) - \int_s^t |C(t,u)|R^2(u,s)du$$
$$\leq \alpha R^2(t,s) + MC^2(t,s) - R^2(t,s)$$
$$= -(1-\alpha)R^2(t,s) + MC^2(t,s).$$

An integration yields the result.

This result is a special case of the next one, but they are worth separating for the following reason. Two things can now be seen. First, if we refer to Theorem 2.6.1.9 we see that we can add conditions to ensure that

$$K(R(t,s) - C(t,s))^2 \leq V(t)$$

for some positive constant K. We could then follow Theorem 2.6.3.2 and construct a new Liapunov functional, W, with the classical properties of Liapunov functions for differential equations.

2. THE REMARKABLE RESOLVENT

Theorem 2.6.3.5. *Let (2.6.5a) and (2.6.5b) hold. Then*

$$\sup_{0 \leq s \leq t < \infty} \int_s^t C^{2^k}(u,s)du < \infty$$

implies that

$$\sup_{0 \leq s \leq t < \infty} \int_0^t R^{2^k}(u,s)du < \infty$$

for any positive integer k.

Proof. This is proved by squaring an inequality in the previous proof to obtain

$$R^4(t,s) \leq M^2 C^4(t,s) + (1+\epsilon)\Big(\int_s^t |C(t,u)|R^2(u,s)du\Big)^2$$

$$\leq M^2 C^4(t,s) + (1+\epsilon)\int_s^t |C(t,u)|du \int_s^t |C(t,u)|R^4(u,s)du$$

$$\leq M^2 C^4(t,s) + \int_s^t |C(t,u)|R^4(u,s)du.$$

Then define

$$V(t) = \int_s^t \int_{t-u}^\infty |C(u+v,u)|dv R^4(u,s)du$$

so that

$$V'(t) = \int_0^\infty |C(t+v,t)|dv R^4(t,s) - \int_s^t |C(t,u)|R^4(u,s)du$$
$$\leq \alpha R^4(t,s) + M^2 C^4(t,s) - R^4(t,s)$$
$$= -(1-\alpha)R^4(t,s) + M^2 C^4(t,s)$$

from which the result follows for $k = 2$. An induction is clear.

In the result below we do not treat the case of $k = 0$, but one may verify that we can conclude also that $x \in L^1[0,\infty)$ under weaker conditions than (2.6.5a). That is, as (2.6.5b) holds, so does (2.6.12). Then ask that $\int_s^t |C(v,s)|dv$ is bounded and refer to (2.6.12) again. Thus, in

$$x(t) = a(t) - \int_0^t R(t,s)a(s)ds$$

we take the absolute value, integrate, and interchange the order of integration and obtain

$$\int_0^t |x(s)|ds \leq \int_0^t |a(s)|ds + \int_0^t \int_s^t |R(u,s)|du|a(s)|ds$$

which is bounded if $a \in L^1[0,\infty)$.

Theorem 2.6.3.6. Let (2.6.5a) and (2.6.5b) hold with $\sup_{0 \leq s \leq t < \infty} \int_s^t C^{2^k}(u,s)du < \infty$ for some positive integer k. If $a \in L^1$ and $a \in L^{2^k}$ then $x \in L^{2^k}$.

Proof. If we square (2.6.3), use the Schwarz inequality, and interchange the order of integration we have

$$x^2(t) \leq 2\left(a^2(t) + \left(\int_0^t R(t,s)a(s)ds\right)^2\right)$$

$$\leq 2\left(a^2(t) + \int_0^t |a(s)|ds \int_0^t |a(s)|R^2(t,s)ds\right).$$

Then by the integrability of a there is an $M > 0$ with

$$(1/2)\int_0^t x^2(s)ds \leq \int_0^t a^2(s)ds + M\int_0^t \int_0^u |a(s)|R^2(u,s)dsdu$$

$$= \int_0^t a^2(s)ds + M\int_0^t \int_s^t R^2(u,s)du|a(s)|ds$$

which is bounded by Theorem 2.6.3.5 when we take $k = 1$. An induction is clear.

2.6.4 Another Liapunov Functional

We will now show that essentially the same results can be obtained by asking conditions on the derivatives of C as were obtained from the integrals of C. The basic assumption here is one of convexity where we ask that

$$C(t,s) \geq 0, \quad C_s(t,s) \geq 0,$$
$$C_t(t,s) \leq 0, \quad C_{st}(t,s) \leq 0. \qquad (2.6.13)$$

2. THE REMARKABLE RESOLVENT

Theorem 2.6.4.1. Let (2.6.13) hold and define

$$V(t) = \int_s^t C_v(t,v) \left(\int_v^t R(u,s)du \right)^2 dv$$

$$+ C(t,s) \left(\int_s^t R(u,s)du \right)^2. \quad (2.6.14)$$

Then along the solution of (2.6.2) we have

$$V'(t) \leq C^2(t,s) - R^2(t,s), \quad 0 \leq s \leq t \quad (2.6.15)$$

so

$$\int_s^t R^2(u,s)du \leq \int_s^t C^2(u,s)du. \quad (2.6.16)$$

If there is an $L > 0$ with $C(t,s) + \int_s^t C_u(t,u)du = C(t,t) \leq L$, $0 \leq s \leq t < \infty$,

(i) then $\dfrac{1}{2L}(R(t,s) - C(t,s))^2 \leq V(t)$.

If there are positive constants M_1, M_2, and M_3 with

$$\int_w^t C(v,w)dv \leq M_1, \quad \int_s^t C(t,u)du \leq M_2, \quad \int_s^t C^2(u,s)du \leq M_3,$$

for $0 \leq s \leq t < \infty$ and $0 \leq w \leq t$ then

(ii) $\displaystyle\int_s^t (R(u,s) - C(u,s))^2 du < M_1 M_2 M_3$.

If $a \in L^1[0,\infty)$ then

(iii) $(x - a) \in L^2[0,\infty)$.

Proof. **Proof of V'.** From (2.6.14) we have

$$V'(t) = \int_s^t C_{vt}(t,v) \left(\int_v^t R(u,s)du \right)^2 dv$$

$$+ C_t(t,s) \left(\int_s^t R(u,s)du \right)^2$$

$$+ 2R(t,s) \int_s^t C_v(t,v) \int_v^t R(u,s)dudv$$

$$+ 2R(t,s)C(t,s) \int_s^t R(u,s)du.$$

2.6. LIAPUNOV FUNCTIONALS FOR RESOLVENTS

If we integrate the next to last term by parts we have

$$2R(t,s)\left[C(t,v)\int_v^t R(u,s)du\Big|_s^t + \int_s^t C(t,v)R(v,s)dv\right]$$

$$= 2R(t,s)\left[-C(t,s)\int_s^t R(u,s)du + \int_s^t C(t,v)R(v,s)dv\right].$$

Hence, taking into account sign conditions we have

$$V'(t) \leq 2R(t,s)\int_s^t C(t,v)R(v,s)dv$$
$$= 2R(t,s)[C(t,s) - R(t,s)] \quad \text{from } (2.6.2^*)$$
$$\leq C^2(t,s) - R^2(t,s) \quad \text{for } s \leq t.$$

Proof of: If there is an $L > 0$ with

$$C(t,s) + \int_s^t C_u(t,u)du \leq L$$

for $0 \leq s \leq t < \infty$ then

(i) $\dfrac{1}{2L}(R(t,s) - C(t,s))^2 \leq V(t).$

Squaring $(2.6.2^*)$ yields

$$\left(R(t,s) - C(t,s)\right)^2 = \left(-\int_s^t C(t,u)R(u,s)du\right)^2$$

$$= \left(C(t,u)\int_u^t R(v,s)dv\Big|_s^t - \int_s^t C_u(t,u)\int_u^t R(v,s)dvdu\right)^2$$

$$= \left(-C(t,s)\int_s^t R(v,s)dv - \int_s^t C_u(t,u)\int_u^t R(v,s)dvdu\right)^2$$

$$\leq 2\left[C^2(t,s)\left(\int_s^t R(v,s)dv\right)^2\right.$$

$$\left. + \int_s^t C_u(t,u)du \int_s^t C_u(t,u)\left(\int_u^t R(v,s)dv\right)^2 du\right]$$

$$\leq 2\left[C(t,s) + \int_s^t C_u(t,u)du\right]V(t)$$

$$= 2C(t,t)V(t) \leq 2LV(t).$$

Notice that this shows how V, itself, is constructed.

Property (ii) is just Theorem 2.6.1.7 when we take (2.6.16) into account.
Property (iii) is just Theorem 2.6.1.5.

REMARK. It is quite remarkable that we obtain (2.6.16) through conditions on the derivatives of C alone, while virtually the identical result is obtained in Theorem 2.6.3.4 through conditions on the integral of C alone.

Theorem 2.6.4.2. *Let (2.6.13) hold and suppose that for fixed $s \geq 0$ there are positive constants J and L with*

$$\sup_{0 \leq s \leq t < \infty} \int_s^t C^2(u,s)du = J \text{ and } C(t,s) + \int_s^t C_u(t,u)du \leq L.$$

Then for $V(t)$ defined in (2.6.14) and for

$$W(t) = [V(t) + 1]e^{-\int_s^t C^2(u,s)du}$$

we have

$$e^{-J}[(1/2L)(R(t,s) - C(t,s))^2 + 1] \leq W(t)$$

and

$$W'(t) \leq -e^{-J}[R^2(t,s) + (R(t,s) - C(t,s))^2].$$

Proof. We have $V(t) \geq (1/2L)(R(t,s) - C(t,s))^2$ so

$$W(t) \geq e^{-J}[(1/2L)(R(t,s) - C(t,s))^2 + 1].$$

Also,

$$\begin{aligned}W'(t) &\leq [V'(t) - C^2(t,s)]e^{-\int_s^t C^2(u,s)du} \\ &\leq [2R(t,s)C(t,s) - 2R^2(t,s) - C^2(t,s)]e^{-\int_s^t C^2(u,s)du} \\ &= [-R^2(t,s) - (R(t,s) - C(t,s))^2]e^{-\int_s^t C^2(u,s)du} \\ &\leq [-R^2(t,s) - (R(t,s) - C(t,s))^2]e^{-J}.\end{aligned}$$

Theorems 2.6.1.6, 2.6.1.7, 2.6.3.1, 2.6.4.1, and 2.6.4.2 now give us conditions under which $C(t,s)$ may be an appropriate approximation to $R(t,s)$.

2.6.5 Large Kernels

So much of our work, both in constructing Liapunov functionals and integrating their derivatives, has involved assumptions of integrability of C with respect to one or both coordinates. If there is a large additive function

2.6. LIAPUNOV FUNCTIONALS FOR RESOLVENTS

of t, possibly a constant, then that is impossible. We solved half of that problem in Section 2.6.2, but we can get around it entirely by differentiating (2.6.2).

Thus, we consider (2.6.2*) and suppose that $C_t(t,s)$ is continuous and write

$$R_t(t,s) = C_t(t,s) - C(t,t)R(t,s) - \int_s^t C_t(t,u)R(u,s)du. \qquad (2.6.17)$$

At times it will be clearer to interchange the notation C_t with C_1.

Theorem 2.6.5.1. *Let $\int_{t-w}^\infty |C_1(u+w,w)|du$ be continuous. Suppose that there is an $\alpha > 0$ with*

$$C(t,t) - \int_0^\infty |C_1(u+t,t)|du \geq \alpha.$$

Then the derivative of

$$V(t) = |R(t,s)| + \int_s^t \int_{t-w}^\infty |C_1(u+w,w)|du|R(w,s)|dw$$

along a solution of (2.6.17) on $[s,\infty)$ satisfies

$$V'(t) \leq -\alpha|R(t,s)| + |C_t(t,s)|.$$

Thus, if, in addition, we have

$$\sup_{0 \leq s \leq t < \infty} \int_s^t |C_1(u,s)|du < \infty$$

and $|C(s,s)|$ is bounded then

$$\sup_{0 \leq s \leq t < \infty} \int_s^t |R(u,s)|du < \infty$$

and $R(t,s)$ is bounded.

Proof. For this function V, fix s and find

$$V'(t) \leq |C_t(t,s)| - C(t,t)|R(t,s)| + \int_s^t |C_t(t,u)R(u,s)|du$$

$$+ \int_0^\infty |C_1(u+t,t)|du|R(t,s)| - \int_s^t |C_t(t,w)||R(w,s)|dw$$

$$\leq |C_t(t,s)| - \alpha|R(t,s)|.$$

But $V(s) = |R(s,s)| = |C(s,s)|$ is bounded so an integration yields the second conclusion. The boundedness of R follows from boundedness of V and $V(t) \geq |R(t,s)|$.

2. THE REMARKABLE RESOLVENT

Theorem 2.6.5.2. *Suppose that $\int_{t-w}^{\infty} |C_1(u+w,w)|du$ is continuous and that there is an $\alpha > 0$ with*

$$2C(t,t) - \int_s^t |C_t(t,u)|du - \int_0^{\infty} |C_1(u+t,t)|du \geq 2\alpha.$$

If

$$\sup_{0 \leq s \leq t < \infty} \int_s^t C_1^2(u,s)du < \infty$$

and $|C(s,s)|$ is bounded then

$$\sup_{0 \leq s \leq t < \infty} \int_s^t R^2(u,s)du < \infty.$$

Also, $R^2(t,s)$ is bounded.

Proof. For fixed s we let $q(t) = R(t,s)$ and define

$$V(t) = q^2(t) + \int_s^t \int_{t-w}^{\infty} |C_t(u+w,w)|du\, q^2(w)dw$$

so that along a solution of (2.6.17) we have

$$V'(t) = 2q(t)C_t(t,s) - 2C(t,t)q^2(t) - 2q(t)\int_s^t C_t(t,u)q(u)du$$

$$+ \int_0^{\infty} |C_t(u+t,t)|du\, q^2(t) - \int_s^t |C_t(t,w)|q^2(w)dw$$

$$\leq 2q(t)C_t(t,s) - 2C(t,t)q^2(t) + \int_s^t |C_t(t,u)|(q^2(t) + q^2(u))du$$

$$+ \int_0^{\infty} |C_t(u+t,t)|du\, q^2(t) - \int_s^t |C_t(t,w)|q^2(w)dw$$

$$= 2q(t)C_t(t,s) + \left[-2C(t,t) + \int_s^t |C_t(t,u)|du\right.$$

$$\left.+ \int_0^{\infty} |C_t(u+t,t)|du\right]q^2(t)$$

$$\leq 2q(t)C_t(t,s) - 2\alpha q^2(t)$$

$$\leq MC_t^2(t,s) - \alpha q^2(t)$$

for some $M > 0$. The conclusion follows from this.

Square both sides of (2.6.2) to see that both sides are integrable with respect to t. We see then that

$$C(t,s) - \int_s^t R(t,u)C(u,s)du$$

is bounded and square integrable. That integral is constructing a copy of $C(t,s)$.

2.7 Integrodifferential Equations

This section is devoted to integrodifferential equations and is parallel to Section 2.2. In several of the theorems it would be possible to simply state that the reader can go back to Section 2.2 and restate those theorems for general functions $A(t)$ and $B(t)$ instead of $C(t,t)$ and $C_t(t,s)$. Instead, we elect to state the correct result and leave the proof as an exercise.

2.7.1 Introduction

In this section we study a scalar equation

$$x'(t) = A(t)x(t) + \int_0^t B(t,s)x(s)ds + a(t) \qquad (2.7.1)$$

where A, B, and a are continuous. There is then Becker's resolvent equation

$$Z_t(t,s) = A(t)Z(t,s) + \int_s^t B(t,u)Z(u,s)du, \quad Z(s,s) = 1, \qquad (2.7.2)$$

and variation of parameters formula

$$x(t) = Z(t,0)x(0) + \int_0^t Z(t,s)a(s)ds. \qquad (2.7.3)$$

We must integrate the resolvent, Z, with respect to the initial time and it will be desirable to show that

$$\sup_{t \geq 0} \int_0^t |Z(t,s)|ds < \infty. \qquad (2.7.4)$$

We see throughout Burton (2005b,c) that it is most useful to have the result that the convolution of an L^1 function with a function tending to zero does, itself, tend to zero. Our first series of results is aimed at presenting an example of a nonconvolution counterpart.

SCHEME

In this section each result is closely connected to the previous one so that an overall scheme is needed.

If A were constant and if B were of convolution type then we could show that Z is of convolution type. Thus, if in addition, $a(t) \to 0$ as $t \to \infty$ and if $\int_0^\infty |Z(u)|du < \infty$ then a classical result would say that in (2.7.3) we have $\int_0^t Z(t-s)a(s)ds \to 0$ as $t \to \infty$. And that is a result which is used seemingly an infinite number of times in both monographs Burton (2005b,c). Our first main result here is to extend that to the case where Z is not of convolution type so that (2.7.4) (together with side conditions) can be substituted for $Z \in L^1$.

As a companion to that result we then give necessary and sufficient conditions for (2.7.4) to hold.

Those necessary and sufficient conditions require that we show that each solution of (2.7.1) is bounded for each bounded and continuous function $a(t)$. And that is hard to prove. We then give several techniques for showing that boundedness. Here, $a(t)$ is an arbitrary bounded and continuous function which makes everything so much more difficult.

Next, to introduce the last section we review Section 2.5 with $A(t+T) = A(t)$ and $B(t+T, s+T) = B(t,s)$ for some $T > 0$ so that those necessary and sufficient conditions need only concern $\int_s^t |Z(u,s)|du$ bounded for $0 \leq s \leq T$; so $a(t)$ is not involved.

Finally, we derive substitute results in the nonperiodic case requiring only that $\int_s^t Z^2(u,s)du$ is bounded. Notice that in both of these cases we are integrating Z with respect to t, another simplification over the integration in (2.7.4) with respect to s.

2.7.2 Uniform Asymptotic Stability

To specify a solution of (2.7.1) we require a $t_0 \geq 0$ and a continuous initial function $\phi : [0, t_0] \to R$. There is then a unique solution $x(t, t_0, \phi)$ satisfying (2.7.1) for $t > t_0$ and with $x(t, t_0, \phi) = \phi(t)$ on $[0, t_0]$. We use the notation $|\phi|_{[0,t_0]} := \sup_{0 \leq t \leq t_0} |\phi(t)|$.

Our main definitions concern the case $a(t) = 0$ so that (2.7.1) becomes

$$x'(t) = A(t)x(t) + \int_0^t B(t,s)x(s)ds. \tag{2.7.5}$$

Definition 2.7.2.1. *The zero solution of* (2.7.5) *is uniformly stable* (US) *if for any $\epsilon > 0$ there exists a $\delta = \delta(\epsilon) > 0$ such that $[t_0 \geq 0, \phi \in C([0,t_0]), |\phi|_{[0,t_0]} < \delta, t \geq t_0]$ imply that $|x(t, t_0, \phi)| < \epsilon$.*

2.7. INTEGRODIFFERENTIAL EQUATIONS

Definition 2.7.2.2. *The zero solution of (2.7.5) is uniformly asymptotically stable (UAS) if it is US and there exists a $\delta_0 > 0$ with the property that for each $\epsilon > 0$ there exists $T = T(\epsilon)$ such that $[t_0 \geq 0, \phi \in C([0,t_0]), |\phi|_{[0,t_0]} < \delta_0, t \geq t_0 + T]$ imply that $|x(t,t_0,\phi)| < \epsilon$.*

Becker (1979) has proved some very important results on UAS. Consider (2.7.5) and the resolvent $Z(t,s)$ defined in (2.7.2). First, define a scalar function

$$d(t,s) := \int_0^s \left| \int_s^t Z(t,\xi) B(\xi,u) d\xi \right| du, \quad 0 \leq s \leq t < \infty.$$

Definition 2.7.2.3. *A continuous function $h(t,s)$ defined for $0 \leq s \leq t < \infty$ is said to promote uniform asymptotic stability if $h(t,s)$ is bounded and if for each $\eta > 0$, there exists a number $T(\eta) > 0$ such that $|h(t+s,s)| \leq \eta$ for all $s \geq 0$ and $t \geq T(\eta)$.*

A main result of Becker (1979; p. 62) is:

Theorem 2.7.2.4. *The zero solution of (2.7.5) is UAS if and only if both $Z(t,s)$ and $d(t,s)$ promote UAS.*

The proposition is important in so many contexts. First, we will quote some results which will show that the zero solution is UAS. This will allow us to say that $Z(t,s)$ promotes UAS. That, in turn, will allow us to obtain a result akin to the classical theorem which states that the convolution of an L^1-function with a function tending to zero does, itself, tend to zero; in that work we use $Z(t,s)$ as the analog of an L^1-function.

There is a major result by Zhang (1997) which shows that (2.7.4) is central to all of our work here. Under very mild assumptions, (2.7.4) characterizes UAS. In the next section we will show how to characterize (2.7.4). Our task then will be to give a variety of conditions under which (2.7.4) holds. The following are Zhang's conditions:

(H_1) $\displaystyle\sup_{t \geq 0}\left(|A(t)| + \int_0^t |B(l,s)| ds\right) < \infty.$

(H_2) for any $\sigma > 0$, there exists an $S = S(\sigma) > 0$ such that

$$\int_0^{t-S} |B(t,u)| du < \sigma \text{ for all } t \geq S.$$

(H_3) $A(t)$ and $B(t, t+s)$ are bounded and uniformly continuous in $(t,s) \in \{(t,s) \in [0,\infty) \times K \mid -t \leq s \leq 0\}$ for any compact set $K \subset (-\infty, 0]$.

Theorem 2.7.2.5. *(Zhang (1997)) Let (H_1), (H_2), (H_3) hold. The zero solution of (2.7.5) is UAS if and only if (2.7.4) holds.*

The chain of reasoning in the next result is as follows. Condition (2.7.4), together with the (H_i), implies UAS. Now UAS implies that $Z(t,s)$ promotes UAS. Thus, (2.7.4) and the property that Z promotes UAS enable us to prove the following convolution-type result.

Theorem 2.7.2.6. *Suppose the resolvent, $Z(t,s)$, satisfies (2.7.4) and promotes UAS. If $\phi : [0,\infty) \to R$ is continuous and satisfies $\phi(t) \to 0$ as $t \to \infty$, then $\int_0^t Z(t,s)\phi(s)ds \to 0$ as $t \to \infty$.*

Proof. By (2.7.4),

$$M := \sup_{t \geq 0} \int_0^t |Z(t,s)|ds < \infty.$$

For a given $\epsilon > 0$ find $L > 0$ with $|\phi(t)| < \epsilon/(2M)$ if $t \geq L$. Then for $t \geq L$, we have

$$\int_L^t |Z(t,s)\phi(s)|ds \leq (\epsilon/(2M)) \int_L^t |Z(t,s)|ds$$

$$\leq (\epsilon/(2M)) \int_0^t |Z(t,s)|ds \leq \epsilon/2.$$

Let $J = \max_{0 \leq t \leq L} |\phi(t)|$. Then

$$\int_0^L |Z(t,s)\phi(s)|ds \leq J \int_0^L |Z(t,s)|ds = J \int_0^L |Z(s+\tau,s)|ds$$

where $\tau := t - s \geq 0$. By Definition 2.7.2.3, we can find $P > 0$ such that $|Z(s+\tau,s)| \leq \epsilon/(2LJ)$ for all $s \geq 0$ and $\tau \geq P$. Now let $T := P + L$. If $t \geq T$, then $\tau \geq P$ because for this integral $0 \leq s \leq L$. Consequently,

$$\int_0^L |Z(t,s)\phi(s)|ds \leq J \int_0^L \frac{\epsilon}{2LJ}ds = \epsilon/2.$$

It follows that

$$\left| \int_0^t Z(t,s)\phi(s)ds \right| \leq \int_0^L |Z(t,s)\phi(s)|ds + \int_L^t |Z(t,s)\phi(s)|ds < \epsilon$$

for $t \geq T$.

This completes the proof.

2.7.3 The Resolvent

In this subsection we offer seven sets of conditions ensuring (2.7.4) and, hence, forming the basis for UAS. This is done by using Liapunov functionals on Volterra equations with bounded continuous forcing functions. In the next section we use one of those Liapunov functionals to obtain a parallel result in the periodic case without that forcing function. It is so much easier without the forcing function. There then emerges the problem of trying to promote all of the following Liapunov functionals to that same framework in which the forcing function is avoided in the periodic case. That problem is stated at the end of the next section.

It turns out that there is a very nice characterization of (2.7.4). We denote by $(\mathcal{BC}, \|\cdot\|)$ the Banach space of bounded continuous functions $\phi : [0, \infty) \to \mathbb{R}$ with the supremum norm.

Theorem 2.7.3.1. *Suppose that $Z(t,0)$ is bounded for $t \geq 0$. Condition (2.7.4) holds if and only if every solution of (2.7.1) on $[0, \infty)$ is bounded for every $a \in \mathcal{BC}$. Moreover, if every solution of (2.7.1) on $[0, \infty)$ is bounded for every $a \in \mathcal{BC}$, then $Z(t,0)$ is bounded and (2.7.4) holds.*

Proof. If $x(t)$ is bounded for all such functions a then so is $\int_0^t Z(t,s)a(s)ds$. By Perron's theorem (1930) (or Burton(2005b; p. 116)) (2.7.4) follows. On the other hand, if (2.7.4) holds and if $Z(t,0)$ is bounded, then the boundedness of x follows for any bounded function a. For the last sentence, take $a(t) \equiv 0$. This completes the proof.

Notice! In this subsection our solutions are always on $[0, \infty)$ so that the solution is $x(t) := x(t, 0, x(0))$. There is no initial function and the solution is differentiable for $t > 0$.

We will now examine four essentially different kinds of proofs showing that every solution of (2.7.1) is bounded for every bounded continuous function $a(t)$. These will range from the pedestrian to the intricate, but they will all illustrate nonconvolution properties. The first is a Razumikhin argument which is very simple and very demanding. But the focus is on the fact that we are integrating B with respect to s. The same result would hold if $B(t,s)$ were replaced by $B(t,s)\lambda(t)$ where $|\lambda(t)| \leq 1$ and there would be no change in the hypotheses or proof.

Theorem 2.7.3.2. *Suppose there is an $\alpha > 0$ such that*

$$\sup_{t \geq 0}\left[A(t) + \int_0^t |B(t,s)|ds\right] \leq -\alpha.$$

Then every solution of (2.7.1) on $[0, \infty)$ is bounded for every $a \in \mathcal{BC}$.

2. THE REMARKABLE RESOLVENT

The proof is almost identical to that of Theorem 2.2.10.

Our next result is based on a contraction mapping and is only one of a class. Here we have $a \in L^\infty$, we use an L^∞ norm, our solution is in L^∞, as are $Z(t,0)$ and $\int_0^t Z(t,s)a(s)ds$. In a similar way, if $a \in L^p$ then we can use a norm with an L^p weight and get our solution and $\int_0^t Z(t,s)a(s)ds$ in L^p. The last result of this section is an example of that process.

Theorem 2.7.3.2 used pointwise conditions, but contractions use averaging conditions. Theorem 2.7.3.2 focused on integration of the second coordinate of $B(t,s)$, but Theorem 2.7.3.5 will focus on integration of both coordinates of B; each will be treated differently. When we use a first order Liapunov functional it will require integration of only the first coordinate of B, but a second order Liapunov functional will require integration of both coordinates. All of these properties are hidden in the convolution case.

Theorem 2.7.3.3. *Suppose there is a $J > 0$ such that $\int_0^t A(u)du \leq J$, $\int_0^t e^{\int_u^t A(s)ds} du \leq J$, and suppose there is an $\alpha < 1$ with*

$$\sup_{t \geq 0} \int_0^t e^{\int_u^t A(s)ds} \int_0^u |B(u,s)| ds\, du < \alpha.$$

Then every solution of (2.7.1) on $[0,\infty)$ is in $\mathcal{BC}$ when $a \in \mathcal{BC}$.

Proof. Use the variation of parameters formula for ordinary differential equations on (2.7.1) and from that define a mapping $P : \mathcal{BC} \to \mathcal{BC}$ by $\phi \in \mathcal{BC}$ implies that

$$(P\phi)(t) = x(0)e^{\int_0^t A(s)ds} + \int_0^t e^{\int_u^t A(s)ds} \left[\int_0^u B(u,s)\phi(s)ds + a(u)\right] du.$$

The reader readily shows it is a contraction.

This result and the next one form what might be called boundaries for a whole set of results of the type which can be found in Burton (2005b; Section 1.6). In the last result we try to derive all the stability from $A(t)$. In the next result we borrow all of $B(t,s)$ to help with the stability. Frequently, we can get nice results by just borrowing some of $B(t,s)$.

Theorem 2.7.3.4. *Let $D(t) := A(t) + \int_0^\infty B(u+t,t)du$ be defined for $t \geq 0$ and suppose there is a $J > 0$ such that $\int_0^t D(u)du \leq J$, $\int_0^t e^{\int_u^t D(s)ds} du \leq J$. Suppose also that there is an $\alpha < 1$ with*

$$\sup_{t \geq 0} \left[\int_0^t \int_{t-s}^\infty |B(v+s,s)| dv\, ds \right.$$
$$\left. + \int_0^t e^{\int_u^t D(s)ds} |D(u)| \int_0^u \int_{u-s}^\infty |B(v+s,s)| dv\, ds\, du\right] \leq \alpha.$$

2.7. INTEGRODIFFERENTIAL EQUATIONS

Then every solution of (2.7.1) on $[0, \infty)$ is in $\mathcal{BC}$ for every $a \in \mathcal{BC}$.

Proof. We can write (2.7.1) as

$$x'(t) = A(t)x(t) + \int_0^\infty B(u+t,t) du\, x(t)$$
$$- \frac{d}{dt} \int_0^t \int_{t-s}^\infty B(u+s,s) du\, x(s)\, ds + a(t)$$
$$= D(t)x(t) - \frac{d}{dt} \int_0^t \int_{t-s}^\infty B(u+s,s) du\, x(s)\, ds + a(t).$$

By the variation of parameters formula followed by integration by parts we have

$x(t)$
$$= x(0) e^{\int_0^t D(s) ds} - \int_0^t e^{\int_u^t D(s) ds} \frac{d}{du} \int_0^u \int_{u-s}^\infty B(v+s,s) dv\, x(s)\, ds\, du$$
$$+ \int_0^t e^{\int_u^t D(s) ds} a(u)\, du$$
$$= x(0) e^{\int_0^t D(s) ds} - e^{\int_u^t D(s) ds} \int_0^u \int_{u-s}^\infty B(v+s,s) dv\, x(s)\, ds \Big|_0^t$$
$$- \int_0^t e^{\int_u^t D(s) ds} D(u) \int_0^u \int_{u-s}^\infty B(v+s,s) dv\, x(s)\, ds\, du$$
$$+ \int_0^t e^{\int_u^t D(s) ds} a(u)\, du$$
$$= x(0) e^{\int_0^t D(s) ds} - \int_0^t \int_{t-s}^\infty B(v+s,s) dv\, x(s)\, ds$$
$$- \int_0^t e^{\int_u^t D(s) ds} D(u) \int_0^u \int_{u-s}^\infty B(v+s,s) dv\, x(s)\, ds\, du$$
$$+ \int_0^t e^{\int_u^t D(s) ds} a(u)\, du.$$

This will define a mapping in the usual way and by the assumptions it will map $\mathcal{BC}$ into $\mathcal{BC}$ and will be a contraction.

The proof of the next result differs in just three places from that of Theorem 2.1.12. As the last step is significantly different, we give the complete proof.

2. THE REMARKABLE RESOLVENT

Theorem 2.7.3.5. *Suppose there is an $\alpha > 0$ with*

$$A(t) + \int_0^\infty |B(u+t,t)|du \leq -\alpha.$$

Suppose also that there is a function $\Phi : [0,\infty) \to (0,\infty)$ which is differentiable, decreasing, and $L^1[0,\infty)$ with

$$\Phi(t-s) \geq \int_{t-s}^\infty |B(u+s,s)|du.$$

If $a \in BC$ so is each solution of (2.7.1) on $[0,\infty)$.

Proof. Define the Liapunov functional

$$V(t,x(\cdot)) = |x(t)| + \int_0^t \int_{t-s}^\infty |B(u+s,s)|du|x(s)|ds$$

with derivative along a solution of (2.7.1) satisfying

$$V'(t,x(\cdot)) \leq A(t)|x(t)| + \int_0^t |B(t,s)x(s)|ds + |a(t)|$$

$$+ \int_0^\infty |B(u+t,t)|du|x(t)| - \int_0^t |B(t,s)x(s)|ds$$

$$\leq -\alpha|x(t)| + |a(t)|.$$

For a fixed solution, we write $V(t,x(\cdot)) = V(t)$. Suppose that there is a $t > 0$ with $V(t) = \max_{0 \leq s \leq t} V(s)$. Then for $0 \leq s \leq t$ we have

$$\frac{dV(s)}{ds} \leq -\alpha|x(s)| + |a(s)|$$

and

$$\frac{dV(s)}{ds}\Phi(t-s) \leq -\alpha|x(s)|\Phi(t-s) + |a(s)|\Phi(t-s).$$

An integration by parts of the left-hand-side yields

$$\int_0^t \frac{dV(s)}{ds}\Phi(t-s)ds = V(s)\Phi(t-s)\Big|_0^t - \int_0^t V(s)\frac{d}{ds}\Phi(t-s)ds$$

$$= V(t)\Phi(0) - V(0)\Phi(t) - \int_0^t V(s)\frac{d}{ds}\Phi(t-s)ds$$

$$\geq V(t)\Phi(0) - V(0)\Phi(t) - V(t)\int_0^t \frac{d}{ds}\Phi(t-s)ds$$

$$= V(t)\Phi(0) - V(0)\Phi(t) - V(t)[\Phi(0) - \Phi(t)]$$

$$= V(t)\Phi(t) - V(0)\Phi(t).$$

2.7. INTEGRODIFFERENTIAL EQUATIONS

Hence, for $|a(t)| \leq a_0$ we then have

$$\begin{aligned}
-|x(0)|\Phi(0) &= -V(0)\Phi(0) \\
&\leq -V(0)\Phi(t) \\
&\leq V(t)\Phi(t) - V(0)\Phi(t) \\
&\leq -\alpha \int_0^t \Phi(t-s)|x(s)|ds + a_0 \int_0^t \Phi(t-s)ds \\
&\leq -\alpha \int_0^t \int_{t-s}^\infty |B(u+s,s)|du|x(s)|ds + a_0 \int_0^\infty \Phi(s)ds.
\end{aligned}$$

From this we see that for that fixed solution there is a positive number K with

$$\int_0^t \int_{t-s}^\infty |B(u+s,s)|du|x(s)|ds \leq K.$$

This means that

$$|x(t)| \leq V(t) \leq |x(t)| + K$$

and

$$V'(t) \leq -\alpha|x(t)| + a_0.$$

From that last pair it follows trivially that there is an $L > 0$ with $|x(t)| \leq L$. Indeed, a simple geometrical argument in the $(V, |x|)$ plane with the lines $V = |x|$ and $V = |x| + K$ shows that $|x(t)| \leq (a_0/\alpha) + K + |x(0)|$ since V' is negative for $|x| > a_0/\alpha$. This completes the proof.

Virtually the same proof would allow us to show that if $a \in L^1[0,\infty)$, then x, $Z(t,0)$, and $\int_0^t Z(t,s)a(s)ds$ are also in $L^1[0,\infty)$. In our next theorem we use a quadratic Liapunov functional and for $a \in L^p$ we can get $x \in L^2$. The next result is interesting in that its main condition averages the two main conditions of Theorems 2.7.3.2 and 2.7.3.5.

Theorem 2.7.3.6. *Suppose there is an $\alpha > 0$ and a $k > 0$ with*

$$\sup_{t \geq 0}\left[2A(t) + \int_0^\infty |B(u+t,t)|du + \int_0^t |B(t,s)|ds\right] + k \leq -\alpha.$$

Suppose also that there is a function $\Phi : [0,\infty) \to (0,\infty)$ which is differentiable, decreasing, and $L^1[0,\infty)$ with

$$\Phi(t-s) \geq \int_{t-s}^\infty |B(u+s,s)|du.$$

If $a \in \mathcal{BC}$, so is the solution of (2.7.1) on $[0,\infty)$.

Proof. For a fixed $a \in \mathcal{BC}$ and a fixed $x(0)$, find a_0 with $a^2(t) \leq a_0$ and define

$$V(t) = x^2(t) + \int_0^t \int_{t-s}^\infty |B(u+s,s)| du\, x^2(s) ds$$

with derivative along the fixed solution of (2.7.1) satisfying

$$V'(t) = 2A(t)x^2(t) + 2x(t)\int_0^t B(t,s)x(s)ds + 2xa(t)$$
$$+ \int_0^\infty |B(u+t,t)| du\, x^2(t) - \int_0^t |B(t,s)| x^2(s) ds$$
$$\leq 2A(t)x^2(t) + \int_0^t |B(t,s)| \left(x^2(t) + x^2(s) \right) ds + kx^2(t)$$
$$+ \frac{1}{k}a^2(t) + \int_0^\infty |B(u+t,t)| du\, x^2(t) - \int_0^t |B(t,s)| x^2(s) ds$$
$$= \left[2A(t) + \int_0^t |B(t,s)| ds + \int_0^\infty |B(u+t,t)| du + k \right] x^2(t) + \frac{a_0}{k}$$
$$\leq -\alpha x^2(t) + a_0/k.$$

We now introduce $\Phi(t-s)$ exactly as in the proof of the last theorem and show that the integral term in V is bounded. We continue and show that $x^2(t)$ is bounded to complete the proof.

The next result and its corollary are proved exactly as in Theorems 2.2.7 and 2.2.8.

Theorem 2.7.3.7. *Suppose that*

$$B(t,s) \leq 0, \quad B_s(t,s) \leq 0, \quad B_{st}(t,s) \geq 0, \quad B_t(t,0) \geq 0.$$

Then the derivative of

$$V(t) = x^2(t) - \int_0^t B_s(t,s) \left(\int_s^t x(u) du \right)^2 ds - B(t,0) \left(\int_0^t x(s) ds \right)^2$$

along any solution of (2.7.1) on $[0, \infty)$ satisfies

$$V'(t) = 2A(t)x^2(t) + 2x(t)a(t) - \int_0^t B_{st}(t,s) \left(\int_s^t x(u) du \right)^2 ds$$
$$- B_t(t,0) \left(\int_0^t x(s) ds \right)^2.$$

Corollary 2.7.3.8. *Let the conditions of Theorem 2.7.3.7 hold and suppose there is an $\alpha > 0$ with $A(t) \leq -\alpha$:*
(i) If $a \in L^2[0,\infty)$, so is any solution of (2.7.1) on $[0,\infty)$, as is $Z(t,0)$ and $\int_0^t Z(t,s)a(s)ds$.
(ii) If $a(t)$ is bounded and if there is an $M > 0$ with

$$\int_0^t |B_s(t,s)|(t-s)^2 ds + t^2|B(t,0)| \leq M$$

then (2.7.4) holds.

Theorem 2.7.3.9. *Let the conditions of Theorem 2.7.3.7 hold. Suppose there is an $\alpha > 0$ and a function $k : [0,\infty) \to (0,\infty)$ with $\int_0^t e^{-\int_u^t k(s)ds} du$ bounded and*

$$2A(t) \leq -k(t) - \alpha, \quad B_{st}(t,s) \geq -k(t)B_s(t,s), \quad B_t(t,0) \geq -k(t)B(t,0).$$

If $a \in \mathcal{BC}$, so is any solution of (2.7.1) on $[0,\infty)$.

Proof. Under these conditions we see that for V defined in Theorem 2.7.3.7 we have

$$V'(t) \leq -k(t)V(t) + (1/\alpha)a^2(t)$$

so that

$$x^2(t) \leq V(t) \leq V(0)e^{-\int_0^t k(s)ds} + (1/\alpha)\int_0^t e^{-\int_u^t k(s)ds} a^2(u)du.$$

and that is bounded by assumption.

In our earlier proofs using Liapunov functionals we could have greatly simplified the proofs by asking a bit more from $A(t)$, using a coefficient larger than 1 in front of the integral, and getting a differential inequality. Such details are left to the reader.

2.7.4 The periodic case

To introduce us to the substitute periodic case of the next section we review the high points of Section 2.5.

We are supposing that

$$B(t+T, s+T) = B(t,s), \quad A(t+T) = A(t), \quad \exists T > 0.$$

Our main result was the following.

2. THE REMARKABLE RESOLVENT

Theorem 2.5.1. Let A and B be continuous, $Z(t,s)$ satisfy (2.5.2), and let (2.5.5) hold.

(i) If there is a $J > 0$ such that

$$\int_s^t |B(u,s)|\, du \leq J \quad \text{for} \quad 0 \leq s \leq t < \infty \tag{2.5.8}$$

and

$$\int_0^T \int_s^\infty |Z(t,s)|\, dt\, ds < \infty, \tag{2.5.9}$$

then

$$\sup_{t \geq 0} \int_0^t |Z(t,s)|\, ds = M \quad \text{for some} \quad M > 0. \tag{2.5.10}$$

(ii) If there is a $K > 0$ such that

$$\int_0^t |B(t,s)|\, ds \leq K \quad \text{for} \quad t \geq 0, \tag{2.5.11}$$

then (2.5.10) implies (2.5.9).

The main utility of the theorem is now stated.

Corollary. Let the conditions of Theorem 2.5.1 hold. If there is an $E > 0$ with

$$\int_s^\infty |Z(t,s)|\, dt \leq E, \quad 0 \leq s \leq T,$$

then

$$\sup_{t \geq 0} \int_0^t |Z(t,s)|\, ds < \infty.$$

With this result we avoided all the difficulties encountered in the last subsection concerning a bounded and continuous function $a(t)$. The function Φ in Theorem 2.7.3.5 is not needed. This was emphasized in our Example 2.5.1 which logically belongs here and will be stated below as Theorem 2.7.5.2.

There is a very nice project here. In the last section we used different Liapunov functionals which would yield $\int_s^\infty |Z(t,s)|^p dt \leq E$ for $0 \leq s \leq T$ and for $p = 2$ when $a(t) = 0$. Can we use those functionals here and change Corollary 4.2 accordingly? In fact, it may be possible to construct a "roundabout theorem" of the type seen in [4; p. 62]. In that case, from the integral condition and properties of the Liapunov functional we can conclude that solutions tend to zero and that may get the result for $p = 1$.

2.7.5 L^p results

If we do not have the periodicity, we can still have some results in the above mentioned direction. Suppose that $a(t) \in L^1[0, \infty)$ and consider the solution of (2.7.1) on $[0, \infty)$ given by

$$x(t) = Z(t, 0)x(0) + \int_0^t Z(t, s)a(s)ds.$$

Could it be that $x \in L^1[0, \infty)$? In the convolution case we would have

$$x(t) = Z(t, 0)x(0) + \int_0^t Z(t-s)a(s)ds$$

and if $Z \in L^1[0, \infty)$, then the convolution of two L^1 functions is an L^1-function. Thus, we would want to show that $Z \in L^1$. But in the nonconvolution case all of this breaks down, while something interesting happens. The easy case is to prove, not (2.7.4), but that there is an $M > 0$ with

$$\sup_{0 \leq s \leq t < \infty} \int_s^\infty |Z(t, s)| dt \leq M. \tag{2.7.6}$$

And that is exactly what is required here.

Theorem 2.7.5.1. *If (2.7.6) holds, then $a \in L^1[0, \infty)$ implies $x \in L^1[0, \infty)$.*

Proof. From the variation of parameters formula we have

$$\int_0^t |x(s)|ds \leq \int_0^t |Z(s,0)x(0)|ds + \int_0^t \int_0^u |Z(u,s)a(s)|dsdu$$

$$= \int_0^t |Z(s,0)x(0)|ds + \int_0^t \int_s^t |Z(u,s)a(s)|duds$$

$$\leq \int_0^t |Z(s,0)x(0)|ds + \int_0^t |a(s)| \int_s^\infty |Z(u,s)|duds$$

$$\leq \int_0^t |Z(s,0)x(0)|ds + M \int_0^t |a(s)|ds.$$

Shortly we will present some results ensuring that (2.7.6) holds. However, this leads us to one of the most tantalizing ideas in all of Volterra equations. We know that if (2.7.4) holds then every solution of (2.7.1) is bounded for every bounded continuous $a(t)$. Using the sequence of inequalities just displayed we can say that if (2.7.6) holds then for every bounded and continuous $a(t)$ it is true that the average of every solution of (2.7.1) with $x(0) = 0$ is bounded:

$$\frac{1}{t}\int_0^t |x(s)|ds \tag{2.7.7}$$

is bounded. This suggests that (2.7.4) and (2.7.6) are very closely related. In many problems it may be that (2.7.7) is as useful as knowing that $x(t)$ is bounded. Here are the details. Note that if (2.7.6) holds and if $|a(t)| \leq K$ for some $K > 0$, then

$$\int_0^t |x(s)|ds \leq \int_0^t |Z(s,0)x(0)|ds + MKt = MKt.$$

We will now give a selection of results proving that (2.7.6) holds, as well as some related forms. Notice how much easier the arguments are when we use the same Liapunov functionals, but take $a(t) = 0$ and strive for (2.7.6) instead of (2.7.4). The natural first result here is Example 2.5.1 which we restate here for reference. Review its proof for the form of the Liapunov functional.

Theorem 2.7.5.2. *Consider (2.7.1) and suppose that $\int_{t-v}^{\infty} |B(u+v,v)|du$ is continuous and that $a(t) \equiv 0$. If there is an $\alpha > 0$ with*

$$\sup_{t \geq 0}\left[A(t) + \int_0^\infty |B(u+t,t)|du\right] \leq -\alpha$$

for $0 \leq s \leq t$, then the resolvent, $Z(t,s)$, satisfies $\int_s^t Z(u,s)du \leq 1/\alpha$ for $0 \leq s \leq t < \infty$.

There is, of course, the parallel result showing the integrability of Z^2 with respect to t.

Theorem 2.7.5.3. *Consider (2.7.1) and suppose that $\int_{t-v}^{\infty} |B(u+v,v)|du$ is continuous and that $a(t) \equiv 0$. If there is an $\alpha > 0$ with*

$$\sup_{t \geq 0}\left[2A(t) + \int_0^\infty |B(u+t,t)|du + \int_s^t |B(t,u)|du\right] \leq -\alpha$$

for $0 \leq s \leq t$, then the resolvent, $Z(t,s)$, satisfies $\int_s^t Z^2(u,s)du \leq 1/\alpha$ for $0 \leq s \leq t < \infty$.

2.7. INTEGRODIFFERENTIAL EQUATIONS

Proof. Use Becker's resolvent equation again and take $z(t) = Z(t, s)$. Define the Liapunov functional

$$V(t) = z^2(t) + \int_s^t \int_{t-v}^\infty |B(u+v, v)| du z^2(v) dv$$

with derivative satisfying

$$V'(t) = 2A(t)z^2(t) + 2z(t) \int_s^t B(t, u) z(u) du$$

$$+ \int_0^\infty |B(u+t, t)| du z^2(t) - \int_s^t |B(t, v)| z^2(v) dv$$

$$\leq 2A(t)z^2(t) + \int_s^t |B(t, u)| \left(z^2(u) + z^2(t) \right) du$$

$$+ \int_0^\infty |B(u+t, t)| du z^2(t) - \int_s^t |B(t, v)| z^2(v) dv$$

$$= \left[2A(t) + \int_0^\infty |B(u+t, t)| du + \int_s^t |B(t, u)| du \right] z^2(t)$$

$$\leq -\alpha z^2(t).$$

Hence,

$$0 \leq V(t) \leq V(s) - \alpha \int_s^t z^2(u) du = 1 - \alpha \int_s^t z^2(u) du$$

from which the result follows.

The interested reader could borrow $B(t, s)$ and form $A(t) + \int_t^\infty B(u, t) du$ in place of $A(t)$ to get a stability result. Details, which are quite lengthy, may be found in Burton (2005c; p. 146).

The reader will recognize that we have already revisited the Liapunov functional in Theorem 2.7.3.5 in our work in Theorem 2.7.4.1 where we integrated Z with respect to t. A side condition there which was not mentioned was that Z is bounded. This means that the integral of Z^2 with respect to t is bounded.

Next, we revisit the Liapunov functional of Theorem 2.7.3.7. Here, the differential inequality needed in the proof of Theorem 2.7.3.9 is totally unnecessary.

2. THE REMARKABLE RESOLVENT

Theorem 2.7.5.4. Consider (2.7.1) with $a(t) \equiv 0$. Suppose there is an $\alpha > 0$ with

$$B(t,s) \leq 0, \ B_s(t,s) \leq 0, \ B_{st}(t,s) \geq 0, \ B_t(t,s) \geq 0, \ 2A(t) \leq -\alpha.$$

Then the derivative of

$$V(t) = z^2(t) - \int_s^t B_u(t,u) \left(\int_u^t z(v)dv \right)^2 du - B(t,s) \left(\int_s^t z(u)du \right)^2$$

along the solution $Z(t,s)$ with $Z(s,s) = 1$ of the resolvent equation

$$z'(t) = A(t)z(t) + \int_s^t B(t,u)z(u)du$$

satisfies

$$V'(t) \leq 2A(t)z^2(t) \leq -\alpha Z^2(t,s).$$

Thus,

$$0 \leq V(t) \leq 1 - \alpha \int_s^t Z^2(u,s)du$$

and

$$\int_s^t Z^2(u,s)du \leq 1/\alpha$$

for $0 \leq s \leq t < \infty$.

Proof. We have

$$V'(t) = 2A(t)z^2(t) + 2z(t)\int_s^t B(t,u)z(u)du$$

$$- \int_s^t B_{ut}(t,u) \left(\int_u^t z(v)dv \right)^2 du$$

$$- 2z(t) \int_s^t B_u(t,u) \int_u^t z(v)dv\,du$$

$$- B_t(t,s) \left(\int_s^t z(u)du \right)^2 - 2z(t)B(t,s)\int_s^t z(u)du.$$

We integrate the third-to-last term by parts and obtain

$$-2z(t)\left[B(t,u) \int_u^t z(v)dv \Big|_s^t + \int_s^t B(t,u)z(u)du \right]$$

$$= -2z(t)\left[-B(t,s) \int_s^t z(v)dv + \int_s^t B(t,u)z(u)du \right].$$

2.7. INTEGRODIFFERENTIAL EQUATIONS

Cancel terms and obtain

$$V'(t) = 2A(t)z^2(t) - \int_s^t B_{ut}(t,u)\left(\int_u^t z(v)dv\right)^2 du$$
$$- B_t(t,s)\left(\int_s^t z(u)du\right)^2.$$

The conclusion follows from this.

Corollary 2.7.5.5. *Let the conditions of Theorem 2.7.5.4 hold and suppose that there is a continuous function $k : [0, \infty) \to [0, \infty)$ with $k(t) \leq \alpha$,*

$$-B_{ut}(t,u) \leq k(t)B_u(t,u)$$

and

$$-B_t(t,s) \leq k(t)B(t,s).$$

Then for V defined in Theorem 2.7.5.4 we have $V' \leq -k(t)V$ and $Z^2(t,s) \leq Z^2(s,s)e^{-\int_s^t k(u)du}$.

We turn now to the fixed point methods of Theorem 2.7.3.3, set $a(t) = 0$, and develop a weighted norm which will yield exponential decay on $Z(t, s)$. The reader will recognize that there are many alternatives to our presentation. Basically, one chooses a norm of the form $|\phi|_r = \sup_{t \geq s} |\phi(t)|/r(t)$ where $r : [0, \infty) \to (0, \infty)$ and $r \in L^2[0, \infty)$. The task then is to show that a certain mapping, which we define below, will map the resulting space into itself and be a contraction. It is a much more flexible technique than some of our Liapunov arguments because it relies on averages instead of pointwise conditions. It is often the case that exponential decay of a solution requires exponential decay of the kernel. Here, we do require such decay on the kernel.

Theorem 2.7.5.6. *Suppose there is a $\gamma > 0$ and an $\alpha < 1$ with $A(v) + \gamma \leq 0$ and*

$$\sup_{0 \leq s \leq t < \infty} \int_s^t e^{\int_u^t [A(v)+\gamma]dv} \int_s^u |B(u,v)e^{\gamma(u-v)}|dv\,du \leq \alpha.$$

Then the unique solution, $Z(t, s)$, of

$$z'(t) = A(t)z(t) + \int_s^t B(t,u)z(u)du, \quad z(s) = 1,$$

tends to zero exponentially.

Proof. Fix $s \geq 0$ and define a norm on the Banach space $(U, |\cdot|_\gamma)$ of continuous functions $\phi : [s, \infty) \to \mathbb{R}$ for which

$$\sup_{t \geq s} |\phi(t)| e^{\gamma(t-s)} =: |\phi|_\gamma$$

exists.

By the variation-of-parameters formula we have

$$z(t) = e^{\int_s^t A(u)du} + \int_s^t e^{\int_u^t A(v)dv} \int_s^u B(u,v) z(v) dv du$$

which we use in the usual way to define a mapping P on U. Then, if $\phi \in U$ we have

$$(P\phi)(t) e^{\gamma(t-s)} = e^{\int_s^t [A(u)+\gamma]du}$$
$$+ \int_s^t e^{\int_u^t [A(v)+\gamma]dv} \int_s^u B(u,v) e^{\gamma(u-v)} \phi(v) e^{\gamma(v-s)} dv du$$

and

$$|(P\phi)(t) e^{\gamma(t-s)}|$$
$$\leq 1 + \int_s^t e^{\int_u^t [A(v)+\gamma]dv} \int_s^u |B(u,v) e^{\gamma(u-v)}| |\phi(v)| e^{\gamma(v-s)} dv du$$

so that $P\phi \in U$. Next, if $\phi, \eta \in U$ then

$$|(P\phi)(t) - (P\eta)(t)| e^{\gamma(t-s)}$$
$$\leq \int_s^t e^{\int_u^t [A(v)+\gamma]dv} \int_s^u |B(u,v) e^{\gamma(u-v)}| |\phi(v) - \eta(v)| e^{\gamma(v-s)} dv du$$
$$\leq \alpha |\phi - \eta|_\gamma$$

a contraction. Thus, there is a fixed point and it is in U, completing the proof.

2.8 A Nonlinear Application

Virtually every result in this entire chapter can be used directly to obtain qualitative properties of solutions of the nonlinear scalar equation

$$x(t) = a(t) - \int_0^t C(t,s) [x(s) + G(s, x(s))] ds \tag{2.8.1}$$

2.8. A NONLINEAR APPLICATION

having the linear part

$$y(t) = a(t) - \int_0^t C(t,s)y(s)ds \tag{2.8.2}$$

with two forms of the resolvent equation

$$R(t,s) = C(t,s) - \int_s^t R(t,u)C(u,s)du$$

$$= C(t,s) - \int_s^t C(t,u)R(u,s)du \tag{2.8.3}$$

generating the variation of parameters formula

$$y(t) = a(t) - \int_0^t R(t,s)a(s)ds. \tag{2.8.4}$$

It is assumed that $a : [0,\infty) \to \Re$, $C : [0,\infty) \times [0,\infty) \to \Re$, and $G : [0,\infty) \times \Re \to \Re$ are all continuous. This will ensure that (2.8.1) has a solution, while (2.8.2) and (2.8.3) have unique solutions.

It is recognized that we have not proved an existence result for (2.8.1). That will be done in the next chapter. However, it will be a significant distraction to stop the work of Chapter 2 and prove an existence result before coming to these applications.

As an appendix at the end of this section we will show that we can write (2.8.1) with another variation of parameters formula as

$$x(t) = a(t) - \int_0^t R(t,u)a(u)du - \int_0^t R(t,u)G(u,x(u))du$$

or

$$x(t) = y(t) - \int_0^t R(t,s)G(s,x(s))ds \tag{2.8.5}$$

where y is the solution of (2.8.2).

We can study (2.8.2) and (2.8.3) separately and apply our conclusions to (2.8.5) which will then be much simpler than (2.8.1). Our goal is to give conditions on a, C, and G to ensure that the solution of (2.8.1) is either bounded, is in $L^p[0,\infty)$, or tends to zero as $t \to \infty$. We ask a variant of

$$|G(t,x)| \le k_1|x| + h(t)|x|^{1/p} \tag{2.8.6}$$

where $p > 1$, $q > 1$, $k_1 > 0$, with $(1/p) + (1/q) = 1$, and $h \in L^q[0,\infty)$.

2. THE REMARKABLE RESOLVENT

Linear preliminaries

We will now go back through Chapter 2 and select out a few of the results which we will need here, including those which yield

$$y \in L^q[0, \infty) \text{ for some } q = 1, 2, 4, ..., 2^r, ..., \infty. \tag{2.8.7}$$

We have mentioned several times that a classical idea is that the solution, y, follows the forcing function, $a(t)$. For example, if $a \in L^p$, so is y. Looking at (2.8.5) we see that we are striving for a nonlinear counterpart: if $y \in L^q$, so is x.

The next two results will be fundamental reminders. The first is Theorem 2.1.3(iii) and is a result of Islam and Neugebauer (2008).

Lemma 2.8.1. *If there is an $\alpha \in (0,1)$ with*

$$\int_0^t |C(t,s)|ds \leq \alpha, \ 0 \leq t < \infty, \tag{2.8.8}$$

then

$$\int_0^t |R(t,s)|ds \leq \frac{\alpha}{1-\alpha}. \tag{2.8.9}$$

Take the absolute values in the first choice in (2.8.3), integrate from 0 to t with respect to s, and finish by interchanging the order of integration.

There is a symmetric result obtained by using the second choice in (2.8.3) and integrating with respect to t; this will yield the first part of (2.8.11) below. It was also seen in Theorem 2.6.3.1 and (2.6.12).

Lemma 2.8.2. *If there is a $\beta \in (0,1)$ with*

$$\int_s^t |C(u,s)|du \leq \beta, \ 0 \leq s \leq t < \infty \tag{2.8.10}$$

then

$$\int_0^\infty |C(v+t,t)|dv \leq \beta \tag{2.8.10*}$$

and

$$\int_s^t |R(u,s)|du \leq \frac{\beta}{1-\beta} \text{ and } \int_0^\infty |R(v+t,t)|dv \leq \frac{\beta}{1-\beta}. \tag{2.8.11}$$

Moreover, if (2.8.10) holds, if $a \in L^1[0,\infty)$, and if $\int_{t-u}^\infty |C(v+u,u)|dv$ is continuous, then $y \in L^1[0,\infty)$.

2.8. A NONLINEAR APPLICATION

Proof. We have indicated how to get the first part of (2.8.11). From (2.8.10) we have $\int_s^\infty |C(u,s)|du \le \beta$ so $\int_0^\infty |C(u+s,s)|du \le \beta$. The last inequality holds for s replaced by t.

In the same way if (2.8.10) holds and we have the first part of (2.8.11) then

$$\int_0^\infty |R(v+t,t)|dv = \int_t^\infty |R(w,t)|dw \le \frac{\beta}{1-\beta}.$$

To prove the last sentence of the lemma, define

$$V(t) = \int_0^t \int_{t-u}^\infty |C(v+u,u)|dv|y(u)|du$$

so that along the solution of (2.8.2) we have

$$V'(t) = \int_0^\infty |C(v+t,t)|dv|y(t)| - \int_0^t |C(t,u)y(u)|du$$
$$\le \beta|y(t)| - |y(t)| + |a(t)|.$$

An integration yields the result since $V(t) \ge 0$ and $\beta < 1$. This completes the proof.

Corollary 2.8.1. *Let (2.8.10) hold. If $a \in L^1[0,\infty)$ then*

$$\int_0^t R(t,s)a(s)ds \in L^1[0,\infty). \tag{2.8.12}$$

Proof. Note that

$$-\int_0^t R(t,s)a(s)ds = y(t) - a(t) \in L^1[0,\infty).$$

There is a companion to that last result which will be needed, found in Theorem 2.1.3(ii).

Proposition 2.8.1. *If (2.8.8) holds, if $a(t) \to 0$ as $t \to \infty$, and if for each $T > 0$ then*

$$\int_0^T |C(t,s)|ds \to 0 \text{ as } t \to \infty, \tag{2.8.13}$$

then both

$$\int_0^t R(t,s)a(s)ds \text{ and } \int_0^T |R(t,s)|ds \to 0 \text{ as } t \to \infty. \tag{2.8.14}$$

Finally, we remark that there are numerous conditions on C which are known to imply that there is a positive constant K such that

$$|R(t,s)| \leq K, \ 0 \leq s \leq t < \infty. \tag{2.8.15}$$

See Section 2.6, particularly Theorems 2.6.3.2, 2.6.3.3, 2.6.4.1, 2.6.5.1.

The nonlinear problem

We begin with two parallel results.

Theorem 2.8.1. *Let(2.8.8) and (2.8.13) hold and suppose there is a continuous function* $h : [0,\infty) \to [0,\infty)$ *with*

$$|G(t,x)| \leq h(t)|x|, \ h(t) \to 0 \text{ as } t \to \infty. \tag{2.8.16}$$

If the solution of (2.8.2), y, is bounded so is x, the solution of (2.8.5).

Proof. Conditions (2.8.8) and (2.8.13) imply (2.8.14) This means that there is a $\gamma < 1$ and $T > 0$ with

$$\int_0^t |R(t,s)|h(s)ds \leq \gamma, \ t \geq T.$$

Let $|y(t)| \leq Y$ for some $Y > 0$. If x is not bounded, then there is a $q > T$ with $|x(t)| \leq |x(q)|$ for $0 \leq t < q$ which yields

$$|x(q)| \leq Y + |x(q)| \int_0^q |R(q,s)|h(s)ds \leq Y + \gamma|x(q)|$$

so $|x(q)| \leq Y/(1-\gamma)$ will be the bound on x for $t \geq q$. As x is continuous, the proof is complete.

Lemma 2.8.3. *Let (2.8.10) hold so that (2.8.12) also holds and let* $y \in L^1[0,\infty)$. *If*

$$|G(t,x)| \leq h(t)|x|, \ h(t) \in L^1[0,\infty), \tag{2.8.17}$$

and if the solution of (2.8.5) is bounded, then it is also in $L^1[0,\infty)$.

Proof. If x is bounded, then $|G(t,x)| \leq h(t)|x| \in L^1[0,\infty)$ so

$$\int_0^t R(t,s)G(s,x(s))ds \in L^1[0,\infty)$$

and the result follows.

2.8. A NONLINEAR APPLICATION

Theorem 2.8.2. *Let (2.8.10), (2.8.15), and (2.8.17) hold and let $y(t)$ be bounded and in $L^1[0, \infty)$. Suppose that there is an $M > 0$ with*

$$|G(t, x_1) - G(t, x_2)| \leq Mh(t)|x_1 - x_2|. \tag{2.8.18}$$

Then the solution of (2.8.5) is bounded and, hence, is in $L^1[0, \infty)$.

Proof. Let $(X, |\cdot|_h)$ be the Banach space of bounded continuous functions $\phi : [0, \infty) \to \Re$ where

$$|\phi|_h := \sup_{t \geq 0} |\phi(t)| e^{-\int_0^t (MK+1)h(s)ds}.$$

Then define $P : X \to X$ by $\phi \in X$ implies that

$$(P\phi)(t) = y(t) - \int_0^t R(t, s)G(s, \phi(s))ds$$

so

$$|(P\phi)(t)| \leq |y(t)| + \left| \int_0^t R(t, s)G(s, \phi(s))ds \right|$$

is bounded, as in the proof of Lemma 2.8.3. To see that P is a contraction, if $\phi, \eta \in X$ then

$$|(P\phi)(t) - (P\eta)(t)| e^{-\int_0^t (MK+1)h(s)ds}$$

$$\leq \int_0^t |R(t, s)| e^{-\int_0^t (MK+1)h(s)ds} Mh(s)|\phi(s) - \eta(s)| ds$$

$$\leq \int_0^t MKe^{-\int_s^t (MK+1)h(u)du} e^{-\int_0^s (MK+1)h(u)du} h(s)|\phi(s) - \eta(s)| ds$$

$$\leq \frac{MK}{MK+1} |\phi - \eta|_h e^{-\int_s^t (MK+1)h(u)du} \Big|_0^t$$

$$\leq \frac{MK}{MK+1} |\phi - \eta|_h,$$

completing the proof.

By changing (2.8.17) we can avoid (2.8.18). The reader may wish to review Lemma 2.8.2 and (2.8.10), (2.8.10*), and (2.8.11). If (2.8.10) holds then we can define

$$V(t) = \int_0^t \int_{t-u}^{\infty} |R(v+u, u)| dv |G(u, x(u))| du \tag{2.8.19}$$

and find that

$$V'(t) = \int_0^\infty |R(v+t,t)|dv|G(t,x)| - \int_0^t |R(t,u)G(u,x(u))|du$$

so that

$$V'(t) \leq \frac{\beta}{1-\beta}|G(t,x)| - |x| + |y|. \tag{2.8.20}$$

Theorem 2.8.3. *Let (2.8.10) hold, let $p > 1$, $q > 1$, $(1/p) + (1/q) = 1$, and suppose there is a function $\lambda : [0, \infty) \to [0, \infty)$, $\lambda \in L^q[0, \infty)$, and with $k_1 > 0$ so that*

$$|G(t,x)| \leq k_1|x| + \lambda(t)|x|^{1/p} \tag{2.8.21}$$

and

$$k_2 := k_1 \frac{\beta}{1-\beta} < 1. \tag{2.8.22}$$

If $y \in L^1[0, \infty)$, so is x.

Proof. From (2.8.20) and (2.8.21) we have

$$V'(t) \leq \frac{\beta}{1-\beta}[k_1|x| + \lambda(t)|x|^{1/p}] - |x| + |y|$$

$$= k_2|x| + \frac{\beta}{1-\beta}\lambda(t)|x|^{1/p} - |x| + |y|$$

$$= -(1-k_2)|x| + |y| + \frac{\beta}{1-\beta}\lambda(t)|x|^{1/p}.$$

An integration yields

$$(1-k_2)\int_0^t |x(s)|ds \leq \int_0^t |y(s)|ds + \frac{\beta}{1-\beta}\left(\int_0^t \lambda^q(s)ds\right)^{1/q}\left(\int_0^t |x(s)|ds\right)^{1/p}.$$

There are then positive constants U and L with

$$\int_0^t |x(s)|ds \leq U + L\left(\int_0^t |x(s)|ds\right)^{1/p},$$

placing a bound on $\int_0^t |x(s)|ds$. This completes the proof.

We have proved a collection of results showing that if $a \in L^{2^n}$ or if $a' \in L^{2^n}$ then $y \in L^{2^n}$. Again, it would be a distraction to repeat those here. We will assume that $y \in L^{2^n}$ and give conditions to ensure that $x \in L^{2^n}$.

2.8. A NONLINEAR APPLICATION

Lemma 2.8.4. *Let (2.8.8) hold and let $\epsilon > 0$ be given. Then there is a positive constant M so that if*

$$\gamma := \frac{\alpha(1+\epsilon)}{1-\alpha} \qquad (2.8.23)$$

then

$$x^2 \leq My^2 + \gamma \int_0^t |R(t,s)|G^2(s,x(s))ds. \qquad (2.8.24)$$

Proof. If we square both sides of (2.8.5) we can readily find M with

$$\begin{aligned}
x^2 &\leq My^2 + (1+\epsilon)\left(\int_0^t R(t,s)G(s,x(s))ds\right)^2 \\
&\leq My^2 + (1+\epsilon)\int_0^t |R(t,s)|ds \int_0^t |R(t,s)|G^2(s,x(s))ds \\
&\leq My^2 + \frac{\alpha(1+\epsilon)}{1-\alpha}\int_0^t |R(t,s)|G^2(s,x(s))ds \\
&= My^2 + \gamma \int_0^t |R(t,s)|G^2(s,x(s))ds,
\end{aligned}$$

as required.

Assume that (2.8.10) holds and for γ defined in (2.8.23) define the Liapunov functional

$$V(t) = \gamma \int_0^t \int_{t-u}^\infty |R(u+v,u)|dv G^2(u,x(u))du. \qquad (2.8.25)$$

and review (2.8.11) to see that $\int_0^\infty |R(v+t,t)|dv \leq \beta/(1-\beta)$.

Thus, the derivative of V satisfies

$$V'(t) = \gamma \int_0^\infty |R(v+t,t)|dv G^2(t,x(t)) - \gamma \int_0^t |R(t,u)|G^2(u,x(u))du$$

so that

$$V'(t) \leq \frac{\gamma\beta}{1-\beta}G^2(t,x(t)) + My^2(t) - x^2(t). \qquad (2.8.26)$$

2. THE REMARKABLE RESOLVENT

Theorem 2.8.4. Let (2.8.8) and (2.8.10) hold, let $p > 1$, $q > 1$, $(1/p) + (1/q) = 1$, and let $y \in L^2[0, \infty)$. Suppose there is a function $\lambda \in L^q[0, \infty)$ and an $\eta > 0$ with

$$\frac{\gamma\eta\beta}{1-\beta} =: \mu < 1 \qquad (2.8.27)$$

and

$$G^2(t, x) \leq \eta x^2 + \lambda(t)|x|^{2/p}. \qquad (2.8.28)$$

Then $x \in L^2[0, \infty)$. If, in addition, y and $C(t, s)$ are bounded, so is x.

Proof. From (2.8.26) and (2.8.27) we have

$$V'(t) \leq \mu x^2(t) + \frac{\mu\lambda(t)}{\eta}|x(t)|^{2/p} + My^2(t) - x^2(t) \qquad (2.8.29)$$

so that

$$(1-\mu)\int_0^t x^2(s)ds \leq \frac{\mu}{\eta}\left(\int_0^t \lambda^q(s)ds\right)^{1/q}\left(\int_0^t x^2(s)ds\right)^{1/p} + M\int_0^t y^2(s)ds.$$

This places a bound on $\int_0^t x^2(s)ds$. If C is bounded, use (2.8.11) in (2.8.3) to see that R is bounded. If, in addition, y is bounded then use the last relation in (2.8.5) to see that x is bounded.

We could continue to square (2.8.5) and (2.8.28) and obtain $x \in L^{2^n}$ when $y \in L^{2^n}$.

RESEARCH PROBLEM. Notice that we need $\beta < 1$, but allow $\beta/(1-\beta) > 1$. In the aforementioned work of Islam and Neugebauer (2008), they ask for an $\alpha < 1$ with $\int_0^t |C(t,s)|ds \leq \alpha$ and obtain $\int_0^t |R(t,s)|ds \leq \alpha/(1-\alpha)$. For $\alpha/(1-\alpha) < 1$, they use contractions to get boundedness when $|G(t,x)| \leq k|x|$. It would be very interesting to extend that work to allow $\alpha/(1-\alpha) > 1$, using a different condition on $G(t,x)$.

We have been working with small C. Now we turn to possibly large C. Differentiate (2.8.3) to obtain

$$R_t(t, s) = C_t(t, s) - C(t, t)R(t, s) - \int_s^t C_t(t, u)R(u, s)du$$

and let $q(t) = R(t, s)$ for fixed s and write

$$q' = C_t(t, s) - C(t, t)q(t) - \int_s^t C_t(t, u)q(u)du.$$

2.8. A NONLINEAR APPLICATION

Theorem 2.8.5. *Let $\int_{t-u}^{\infty} |C_1(v+u, u)| dv$ be continuous and suppose there is a constant $\beta > 0$ with*

$$C(t,t) - \int_0^{\infty} |C_1(v+t,t)| dv \geq \beta. \tag{2.8.30}$$

Then

$$|R(t,s)| + \beta \int_s^t |R(u,s)| du \leq |C(s,s)| + \int_s^t |C_1(u,s)| du. \tag{2.8.31}$$

Proof. Use

$$V(t) = |q(t)| + \int_s^t \int_{t-u}^{\infty} |C_1(v+u, u)| dv |q(u)| du,$$

note that $V(s) = |C(s,s)|$, and obtain

$$V'(t) \leq |C_t(t,s)| - C(t,t)|q(t)| + \int_s^t |C_t(t,u)q(u)| du$$
$$+ \int_0^{\infty} |C_1(v+t,t)| dv |q(t)| - \int_s^t |C_1(t,u)q(u)| du$$
$$\leq |C_t(t,s)| - \beta |q(t)| = |C_t(t,s)| - \beta |R(t,s)|$$

or

$$|R(t,s)| \leq V(t) \leq V(s) + \int_s^t |C_t(u,s)| du - \beta \int_s^t |R(u,s)| du,$$

yielding the result.

Corollary 2.8.2. *Let the conditions of Theorem 2.8.5 hold and suppose there exists $M > 0$ with*

$$|C(s,s)| + \int_s^t |C_1(u,s)| du \leq M. \tag{2.8.32}$$

Then

$$|R(t,s)| + \beta \int_s^t |R(u,s)| du \leq M. \tag{2.8.33}$$

If, in addition, $a \in L^1[0, \infty)$ then the unique solution y of (2.8.2) is in $L^1[0, \infty)$; moreover, if $a \in L^1$ and bounded, so is y.

Proof. Since $|R(t,s)|$ is bounded and $a \in L^1$ and bounded, y is bounded by (2.8.4). If $\int_0^t |a(u)|du \leq A$ then

$$\int_0^t |y(u)|du \leq \int_0^t |a(u)|du + \int_0^t \int_0^u |R(u,s)a(s)|ds\,du$$

$$= \int_0^t |a(u)|du + \int_0^t \int_s^t |R(u,s)|du|a(s)|ds$$

$$\leq A + (M/\beta)A,$$

as required.

Corollary 2.8.3. *Let the conditions of Theorem 2.8.5 and (2.8.32) hold. Suppose there are constants $p > 1$, $q > 1$, with $(1/p) + (1/q) = 1$ and a function $h \in L^q[0,\infty)$ with*

$$|G(t,x)| \leq h(t)|x|^{1/p}.$$

Then the solution of (2.8.5) satisfies $x \in L^1[0,\infty)$. If, in addition, $a(t)$ is bounded then x is bounded.

Proof. Notice that

$$\int_0^\infty |R(v+t,t)|dv = \int_t^\infty |R(u,t)|du \leq M/\beta$$

by (2.8.33). Thus, from the derivative of

$$V(t) = \int_0^t \int_{t-u}^\infty |R(v+u,u)|dv|G(u,x(u))|du$$

we have

$$V'(t) \leq (M/\beta)|G(t,x)| + |y| - |x|$$
$$\leq (M/\beta)h(t)|x|^{1/p} + |y| - |x|.$$

An integration and use of Hölder's inequality will give $x \in L^1$ since Corollary 2.8.2 yields $y \in L^1$. But Corollary 2.8.2 also gives y bounded and R bounded so boundedness of x will now readily follow from (2.8.5).

We have given two ways of showing $\int_0^\infty |R(v+t,t)|dv$ bounded and there are many ways to show $y \in L^1$. Every result in Sections 2.6.3, 2.6.4, and 2.6.5 with R^{2n} integrable can be used with Hölder's inequality in the second choice in (2.8.3) to yield R bounded. We now offer a result which allows flexibility for other methods.

2.8. A NONLINEAR APPLICATION

Theorem 2.8.6. *Let $y \in L^1[0, \infty)$ satisfy (2.8.2) and suppose that there is a constant μ with $\int_0^\infty |R(v+t,t)|dv \leq \mu$. If there is a constant $k > 0$ with $|G(t,x)| \leq k|x|$ and $k\mu < 1$ then the solution x of (2.8.5) is in $L^1[0,\infty)$. If, in addition, y is bounded and $R(t,s)$ is bounded then x is bounded.*

Proof. Define
$$V(t) = \int_0^t \int_{t-u}^\infty |R(v+u,u)|dv|G(u,x(u))|du$$
and obtain
$$V'(t) = \int_0^\infty |R(v+t,t)|dv|G(t,x)| - \int_0^t |R(t,u)G(u,x(u))|du$$
$$\leq \mu k|x| - |x| + |y|.$$

This yields $x \in L^1[0,\infty)$. Next, using these conditions in (2.8.5) yields x bounded.

Remark. Our work is stated for scalar equations, but much of it is valid also for systems. Miller's work in the next subsection is definitely for systems so one needs to be careful about commuting terms. If we go back to the early parts of the chapter we see that some of our Liapunov functionals used here also work for systems. That is true for the contraction result, as well. It fails in all the cases where we square the equation. It would be very interesting to advance Miller's variation of parameters formula to the case of $G(t,s,x(s))$.

2.8.1 The Nonlinear Variation of Parameters

Here, we follow the presentation of Miller (1971a; pp. 190-2). Given a linear system
$$X(t) = f(t) - \int_0^t C(t,s)X(s)ds, \tag{2.8.34}$$
we can write
$$X(t) = f(t) - \int_0^t R(t,u)f(u)du \tag{2.8.35}$$
and we suppose that properties of X and R are known. Then consider
$$x(t) = f(t) - \int_0^t C(t,s)[x(s) + G(s,x(s))]ds$$
$$=: F(t) - \int_0^t C(t,s)x(s)ds \tag{2.8.36}$$

where

$$F(t) := f(t) - \int_0^t C(t,s)G(s,x(s))ds.$$

Thus, we apply (2.6.3) to (2.8.36) and obtain

$$\begin{aligned} x(t) &= F(t) - \int_0^t R(t,s)F(s)ds \\ &= \{f(t) - \int_0^t C(t,s)G(s,x(s))ds\} - \int_0^t R(t,s)f(s)ds \\ &\quad + \int_0^t R(t,s) \int_0^s C(s,u)G(u,x(u))du\, ds \\ &= \{f(t) - \int_0^t R(t,s)f(s)ds\} - \int_0^t C(t,s)G(s,x(s))ds \\ &\quad + \int_0^t \{\int_u^t R(t,s)C(s,u)ds\}G(u,x(u))du \\ &= \{f(t) - \int_0^t R(t,s)f(s)ds\} \\ &\quad - \int_0^t \{C(t,u) - \int_u^t R(t,s)C(s,u)ds\}G(u,x(u))du. \end{aligned}$$

Using (2.6.2) and (2.8.35) yields

$$\begin{aligned} x(t) &= X(t) - \int_0^t R(t,u)G(u,x(u))du \\ &= f(t) - \int_0^t R(t,u)[f(u) + G(u,x(u))]du \end{aligned} \quad (2.8.37)$$

which Miller calls the variation of constants form of (2.8.36).

2.9 Notes

All of the chapter was strongly motivated by the work of Miller displayed in subsection 2.8.1. The material in the first four sections of this chapter is taken from various parts of Burton (2006a) and (2007a,b,c). That in (2006a) was published first in the indicated journal which is published by the University of Szeged, Hungary; it formed the basis of most of the material to follow. That in (2007a,b) was published first in the indicated journal by Elsevier. The material in (2007c) was first published in the indicated journal by the University of Baia Mare, Romania. The material in Section 5 was taken mainly from Becker, Burton, and Krisztin (1988). The material in Section 2.8.1 is taken directly from indicated work of Miller. That in Section 2.7 rests entirely on the indicated work of Becker and Zhang.

Chapter 3

Existence Properties

3.1 A Global Existence Theorem

In this section we assume continuity of the functions in the integral equation and show that there is global existence of solutions if either the growth of the functions involved is not too fast or if there is a certain type of Liapunov function.

We begin by reviewing some of the history concerning an ordinary differential equation

$$x' = f(t, x) \tag{3.1.1}$$

where $f : [0, \infty) \times \Re^n \to \Re^n$ is continuous. Then we construct a global existence theorem for

$$x(t) = a(t) + \int_0^t D(t, s, x(s))ds \tag{3.1.2}$$

where $D : [0, \infty) \times [0, \infty) \times \Re^n \to \Re^n$ and $a : [0, \infty) \to \Re^n$ are both continuous. If we extend the domain of D to $-\infty < s \leq t < \infty$, that will contain results for the infinite delay equation

$$x(t) = b(t) - \int_{-\infty}^t D(t, s)g(s, x(s))ds \tag{3.1.3}$$

where b, D, and g are continuous, while $xg(t,x) \geq 0$. To specify a solution of (3.1.3) we require a continuous initial function $\varphi : (-\infty, 0] \to R$ such that

$$a(t) := b(t) - \int_{-\infty}^0 D(t,s)g(s, \varphi(s))ds \text{ is continuous}$$

3.1. A GLOBAL EXISTENCE THEOREM

so that

$$x(t) = a(t) - \int_0^t D(t,s)g(s,x(s))ds$$

is essentially of the form of (3.1.2) when φ is chosen so that $\varphi(0) = a(0)$. Fulfillment of that condition will be prominent in Chapter 7.

Existence theory for (3.1.1) usually rests on limiting arguments with ε-approximate solutions or on careful application of Schauder's fixed point theorem after constructing an appropriate set and a mapping of that set into itself; this is usually an intricate and tedious task. Schaefer's theorem is mainly Schauder's theorem followed by a simple retract argument. Its great advantage over Schauder's theorem is that a self-mapping set need not be found.

The reader will find standard developments of existence theory for (3.1.1) in any treatment of differential equations such as Hartman (1964). Existence theory for (3.1.2) is found in the books of Corduneanu (1991), Gripenberg, Londen, and Staffans (1990), and Miller (1971a), while such results for both (3.1.2) and (3.1.3) may be found in Burton (2005 b,c), for example.

We next consider some background and motivation. Classical existence theory for (3.1.1) begins with a local result. It is shown that for each $(t_0, x_0) \in [0, \infty) \times R^n$, there is at least one solution $x(t) = x(t, t_0, x_0)$ with $x(t_0, t_0, x_0) = x_0$ and satisfying (3.1.1) on an interval $[t_0, t_1]$, where t_1 is computed from a bound on f in a closed neighborhood of (t_0, x_0). This yields a new point $(t_1, x(t_1))$ and we begin once more computing bounds on f in a closed neighborhood of $(t_1, x(t_1))$ and obtain a continuation of the solution on $[t_1, t_2]$.

If we continue this process on intervals $\{[t_n, t_{n+1}]\}$, can we say that $t_n \to \infty$ as $n \to \infty$? It turns out that we can unless there is an α such that $|x(t)|$ tends to infinity as t tends to α from the left. Either implicitly or explicitly we invoke Zorn's lemma to claim that there is a solution on $[t_0, \infty)$ or one on $[t_0, \alpha)$ which can not be continued to α. Some authors call a solution on $[t_0, \infty)$ noncontinuable. We do not; for our purposes, if a solution is defined on $[t_0, \alpha)$ with $\alpha < \infty$, and if it can not be extended to α, it is said to be noncontinuable. An example of the latter case is

$$x' = x^2, \quad (t_0, x_0) = (0, 1)$$

which has the solution $x(t) = \frac{1}{1-t}$ on $[0,1)$ and is noncontinuable. In fact, two more examples complete the range of possibilities. Solutions of $x' = -x^3$ are all continuable to $+\infty$ because they are bounded, even though the right-hand-side grows faster than in the first example. Finally, solutions of $x' = t^3 x \ell n(1 + |x|)$ are unbounded, but continuable to $+\infty$ because the right-hand-side does not grow too fast.

We come then to the question of how to rule out noncontinuable solutions of the kind mentioned above. In principle, there is a fine way of doing so. Kato and Strauss (1967) prove that it always works.

Definition 3.1.1. *A continuous function $V : [0, \infty) \times R^n \to [0, \infty)$ which is locally Lipschitz in x is said to be mildly unbounded if for each $T > 0$, $\lim_{|x| \to \infty} V(t, x) = \infty$ uniformly for $0 \leq t \leq T$.*

Notice that there is no mention of the differential equation or its solution. That will necessarily change when we advance the theory to integral equations.

If there is a mildly unbounded V which is differentiable, then we invoke the local existence theory and consider a solution $x(t)$ of (3.1.1) on $[t_0, \alpha)$ so that $V(t, x(t))$ is an unknown but well-defined function. The chain rule then gives

$$\frac{dV}{dt}(t, x(t)) = \sum_{i=1}^{n} \frac{\partial V}{\partial x_i} \frac{dx_i}{dt} + \frac{\partial V}{\partial t}$$

$$= \text{grad } V \cdot f + \frac{\partial V}{\partial t}.$$

We can also compute V' when V is only locally Lipschitz in x (cf. Yoshizawa (1966; p. 3)) and we will display such an example in a moment; in that case, one uses the upper right-hand derivative.

If V is so shrewdly chosen that it is mildly unbounded and $V' \leq 0$, then there can be no $\alpha < \infty$ with $\lim_{t \to \alpha-} |x(t)| = \infty$ because $V(t, x(t)) \leq V(t_0, x_0)$.

There is a converse theorem: If f is continuous and locally Lipschitz in x for each fixed t, then Kato and Strauss (1967) show that there is a mildly unbounded V with $V' \leq 0$ if and only if all solutions can be continued for all future time. Their result is not constructive, but investigators have constructed suitable V for many important systems without any growth condition on f. In the example $x' = -x^3$ mentioned above, $V = x^2$ yields $V' = -2x^4 \leq 0$, showing global existence.

These remarks for (3.1.1) apply in large measure to (3.1.2). In those cases we require a functional $V(t, x(\cdot))$. More importantly, as mentioned above, for (3.1.1) the only way a solution can fail to be continuable to $+\infty$ is for there to exist an α with $\lim_{t \to \alpha-} |x(t)| = +\infty$; but for (3.1.2) we must take the limit supremum.

Wintner derived conditions on the growth of f to ensure that solutions of (3.1.1) could be continued to $+\infty$ and Conti used these to construct a

3.1. A GLOBAL EXISTENCE THEOREM

suitable V. These results are most accessible in Hartman (1964; pp. 29–30) for the Wintner condition and Sansone and Conti (1964; p. 6) for V. A proof here will show how it works. The interested reader should also consult Kato and Strauss (1967) concerning the concept of mildly unbounded.

Theorem 3.1.1 (Conti-Wintner). *If there are continuous functions $\Gamma : [0, \infty) \to [0, \infty)$ and $W : [0, \infty) \to [1, \infty)$ with*

$$|f(t,x)| \leq \Gamma(t) W(|x|) \text{ and } \int_0^\infty \frac{ds}{W(s)} = \infty, \text{ then}$$

$$V(t,x) = \left\{ \int_0^{|x|} \frac{ds}{W(s)} + 1 \right\} \exp - \int_0^t \Gamma(s) ds$$

is mildly unbounded and $V'(t, x(t)) \leq 0$ along any solution of (3.1.1) and that solution can be continued on $[t_0, \infty)$.

Proof. Let $x(t)$ be a noncontinuable solution of (3.1.1) on $[t_0, \alpha)$. By examining the difference quotient we see that $|x(t)|' \leq |x'(t)|$. Thus,

$$V'(t,x(t)) \leq \frac{|x(t)|'}{W(|x(t)|)} \left[\exp - \int_0^t \Gamma(s) ds \right] - \Gamma(t) V(t, x(t)) \leq 0$$

when we use $|x(t)|' \leq |x'(t)| \leq \Gamma(t) W(|x(t)|)$. This means that $V(t, x(t)) \leq V(t_0, x_0)$; since V is mildly unbounded, $\lim_{t \to \alpha^-} |x(t)| \neq \infty$. This completes the proof.

Remark 3.1.1. Notice here and throughout the section that we only used $|f(t,x)| \leq \Gamma(t) W(|x|)$ for $0 \leq t \leq \alpha$. In many problems it turns out that the inequality fails for t greater than some α, but holds for all smaller α. In such cases we have proved that the solution can be continued to the latter α and that may be all that is needed.

The main result of this section is based on the following modified theorem of Schaefer (1955) which is discussed and proved also in Smart (1980; p. 29). Schaefer's result was for a locally convex topological vector space.

Theorem 3.1.2 (Schaefer). *Let $(C, \|\cdot\|)$ be a normed space, H a continuous mapping of C into C which is compact on each bounded subset of C. Then either*

(i) *the equation $x = \lambda H x$ has a solution for $\lambda = 1$, or*

(ii) *the set of all such solutions x, for $0 < \lambda < 1$, is unbounded.*

Definition 3.1.2. *Let $\{f_n(t)\}$ be a sequence of functions from an interval $[a, b]$ to real numbers.*

(a) $\{f_n(t)\}$ is uniformly bounded on $[a,b]$ if there exists M such that

$$\left[n \text{ a positive integer and } t \in [a,b]\right]$$

imply $|f_n(t)| \leq M$.

(b) $\{f_n(t)\}$ is equicontinuous if for any $\varepsilon > 0$ there exists $\delta > 0$ such that

$$\left[n \text{ a positive integer, } t_1 \in [a,b], t_2 \in [a,b], \text{ and } |t_1 - t_2| < \delta\right]$$

imply $|f_n(t_1) - f_n(t_2)| < \varepsilon$.

Theorem 3.1.3 (Ascoli-Arzela). *If $\{f_n(t)\}$ is a uniformly bounded and equicontinuous sequence of real functions on an interval $[a,b]$, then there is a subsequence that converges uniformly on $[a,b]$ to a continuous function.*

The lemma is, of course, also true for vector functions. Suppose that $\{\mathbf{F}_n(t)\}$ is a sequence of functions from $[a,b]$ to R^p, for instance, $\mathbf{F}_n(t) = (f_n(t)_1, \ldots, f_n(t)_p)$. [The sequence $\{\mathbf{F}_n(t)\}$ is uniformly bounded and equicontinuous if all the $\{f_n(t)_j\}$ are.] Pick a uniformly convergent subsequence $\{f_{kj}(t)_1\}$ using the lemma. Consider $\{f_{kj}(t)_2\}$ and use the lemma to obtain a uniformly convergent subsequence $\{f_{kjr}(t)_2\}$. Continue and conclude that $\{\mathbf{F}_{kjr\cdots s}(t)\}$ is uniformly convergent.

We now turn to (3.1.2) and our main result for this section and in preparation we refer to Definition 3.1.1. There, the function V is mildly unbounded without any reference to (3.1.2). A typical example would be $V(t,x) = x^2/(t+1)$. Now things are going to change radically. For example, in (3.1.2) we will prepare to use Schaefer's theorem, introduce λ, and write

$$x(t) = \lambda\left[a(t) - \int_0^t D(t,s)g(s,x(s))ds\right]$$
$$=: \lambda(Hx)(t),$$
$$0 \leq \lambda \leq 1, \tag{3.1.2$_\lambda$}$$

retaining all of the continuity conditions.

We will see this several times in Chapter 5 and significant preparation is needed. First, we require a continuous function M with

$$-2g(t,x)[x - \lambda a(t)] \leq M(t),$$
$$g(t,x) \text{ bounded for } t \geq 0 \text{ if } x \text{ is bounded}. \tag{3.1.4}$$

But the defining property of this example is that the kernel is convex:

$$D(t,s) \geq 0, \quad D_s(t,s) \geq 0, \quad D_{st}(t,s) \leq 0, \quad D_t(t,0) \leq 0. \tag{3.1.5}$$

3.1. A GLOBAL EXISTENCE THEOREM

Now, define

$$V(\lambda, t, x(\cdot)) = \left\{ \int_0^t D_s(t,s) \left(\int_s^t \lambda g(v, x(v)) dv \right)^2 ds \right.$$
$$\left. + D(t,0) \left(\int_0^t \lambda g(v, x(v)) dv \right)^2 + 1 \right\} \exp - \int_0^t M(s) ds. \quad (3.1.6)$$

It seems very clear that this function is not necessarily mildly unbounded as it stands. But we will see that it definitely is mildly unbounded along a solution and that is all that is needed. To avoid too much repetition, we ask the reader to refer back to the proof of Theorem 2.1.10 in the linear case and forward to Theorem 5.2.1 for this nonlinear case. Here are the steps: Differentiate V; integrate by parts the term in V' obtained from differentiating the inner integral in the first integral in V; substitute $\lambda a(t) - x(t)$ from $(3.1.2_\lambda)$ into the expression just obtained.

Since $D_{st} \leq 0$ if there is a solution of $(3.1.2_\lambda)$ then we will have

$$V'(\lambda, t, x(\cdot)) \leq \{-2\lambda g(t, x(t))[x(t) - \lambda a(t)]] - M(t)\} e^{-\int_0^t M(s) ds} \quad (3.1.7)$$

and this is not positive by the assumptions on $M(t)$ and the fact that $xg(t,x) \geq 0$. Hence

$$V(\lambda, t, x(\cdot)) \leq V(\lambda, 0, x(\cdot)) = 1, \quad (3.1.8)$$

a bound on V along that supposed solution, and the bound is uniform in λ.

We now show that V is mildly unbounded along that supposed solution. We have

$$(\lambda a(t) - x(t))^2 = \left(\int_0^t \lambda D(t,s) g(s, x(s)) ds \right)^2$$

(from $(3.1.2_\lambda)$)

$$= \left(D(t,0) \int_0^t \lambda g(v, x(v)) dv + \int_0^t D_s(t,s) \int_s^t \lambda g(v, x(v)) dv \, ds \right)^2$$

(upon integration by parts)

$$\leq 2 \int_0^t D_s(t,s) ds \int_0^t D_s(t,s) \left(\int_s^t \lambda g(v, x(v)) dv \right)^2 ds$$
$$+ 2 D^2(t,0) \left(\int_0^t \lambda g(v, x(v)) dv \right)^2$$

(by Schwarz's inequality)

$$\leq 2[D(t,0) + D(t,t) - D(t,0)]V(\lambda, t, x(\cdot))e^{\int_0^t M(s)ds}$$
$$= 2D(t,t)V(\lambda, t, x(\cdot))e^{\int_0^t M(s)ds}.$$

But from (3.1.8) we have

$$(\lambda a(t) - x(t))^2 \leq 2D(t,t)e^{\int_0^t M(s)ds}.$$

Here is the objective which we have achieved. Independent of $\lambda \in [0,1]$, for each $T > 0$ we can find $K > 0$ with $|x(t)| \leq K$ if $0 \leq t \leq T$ and $0 \leq \lambda \leq 1$. Of course, there are many different Liapunov functions which could give us the same result so we formulate the end result as a definition.

Definition 3.1.3. *Let Q be the set of continuous solutions $\phi : [0,\infty) \to \Re^n$ of $(3.1.2_\lambda)$ and let $V : [0,1] \times [0,\infty) \times Q \to [0,\infty)$. Then $V(\lambda, t, x(\cdot))$ is said to be mildly unbounded along any solution of $(3.1.2_\lambda)$ if there is an $L > 0$ with $V(\lambda, 0, x(\cdot)) \leq L$ and for each $T > 0$ there is a $K > 0$ such that $V(\lambda, t, x(\cdot)) \leq L$ for $t \geq 0$ implies that $|x(t)| \leq K$ for $0 \leq t \leq T$.*

Note that $\lambda = 0$ yields $x = 0 \in Q$.

Theorem 3.1.4. *If either of the following conditions hold, then (3.1.2) has a solution on $[0,\infty)$:*

(I) There are continuous increasing functions $\Gamma : [0,\infty) \to [0,\infty)$ and $W : [0,\infty) \to [1,\infty)$ with

$$\int_0^\infty \frac{ds}{W(s)} = \infty \text{ and } |D(t,s,x)| \leq \Gamma(t)W(|x|) \text{ for } 0 \leq s \leq t. \quad (3.1.9)$$

(II) There is a differentiable scalar functional $V(\lambda, t, x(\cdot))$ which satisfies Definition 3.1.3 with $V'(\lambda, t, x(\cdot)) \leq 0$ along any solution of $(3.1.2_\lambda)$.

Proof. Let $T > 0$ and $(C, \|\cdot\|)$ be the Banach space of continuous functions $\varphi : [0,T] \to R^n$ with the supremum norm. We will show that there is a solution $x(t)$ of (3.1.2) on $[0,T]$. These lemmas are for $0 \leq t \leq T$.

Lemma 1. *If H is defined by $(3.1.2_\lambda)$ then $H : C \to C$ and H maps bounded sets into compact sets.*

Proof. If $\varphi \in C$, then $D(t, s, \varphi(s))$ is continuous and so $H\varphi$ is a continuous function of t. Let $J > 0$ be given and let $B = \{\varphi \in C | \|\varphi\| \leq J\}$. Now $a(t)$ is uniformly continuous on $[0,T]$ and $D(t,s,x)$ is uniformly continuous on $\Delta = \{(t,s,x) | 0 \leq s \leq t \leq T, |x| \leq J\}$. Thus, for each $\varepsilon > 0$ there is a $\delta > 0$

3.2. A GLOBAL EXISTENCE THEOREM

such that for $(t_i, s_i, x_i) \in \Delta$, $i = 1, 2$, then $|(t_1, s_1, x_1) - (t_2, s_2, x_2)| < \delta$ implies that $|D(t_1, s_1, x_1) - D(t_2, s_2, x_2)| < \varepsilon$; a similar statement holds for $a(t)$. If $\varphi \in B$ then $0 \leq t_i \leq T$ and $|t_1 - t_2| < \delta$ imply that

$$|(H\varphi)(t_1) - (H\varphi)(t_2)| \leq |a(t_1) - a(t_2)|$$
$$+ \left| \int_0^{t_1} \left[D(t_1, s, \varphi(s)) - D(t_2, s, \varphi(s)) \right] ds \right|$$
$$+ \left| \int_0^{t_1} D(t_2, s, \varphi(s)) ds - \int_0^{t_2} D(t_2, s, \varphi(s)) ds \right|$$
$$\leq \varepsilon + t_1 \varepsilon + |t_1 - t_2| M \leq \varepsilon(1 + T) + \delta M$$

where $M = \max_\Delta |D(t, s, x)|$. Hence, the set $A = \{H\varphi | \varphi \in B\}$ is equicontinuous. Moreover, $\varphi \in B$ implies that $\|H\varphi\| \leq \|a\| + TM$. Thus, A is contained in a compact set by Ascoli's theorem.

Lemma 2. *H is continuous in φ.*

Proof. Let $J > 0$ be given, $\|\varphi_i\| \leq J$ for $i = 1, 2$, and for a given $\varepsilon > 0$ find the δ of uniform continuity on the region Δ of the proof of Lemma 1 for D. If $\|\varphi_1 - \varphi_2\| < \delta$, then

$$|(H\varphi_1)(t) - (H\varphi_2)(t)| \leq \int_0^t |D(t, s, \varphi_1(s)) - D(t, s, \varphi_2(s))| ds$$
$$\leq T\varepsilon$$

so $\|(H\varphi_1) - (H\varphi_2)\| \leq T\varepsilon$.

Lemma 3. *There is a $K > 0$ such that any solution of $(3.1.2_\lambda)$ on $[0, T]$, satisfies $\|\varphi\| \leq K$.*

Proof. Let (I) hold. If φ satisfies (3.1.5) on $[0, T]$, then

$$|\varphi(t)| \leq \lambda \left[A(T) + \int_0^t \Gamma(T) W(|\varphi(s)|) ds \right] \text{ for } 0 \leq t \leq T$$

where $A(T) = \max_{0 \leq t \leq T} |a(t)|$. If we define $y(t)$ by

$$y(t) = \lambda \left[A(T) + 1 + \Gamma(T) \int_0^t W(|y(s)|) ds \right]$$

then $y(t) \geq |\varphi(t)|$ on $[0, T]$; clearly, $y(0) > |\varphi(0)|$ so if there is a first t_1 with $y(t_1) = |\varphi(t_1)|$, then a contradiction is clear. But the Conti-Wintner result gives a bound K on $\|y\|$ and so the lemma is true for (I). The argument when (II) holds follows directly from Definition 3.1.3 and $V' \leq 0$.

Interesting global existence is also found in Lakshmikantham and Leela (1969; p. 46).

3.2 Classical Theory

When we consider

$$x(t) = f(t) + \int_0^t g(t,s,x(s))\,ds, \quad t \geq 0, \tag{3.2.1}$$

it is to be understood that $x(0) = f(0)$ and we are looking for a continuous solution $x(t)$ for $t \geq 0$. However, it may happen that $x(t)$ is specified to be a certain *initial function* on an *initial interval*, say,

$$x(t) = \phi(t) \quad \text{for} \quad 0 \leq t \leq t_0$$

(see Fig. 3.1). We are then looking for a solution of

$$x(t) = f(t) + \int_0^{t_0} g(t,s,\phi(s))\,ds + \int_{t_0}^t g(t,s,x(s))\,ds, \quad t \geq t_0.$$

Notice that at $t = t_0$ we have

$$x(t_0) = f(t_0) + \int_0^{t_0} g(t,s,\phi(s))\,ds$$

which may not equal $\phi(t_0)$ so the graph in Fig. 3.1 could have a discontinuity, not just the indicated cusp.

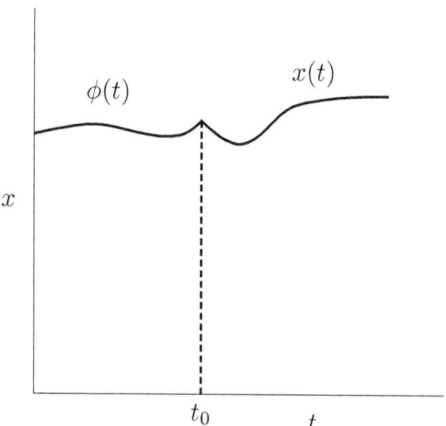

Figure 3.1: The cusp (or discontinuity).

3.2. CLASSICAL THEORY

As an example, (3.2.1) may describe the population density $x(t)$. A given population is observed over a time period $[0, t_0]$ and is given by $\phi(t)$. The subsequent behavior of that density may depend greatly on $\phi(t)$.

A change of variable will reduce the problem back to one of form (3.2.1). Let $x(t + t_0) = y(t)$, so that we have

$$x(t+t_0) = f(t+t_0) + \int_0^{t_0} g(t+t_0, s, \phi(s))\, ds$$
$$+ \int_{t_0}^{t_0+t} g(t_0+t, s, x(s))\, ds$$
$$= f(t+t_0) + \int_0^{t_0} g(t+t_0, s, \phi(s))\, ds$$
$$+ \int_0^t g(t_0+t, u+t_0, x(u+t_0))\, du$$

or

$$y(t) = h(t) + \int_0^t g(t_0+t, u+t_0, y(u))\, du$$

where

$$h(t) = f(t+t_0) + \int_0^{t_0} g(t+t_0, s, \phi(s))\, ds$$

and we want the solution for $t \geq 0$.

Thus, the initial function on $[0, t_0]$ is absorbed into the forcing function, and hence, it always suffices to consider (3.2.1) with the simple condition $x(0) = f(0)$.

Exactly the same may happen for an integrodifferential equation

$$x'(t) = A(t)x(t) + \int_0^t B(t,s)x(s)\, ds + f(t).$$

The natural initial condition is $x(0) = x_0$ and we seek a solution for $t \geq 0$. But there may be a given continuous initial function $\phi : [0, t_0] \to \Re^n$. Make the same change of variable as was done for the integral equation and we obtain a new equation of this same form where the initial function is absorbed into the forcing function. There is one large difference. This time we will get a continuous solution, but it will likely have a corner at t_0 just as our figure indicates. The solution does smooth as time goes on.

We now proceed to obtain two of the most standard existence results for integral equations.

3. EXISTENCE PROPERTIES

Definition 3.2.1. *Let $U \subset \Re^{n+1}$ and $G : U \to \Re^n$. We say that G satisfies a local Lipschitz condition with respect to x if for each compact subset M of U there is a constant K such that (t, x_i) in M implies*

$$|G(t, x_1) - G(t, x_2)| \leq K|x_1 - x_2|.$$

Frequently we need to allow t on only one side of t_0. Let x, f, and g be n vectors and consider again (3.2.1)

$$x(t) = f(t) + \int_0^t g(t, s, x(s))\, ds.$$

From the proof of Theorem 1.2.5 and the introductory remarks of this section we know that an integro-differential equation with initial conditions can be put into the form of (3.2.1).

Theorem 3.2.1. *Let a, b, and L be positive numbers, and for some fixed $\alpha \in (0, 1)$ define $c = \alpha/L$. Suppose*

(a) *f is continuous on $[0, a]$,*

(b) *g is continuous on*

$$U = \{(t, s, x) : 0 \leq s \leq t \leq a \text{ and } |x - f(t)| \leq b\},$$

(c) *g satisfies a Lipschitz condition with respect to x on U of the form*

$$|g(t, s, x) - g(t, s, y)| \leq L|x - y|$$

if $(t, s, x), (t, s, y) \in U$.

If $M = \max_U |g(t, s, x)|$, then there is a unique solution of (3.2.1) on $[0, T]$, where $T = \min[a, b/M, c]$.

Proof. Let $\mathcal{S}$ be the space of continuous functions from $[0, T] \to \Re^n$ with $\psi \in \mathcal{S}$ if

$$\|\psi - f\| \stackrel{\text{def}}{=} \max_{0 \leq t \leq T} |\psi(t) - f(t)| \leq b.$$

Define an operator $A : \mathcal{S} \to \mathcal{S}$ by

$$(A\psi)(t) = f(t) + \int_0^t g(t, s, \psi(s))\, ds.$$

3.2. CLASSICAL THEORY

To see that $A : \mathcal{S} \to \mathcal{S}$ notice that ψ continuous implies $A(\psi)$ continuous and that
$$\|A(\psi) - f\| = \max_{0 \le t \le T} |(A\psi)(t) - f(t)|$$
$$= \max_{0 \le t \le T} \left| \int_0^t g(t, s, \psi(s)) \, ds \right| \le MT \le b \,.$$

To see that A is a contraction mapping, notice that if ϕ and $\psi \in \mathcal{S}$, then
$$\|A\phi - A\psi\| = \max_{0 \le t \le T} \left| \int_0^t g(t, s, \phi(s)) \, ds - \int_0^t g(t, s, \psi(s)) \, ds \right|$$
$$\le \max_{0 \le t \le T} \int_0^t |g(t, s, \phi(s)) - g(t, s, \psi(s))| \, ds$$
$$\le \max_{0 \le t \le T} L \int_0^t |\phi(s) - \psi(s)| \, ds$$
$$\le T \max_{0 \le t \le T} L |\phi(s) - \psi(s)|$$
$$= TL\|\phi - \psi\| \le cL\|\phi - \psi\| = \alpha\|\phi - \psi\| \,.$$

Thus, by the contraction mapping principle, there is a unique function $x \in \mathcal{S}$ with
$$(Ax)(t) = x(t) = f(t) + \int_0^t g(t, s, x(s)) \, ds \,.$$

This completes the proof.

Remark 3.2.1. We can certainly get by with less than continuity of g in t and s. We need the Lipschitz condition and we need to say that if $\psi \in \mathcal{S}$ then $A\psi \in \mathcal{S}$. This will allow for mild singularities. One may find this throughout Miller (1971a), Corduneanu (1991), and many other places. If we focus on such problems, then it will change the direction of the book entirely and we will be led away from our main point of showing qualitative properties based on Liapunov functions.

The interval $[0, T]$ can be improved by using a different norm, as we saw in Chapter 1. The conclusion can be strengthened to $T = \min[a, b/M]$. On an interval $[0, T]$ one may ask that
$$\|\phi\| = \max_{0 \le s \le T} Ae^{-Bs} |\phi(s)|$$
for A and B positive constants.

According to C. Corduneanu, the next result is by L. Tonelli and appeared in the Bull. Calcutta Math. Soc. in 1928.

Theorem 3.2.2. Let a and b be positive numbers and let $f : [0, a] \to \Re^n$ and $g : U \to \Re^n$ both be continuous, where

$$U = \{(t, s, x) : 0 \leq s \leq t \leq a \text{ and } |x - f(t)| \leq b\}.$$

Then there is a continuous solution of

$$x(t) = f(t) + \int_0^t g(t, s, x(s)) \, ds \qquad (3.2.1)$$

on $[0, T]$, where $T = \min[a, b/M]$ and $M = \max_U |g(t, s, x)|$.

Proof. We construct a sequence of continuous functions on $[0, T]$ such that

$$x_1(t) = f(t),$$

and if j is a fixed integer, $j > 1$, then $x_j(t) = f(t)$ when $t \in [0, T/j]$ and

$$x_j(t) = f(t) + \int_0^{t-(T/j)} g(t - (T/j), s, x_j(s)) \, ds$$

for $T/j \leq t \leq T$.

Let us examine this definition. If j is a fixed integer, $j > 1$, then

$$x_j(t) = f(t) \quad \text{on} \quad [0, T/j]$$

and

$$x_j(t) = f(t) + \int_0^{t-(T/j)} g(t - (T/j), s, f(s)) \, ds$$

for $t \in [T/j, 2T/j]$. At $t = 2T/j$, the upper limit is T/j, so the integrand was defined on $[T/j, 2T/j]$, thereby defining $x_j(t)$ on $[T/j, 2T/j]$. Now, for $t \in [2T/j, 3T/j]$ we still have

$$x_j(t) = f(t) + \int_0^{t-(T/j)} g(t - (T/j), s, x_j(s)) \, ds,$$

and the upper limit goes to $2T/j$ on that interval, so $x_j(s)$ is defined, and hence the integral is well defined. This process is continued to $[(j-1)T/j, T]$, obtaining $x_j(t)$ on $[0, T]$ for each j. A Lipschitz condition will be avoided because $x_j(t)$, $j > 1$, is independent of all other $x_k(t)$, $j \neq k$.

3.2. CLASSICAL THEORY

Notice that

$$|x_j(t) - f(t)| \leq \int_0^{t-(T/j)} |g(t - (T/j), s, x_j(s))| \, ds$$
$$\leq M(t - (T/j)) \leq M(b/M) = b.$$

This sequence $\{x_j(t)\}$ is uniformly bounded because

$$|x_j(t)| \leq |f(t)| + b \leq \max_{0 \leq t \leq T} |f(t)| + b.$$

To see that $\{x_j(t)\}$ is an equicontinuous sequence, let $\varepsilon > 0$ be given, let n be an arbitrary integer, let t and v be in $[0, T]$ with $t > v$, and consider

$$|x_n(t) - x_n(v)| \leq |f(t) - f(v)|$$
$$+ \left| \int_0^{v-(T/n)} [g(t - (T/n), s, x_n(s)) - g(v - (T/n), s, x_n(s))] \, ds \right|$$
$$+ \left| \int_{v-(T/n)}^{t-(T/n)} g(t - (T/n), s, x_n(s)) \, ds \right|$$
$$\leq |f(t) - f(v)| + M|t - v|$$
$$+ \int_0^{v-(T/n)} |g(t - (T/n), s, x_n(s)) - g(v - (T/n), s, x_n(s))| \, ds.$$

By the uniform continuity of f, there is a $\delta_1 > 0$ such that $|t - v| < \delta_1$ implies $|f(t) - f(v)| < \varepsilon/3$.

By the uniform continuity of g on U, there is a $\delta_2 > 0$ such that $|t - v| < \delta_2$ implies

$$T|g(t - (T/n), s, x_n(s)) - g(v - (T/n), s, x_n(s))| < \varepsilon/3.$$

Let $\delta = \min[\delta_1, \delta_2, \varepsilon/3M]$. This yields equicontinuity. The conclusion now follows from Ascoli's theorem.

Theorems 3.2.1 and 3.2.2 are local existence results. They guarantee a solution on an interval $[0, T]$. There is extensive discussion of existence in Burton (1983b). It is shown in detail how the solution can be continued until it approaches the boundary of the set where f and g are defined or there is a number L with $\limsup_{t \uparrow L} |x(t)| = +\infty$. This is finite escape time.

One can construct examples of equations whose solutions tend to infinity in finite time in several ways. Perhaps the easiest way is as follows. Let

$g : \Re \to \Re$, $xg(x) > 0$ if $x \neq 0$, g be continuous, and let $\int_0^\infty [dx/g(x)] < \infty$. Suppose that C and C_t are continuous. If, in addition, $C(t,s) \geq c_0 > 0$ and $C_t(t,s) \geq 0$, then for $x_0 > 0$ the solution of

$$x(t) = x_0 + \int_0^t C(t,s)g(x(s))ds$$

tends to infinity in finite time. Just note that $x' \geq c_0 g(x)$, divide by $g(x)$ and integrate both sides form 0 to t.

3.3 Equations of Chapter 2

In Chapter 2 we focused on a variety of problems illustrating differences between the convolution and nonconvolution cases as shown by Liapunov's direct method. A rich field for research was introduced. In much the same way we now re-examine those equations with x replaced by $g(x)$ and see the richness for research emerge. In particular, one is always confronted with the problem of removing the growth condition on g. There is always the possibility that the growth is merely a function of the method used to study the problem so that a slightly modified method would completely remove the growth condition.

This section has four distinct parts. First, we differentiate the nonlinear equation and use several different Liapunov functionals to prove qualitative properties. Next, we use a Liapunov functional directly on the integral equation to obtain several qualitative properties of solutions, sometimes under the assumption that the nonlinearity is not too large and then under the assumption that it is not too small.

In Chapter 2 we studied

$$x(t) = a(t) - \int_0^t C(t,s)x(s)ds, \tag{3.3.1}$$

together with the resolvent equation

$$R(t,s) = C(t,s) - \int_s^t R(t,u)C(u,s)du, \tag{3.3.2}$$

and variation of parameters formula

$$x(t) = a(t) - \int_0^t R(t,s)a(s)ds. \tag{3.3.3}$$

The objective was to prove that solutions are bounded, L^p, or tend to zero. Here, we change the equation in a minimal way and use Sections 3.1

3.3. EQUATIONS OF CHAPTER 2

and 3.2 to study what can be said about solutions concerning the same properties, as well as continuation.

Early on we motivated much of the work by the following question. Under "reasonable" conditions on the kernel $C(t,s)$, what is the qualitative difference between the solutions of

$$x(t) = t + (t+1)^{1/2} \sin(t+1)^{1/3} + \sin t - \int_0^t C(t,s)x(s)ds \qquad (3.3.4)$$

and

$$x(t) = \sin t - \int_0^t C(t,s)x(s)ds? \qquad (3.3.5)$$

In spite of the magnitude of the difference in forcing functions, the difference in solutions is merely an L^p-function. Moreover, there are vector space of unbounded functions, $a(t)$, for which $\int_0^t R(t,s)a(s)ds$ differs from $a(t)$ only by an L^p-function, while there are also vector spaces of small functions which that integral is totally unable to copy within an error of an L^p-function. The wary investigator certainly suspects that all of this is mainly a result of linearity and the principle of superposition. But that is not true. Here, we show that it is a fully nonlinear situation.

The application to physical problems is obvious. There will be large disturbances which may have little effect on problems of a certain nature, while some small disturbances will have enormous effect on these same problems. Ignoring small perturbations can lead to disaster, while worrying over large perturbations may be totally unnecessary.

Here, we begin the study of similar properties in a nonlinear case, trying to see exactly how the techniques must be changed to accommodate the nonlinearities. Our work will center on the scalar equation

$$x(t) = a(t) - \int_0^t C(t,s)g(x(s))ds \qquad (3.3.6)$$

where $a : [0, \infty) \to \Re$, $C : [0, \infty) \times [0, \infty) \to \Re$, $g : \Re \to \Re$, with a, C, and g continuous and

$$xg(x) > 0 \text{ for } x \neq 0. \qquad (3.3.7)$$

These conditions are sufficient to ensure that (3.3.6) has at least one solution which, if it remains bounded, can be continued to $[0, \infty)$. We will now offer two results to the general effect that if $a' \in L^p$ then any solution of (3.3.6) satisfies $g(x(t)) \in L^p$. Here, C_1 or C_t means the partial of $C(t,s)$ with respect to t. The first has appeared many places in the linear case,

3. EXISTENCE PROPERTIES

but notice how (3.3.8) integrates only the first coordinate and how well $a(t)$ and $x(t)$ separate in V'; that will be a major concern here.

In Chapter 2 we had exercises of combining theorems on boundedness. Continue that idea in all of the theorems here. For each of these theorems, find another theorem in this set or in Chapter 2 which can be combined into one theorem.

Notice below that there is no growth condition on g.

Theorem 3.3.1. *Suppose that $\int_{t-s}^{\infty} |C_1(u+s,s)| du$ is continuous, (3.3.7) holds, $a' \in L^1[0,\infty)$, and that there is an $\alpha > 0$ with*

$$-C(t,t) + \int_0^{\infty} |C_1(u+t,t)| du \leq -\alpha. \tag{3.3.8}$$

If x is a solution of (3.3.6) then $g(x) \in L^1[0,\infty)$ and x is bounded so $g(x) \in L^p$ for $p \geq 1$.

Proof. If $x(t)$ solves (3.3.6) then it also solves

$$x'(t) = a'(t) - C(t,t)g(x(t)) - \int_0^t C_1(t,s)g(x(s))ds \tag{3.3.9}$$

and we define a Liapunov functional by

$$V(t) = |x(t)| + \int_0^t \int_{t-s}^{\infty} |C_1(u+s,s)| du |g(x(s))| ds. \tag{3.3.10}$$

We find that

$$V'(t) \leq |a'(t)| - C(t,t)|g(x(t))| + \int_0^t |C_1(t,s)g(x(s))| ds$$
$$+ \int_0^{\infty} |C_1(u+t,t)| du |g(x(t))| - \int_0^t |C_1(t,s))g(x(s))| ds$$
$$\leq |a'(t)| - \alpha |g(x(t))|.$$

Thus, so long as the solution is defined then

$$|x(t)| \leq V(t) \leq V(0) + \int_0^t |a'(s)| ds - \alpha \int_0^t |g(x(s))| ds.$$

Hence, $|x(t)| \leq V(0) + \int_0^{\infty} |a'(s)| ds$ so x can be continued on $[0,\infty)$. Moreover, $g(x) \in L^1[0,\infty)$.

3.3. EQUATIONS OF CHAPTER 2

Exercise 3.3.1. For each of these theorems we will want to drop the smallness condition on a or a' and get a mildly unbounded Liapunov functional which will prove that the solution can be continued for all future time. Thus, in Theorem 3.3.1, let a' be an arbitrary continuous function and from V construct a new functional $W(t) = [V(t)+1]e^{-\int_0^t |a'(s)|ds}$. Find the derivative of W and give a careful statement about continuability of solutions.

One may note that we do not need a' integrable on the half-line to parlay the above proof into existence of the solution for $t > 0$ using only (3.3.8). It is also a curious fact that we did not need $\int_0^x g(s)ds$ to diverge with x in order to obtain boundedness. This will change in the next result.

In this result a' and $g(x)$ separate completely in V' so that we can integrate them separately and achieve the conclusion. In the next result our main problem is to contrive a Liapunov functional so that a' and $g(x)$ will separate in V' in such a way that no assumption on the nonlinearity need be made. That is a significant challenge, but it can be conquered exactly as desired. In a later result we will be forced to use Young's inequality to achieve a separation, and that will require monotonicity of g.

The reader should note that it is $a' \in L^{2n}$, resulting in $x \in L^{2n}$. Notice also how both coordinates of C are integrated, but the burden falls on the second coordinate as n becomes large in marked contrast to the last result. These properties are so important to help us understand the richness of the nonconvolution case.

Exercise 3.3.2. Refer back to Chapter 2 and find an example where we used a Liapunov functional beginning with $V = |x| + ...$ Combine that theorem with Theorem 3.3.1.

Exercise 3.3.3. Combine any theorem in Chapter 2 which used a Liapunov functional, say W, on the integral equation with the Liapunov functional, V, used in the proof of Theorem 3.3.1. Form the new functional $W + V$ on (3.3.9) and then bring in the integral equation (3.3.6) for use with W'.

Theorem 3.3.2. *Let $\int_{t-s}^\infty |C_1(u+s,s)|du$ be continuous and (3.3.7) hold. Suppose there is a positive integer n with $a'(t) \in L^{2n}[0,\infty)$, a constant $\alpha > 0$, and a constant $N > 0$ with*

$$\frac{2n-1}{2nN^{\frac{2n}{2n-1}}} - C(t,t) + \frac{2n-1}{2n}\int_0^t |C_1(t,s)|ds$$
$$+ \frac{1}{2n}\int_0^\infty |C_1(u+t,t)|du \leq -\alpha. \qquad (3.3.11)$$

Then the solution x of (3.3.6) is defined on $[0, \infty)$ and $g(x) \in L^{2n}[0, \infty)$. If $\int_0^x g^{2n-1}(s)ds \to \infty$ as $|x| \to \infty$ then any solution x of (3.3.6) is bounded.

Proof. Obtain (3.3.9) and define

$$V(t) = \int_0^x g^{2n-1}(s)ds + \frac{1}{2n}\int_0^t \int_{t-s}^\infty |C_t(u+s,s)|dug^{2n}(x(s))ds.$$

Then we have

$$V'(t) = a'(t)g^{2n-1}(x) - C(t,t)g^{2n}(x) - g^{2n-1}(x)\int_0^t C_1(t,s)g(x(s))ds$$

$$+ \frac{1}{2n}\int_0^\infty |C_1(u+t,t)|dug^{2n}(x(t)) - \frac{1}{2n}\int_0^t |C_1(t,s)|g^{2n}(x(s))ds$$

$$\leq \frac{(2n-1)g^{2n}(x)}{2nN^{\frac{2n}{2n-1}}} + \frac{(Na'(t))^{2n}}{2n} - C(t,t)g^{2n}(x)$$

$$+ \int_0^t |C_t(t,s)|\left[\frac{(2n-1)g^{2n}(x(t))}{2n} + \frac{g^{2n}(x(s))}{2n}\right]ds$$

$$+ \frac{1}{2n}\int_0^\infty |C_1(u+t,t)|dug^{2n}(x(t)) - \frac{1}{2n}\int_0^t |C_1(t,s)|g^{2n}(x(s))ds$$

$$\leq \frac{(Na'(t))^{2n}}{2n} - \alpha g^{2n}(x(t)).$$

If we integrate the first and last terms we see that there is a positive number, K, such that so long as $x(t)$ is defined, say on $[0,T)$, then $\int_0^t g^{2n}(x(s))ds \leq K$. Let $2n = p$ and $(1/p) + (1/q) = 1$. Then for such t we have

$$|x(t)| \leq |a(t)| + \int_0^t |C(t,s)||g(x(s))|ds$$

$$\leq |a(t)| + \left(\int_0^t |C(t,s)|^q ds\right)^{1/q}\left(\int_0^t g^{2n}(x(s))ds\right)^{1/2n}$$

$$\leq |a(t)| + \left(\int_0^t |C(t,s)|^q ds\right)^{1/q} K^{1/2n};$$

thus, $x(t)$ is bounded if $T < \infty$, and this means that we can continue the solution to ∞. We now have

$$\int_0^x g^{2n-1}(s)ds \leq V(t) \leq V(0) + \int_0^t \frac{(Na'(s))^{2n}}{2n}ds - \alpha\int_0^t g^{2n}(x(s))ds.$$

As the solution can be defined for all future time, $g(x) \in L^{2n}$. If the integral on the left diverges with $|x|$, then x is bounded.

3.3. EQUATIONS OF CHAPTER 2

This result shows that when (3.3.11) holds then for $0 < \beta < 1$ and

$$a(t) = (t+1)^\beta + \sin(t+1)^\beta + (t+1)^{1/2}\sin(t+1)^{1/3} \qquad (3.3.12)$$

then $g(x) \in L^{2n}[0,\infty)$ for some $n > 0$. Now, under a compatible set of conditions on C, we let $\beta = 1$ and notice that the solution is asymptotically periodic. The big functions are largely absorbed into the equation, while the little function takes permanent control.

The fact that no condition on divergence of the integral of g is needed for boundedness in Theorem 3.3.1, but divergence is needed in Theorem 3.3.2 is a common problem usually traceable to method of proof rather than a property of the equation. Even if $\int_0^x g(s)ds$ fails to diverge, one may use $g \in L^{2n}$ in several ways to obtain boundedness of the solution. For example, by squaring (3.3.1) we get $(1/2)x^2(t) \leq a^2(t) + \int_0^t C^2(t,s)ds \int_0^t g^2(x(s))ds$.

Exercise 3.3.4. Assume that a' is an arbitrary continuous function, while $\int_0^x g^{2n-1}(s)ds \to \infty$ as $|x| \to \infty$. Then, construct a mildly unbounded functional $W = [V+1]e^{-\int_0^t \frac{(Na'(s))^2}{2n}ds}$ with non-positive derivative. Give a careful statement about continuability of solutions of (3.3.6).

Exercise 3.3.5. For scalar equations it is frequently too much to ask that V be radially unbounded. In the proof of Theorem 3.3.2 suppose that g is odd and notice that

$$\int_0^x g^{2n-1}(s)ds \leq V(0, x(\cdot)) \leq \int_0^{x_0} g^{2n-1}(s)ds + \int_0^t \frac{(Na'(s))^{2n}}{2n}ds.$$

Carefully state the conditions to ensure that the solution is bounded for that x_0.

Let $\mathcal{P}_T$ be the set of continuous T-periodic scalar functions and let Q be the set of continuous functions $q : [0,\infty) \to R$ such that $q(t) \to 0$ as $t \to \infty$. Denote by $(Y, \|\cdot\|)$ the Banach space of functions $\phi : [0,\infty) \to R$ where $\phi \in Y$ implies that $\phi = p + q$ with $p \in \mathcal{P}_T$ and $q \in Q$ and where $\|\cdot\|$ is the supremum norm.

In our work here, we will use a contraction mapping and that will require that g have a bounded derivative, as required in (3.3.15) below. But that is a requirement based only on the method of proof and it can likely be removed using a different kind of proof, possibly by the method of *a priori* bounds.

3. EXISTENCE PROPERTIES

Suppose that $C: \Re \times \Re \to \Re$ is continuous with

$$C(t+T, s+T) = C(t,s), \qquad (3.3.13)$$

that

$$\int_{-\infty}^{t} C(t,s)ds \text{ is bounded and continuous}, \qquad (3.3.14)$$

and that g has a continuous derivative, denoted by g^*, with

$$|g^*(x)| \leq 1. \qquad (3.3.15)$$

For a fixed $\phi = p + q \in Y$ by the mean value theorem for derivatives we have

$$g(\phi(t)) = g(p(t)) + g(p(t) + q(t)) - g(p(t)) = g(p(t)) + g^*(\xi(t))q(t)$$

where $\xi(t)$ is between $p(t) + q(t)$ and $q(t)$. This means that $g(\phi) \in Y$. Note that this did not require (3.3.15) since g^* would be bounded along the fixed function, ϕ. Thus, the ideas seem fully nonlinear.

We will also require

$$\int_{-\infty}^{0} |C(t,s)|ds \to 0 \text{ as } t \to \infty \qquad (3.3.16)$$

and for $q \in Q$ then

$$\int_{0}^{t} C(t,s)q(s)ds \to 0 \text{ as } t \to \infty. \qquad (3.3.17)$$

In order to have a contraction we ask that there exists an $\alpha < 1$ with

$$\int_{0}^{t} |C(t,s)|ds \leq \alpha. \qquad (3.3.18)$$

To prove that (3.3.6) has an asymptotically periodic solution, we begin by defining a mapping from (3.3.6) by $\phi = p + q \in Y$ implies that

$$(P\phi)(t) = a(t) - \int_{0}^{t} C(t,s)g(\phi(s))ds. \qquad (3.3.19)$$

Notice that some growth of g is needed to bound the solution. Much later, in Chapter 5, we will argue that the growth condition is probably not needed because the behavior of the solution so obtained is bounded and, hence, its behavior could hardly be influenced by large values of x.

Notice also that (3.3.7) is not needed.

3.3. EQUATIONS OF CHAPTER 2

Theorem 3.3.3. *If (3.3.13 - 3.3.18) hold and if $a \in Y$, so is the unique solution of (3.3.6) on $[0, \infty)$.*

Proof. Clearly, (3.3.19) is a contraction, but we must show that $P : Y \to Y$. Write $a = p^* + q^* \in Y$ and for $\phi = p + q \in Y$ define

$$(P\phi)(t) = p(t) - \int_{-\infty}^{t} C(t,s)g(p(s))ds - \int_{0}^{t} C(t,s)g'(\xi(s))q(s)ds$$
$$+ \int_{-\infty}^{0} C(t,s)g(p(s))ds.$$

Clearly, the first two terms on the right are periodic, while the remainder is in Q. Thus, $P : Y \to Y$ and there is a fixed point.

It may be seen that (3.3.8) and (3.3.18) are closely related, but for large n then (3.3.11) and (3.3.18) are very different. We come now to a Liapunov functional which will require much more about the behavior of C, but it is a naturally nonlinear functional. It is closely adapted from work of Levin (1963) and we have used it several times in our linear work.

Notice that growth of g is required for boundedness of the solution.

Theorem 3.3.4. *Let (3.3.7) hold, $H(t,s) := C_t(t,s)$, and suppose there is an $\alpha > 0$ with $C(t,t) \geq \alpha$ and*

$$H(t,s) \geq 0, \; H_t(t,s) \leq 0, \; H_s(t,s) \geq 0, \; H_{st}(t,s) \leq 0. \tag{3.3.23}$$

(i) If $a' \in L^2[0, \infty)$, then any solution of (3.3.6) or (3.3.9) defined on $[0, \infty)$ satisfies $g(x) \in L^2[0, \infty)$.

(ii) If a' is bounded, if $g^2(x) \to \infty$ as $|x| \to \infty$, and if there is an $M > 0$ with

$$\int_{0}^{t} H_s(t,s)(t-s) \int_{s}^{t} |a'(u)|^2 du\, ds + H(t,0)t \int_{0}^{t} |a'(u)|^2 du \leq M, \tag{3.3.24}$$

then any solution of (3.3.6) or (3.3.9) defined on $[0, \infty)$ has $\int_{0}^{x} g(s)ds$ bounded.

3. EXISTENCE PROPERTIES

Proof. Define
$$V(t) = \int_0^x g(s)ds + (1/2)\int_0^t H_s(t,s)\left(\int_s^t g(x(u))du\right)^2 ds$$
$$+ (1/2)H(t,0)\left(\int_0^t g(x(s))ds\right)^2$$
$$= \int_0^x g(s)ds + (1/2)\int_0^t C_{st}(t,s)\left(\int_s^t g(x(u))du\right)^2 ds$$
$$+ (1/2)C_t(t,0)\left(\int_0^t g(x(s))ds\right)^2 \qquad (3.3.25)$$

so that if $V(t)$ is bounded, so is $\int_0^x g(s)ds$. Next, write
$$x' = a'(t) - C(t,t)g(x) - \int_0^t H(t,s)g(x(s))ds.$$
Then the derivative of V along a solution is
$$V'(t) = a'(t)g(x) - C(t,t)g^2(x) - g(x)\int_0^t C_t(t,s)g(x(s))ds$$
$$+ (1/2)\int_0^t H_{st}(t,s)\left(\int_s^t g(x(u))du\right)^2 ds$$
$$+ g(x(t))\int_0^t H_s(t,s)\int_s^t g(x(u))duds$$
$$+ (1/2)H_t(t,0)\left(\int_0^t g(x(s))ds\right)^2$$
$$+ H(t,0)\int_0^t g(x(s))dsg(x(t)).$$

We integrate the fifth term on the right by parts and obtain
$$g(x(t))[H(t,s)\int_s^t g(x(u))du\Big|_0^t + \int_0^t H(t,s)g(x(s))ds]$$
$$= -g(x(t))H(t,0)\int_0^t g(x(u))du + \int_0^t H(t,s)g(x(s))dsg(x(t)).$$

Cancelling terms and taking into account sign conditions yields
$$V'(t) \leq a'(t)g(x) - C(t,t)g(x)^2$$
$$\leq (1/2\alpha)|a'(t)|^2 + (\alpha/2)g^2(x) - \alpha g^2(x)$$
$$\leq (1/2)(|a'(t)|^2)/\alpha - \alpha g^2(x)).$$

Hence,
$$2\int_0^x g(s)ds \leq 2V(t) \leq 2V(0) + (1/\alpha)\int_0^t |a'(s)|^2 ds - \alpha\int_0^t g^2(x(s))ds$$

so (i) follows. Note that $\int_0^x g(s)ds$ is bounded if V is bounded.

Now, assume $a'(t)$ bounded and let (3.3.24) hold; we will bound V and, hence, $\int_0^x g(s)ds$. From V' we see that there is a $\mu > 0$ such that if $V'(t) > 0$ then $|x(t)| < \mu$. Suppose, by way of contradiction, that V is not bounded. Then there is a sequence $\{t_n\} \uparrow \infty$ with $V'(t_n) \geq 0$ and $V(t_n) \geq V(s)$ for $0 \leq s \leq t_n$; thus, $|x(t_n)| \leq \mu$. If $0 \leq s \leq t_n$ then

$$0 \leq 2V(t_n) - 2V(s) \leq -\alpha\int_s^{t_n} g^2(x(u))du + (1/\alpha)\int_s^{t_n} |a'(u)|^2 du.$$

Using these values in the formula for V, taking $|x(t_n)| \leq \mu$, $t = t_n$, and applying the Schwarz inequality yields

$$V(t) \leq \int_0^{\pm\mu} g(s)ds + \int_0^t H_s(t,s)(t-s)\int_s^t (1/\alpha^2)|a'(u)|^2 du\, ds$$
$$+ H(t,0)t(1/\alpha^2)\int_0^t |a'(u)|^2 du \leq \int_0^{\pm\mu} g(s)ds + (1/\alpha^2)M.$$

Thus, $V(t)$ and $\int_0^x g(s)ds$ are bounded.

We have constructed a very intricate Liapunov functional and it has exactly the derivative we sought. But it is crippled by the inconclusive requirement that the solution be defined on $[0, \infty)$. That can be cured, of course, by asking that $\int_0^x g(s)ds \to \infty$ as $|x| \to \infty$, but we would like to do more.

Exercise 3.3.6. Review the proof of Theorem 2.1.10 and conclusion (2.1.15). Write the equation as

$$(x' - a'(t) + C(t,t)g(x))^2 \leq \left(\int_0^t H(t,s)g(x(s))ds\right)^2.$$

Integrate the right-hand-side by parts and, under the appropriate hypothesis patterned after (2.1.15) conclude that

$$\left(x' - a'(t) + C(t,t)g(x)\right)^2 \leq K(t)V(t)$$

for an appropriate function $K(t)$. Thus, on any fixed interval $[0, T]$ we have

$$x' - a'(t) + C(t,t)g(x) = r(t),$$

where $r(t)$ is a bounded function on $[0, T]$. Use the Razumikhin function $W(x) = |x|$ to show that the last equation has a solution bounded on

$[0, T]$. Argue from this that V is mildly unbounded. Define a new Liapunov functional by

$$[V(t) + 1]e^{-Lt}$$

which is mildly unbounded and has a non-positive derivative.

In working with our next Liapunov functional we find

$$V'(t) \leq 2a(t)g(x) - 2xg(x) \tag{3.3.26}$$

and we need to separate $a(t)g(x)$. There are many *ad hoc* ways of doing that. We could simply ask for positive constants c_i with

$$2a(t)g(x) - 2xg(x) \leq c_1|a(t)| - c_2 xg(x) \tag{3.3.26*}$$

and walk away. See Lemma 5.3.2. Under certain conditions there is a very exact result which we now develop.

Notice that in Theorem 3.3.5 we use the lemma below which asks strong growth of $g(x)$ in order to separate two functions in the derivative of the Liapunov functional. Compare this with the growth condition required in Theorem 3.3.4(ii).

Lemma 3.3.1. *Let $g(x) = -g(-x)$, g be strictly increasing, for $x \geq 0$ let $\phi(x) := \frac{d}{dx} xg^{-1}(x)$ be monotone increasing to infinity. Then*

$$2|a(t)g(x)| \leq xg(x) + \int_0^{2|a(t)|} \phi^{-1}(s) ds. \tag{3.3.27}$$

Proof. Young's inequality (Hewitt and Stromberg (1971; p. 189)) states that if $\phi : [0, \infty) \to [0, \infty)$ is continuous, strictly increasing to ∞, satisfies $\phi(0) = 0$, and if $\psi = \phi^{-1}$ then for $\Phi(x) = \int_0^x \phi(u) du$ and $\Psi(x) = \int_0^x \psi(u) du$ we have

$$2|a(t)g(x)| \leq \Phi(g(x)) + \Psi(2|a(t)|). \tag{3.3.28}$$

But

$$\Phi(g(x)) = \int_0^{g(x)} \frac{d}{ds} sg^{-1}(s) ds = g(x) g^{-1}(g(x)) = xg(x),$$

as required.

There are two pertinent remarks concerning this material. The technique of Section 8.2 can avoid some of these inequality challenges, as seen in Burton (2009). A vector extension of the next result is found in Zhang (2009).

3.3. EQUATIONS OF CHAPTER 2

Theorem 3.3.5. If $a : [0, \infty) \to R$ is continuous, if (3.3.7) holds, and if

$$C(t,s) \geq 0, \quad C_s(t,s) \geq 0, \quad C_t(t,s) \leq 0, \quad C_{st}(t,s) \leq 0 \quad (3.3.29)$$

then along the solution of (3.3.6) the functional

$$V(t) = \int_0^t C_s(t,s) \left(\int_s^t g(x(u))du \right)^2 ds + C(t,0) \left(\int_0^t g(x(s))ds \right)^2 \quad (3.3.30)$$

satisfies

$$V'(t) \leq 2a(t)g(x) - 2xg(x).$$

(i) If there are constants B and K with

$$\sup_{t\geq 0} \int_0^t C_s(t,s)ds = B < \infty \text{ and } \sup_{t\geq 0} C(t,0) = K < \infty \quad (3.3.31)$$

then along any solution of (3.3.6) on $[0, \infty)$ we have

$$(a(t) - x(t))^2 \leq 2(B+K)V(t). \quad (3.3.32)$$

(ii) If the conditions of Lemma 3.3.1 hold then along the solution of (3.3.6) we have

$$V'(t) \leq -x(t)g(x(t)) + \int_0^{2|a(t)|} \phi^{-1}(s)ds.$$

Hence, if the last term is $L^1[0,\infty)$ then so is $x(t)g(x(t))$. Moreover, V is then bounded so if (3.3.31) holds then $|a(t) - x(t)|$ is bounded.

Proof. The details are like the linear case in Theorem 2.1.10. We have

$$V(t) = \int_0^t C_s(t,s) \left(\int_s^t g(x(u))du \right)^2 ds + C(t,0) \left(\int_0^t y(x(s))ds \right)^2$$

and differentiate along any solution of (3.3.6) to obtain

$$V'(t) =$$

$$\int_0^t C_{st}(t,s) \left(\int_s^t g(x(u))du \right)^2 ds + 2g(x) \int_0^t C_s(t,s) \int_s^t g(x(u))du\,ds$$

$$+ C_t(t,0) \left(\int_0^t g(x(s))ds \right)^2 + 2g(x)C(t,0) \int_0^t g(x(s))ds.$$

We now integrate the third-to-last term by parts to obtain

$$2g(x)\left[C(t,s)\int_s^t g(x(u))du\bigg|_0^t + \int_0^t C(t,s)g(x(s))ds\right]$$

$$= 2g(x)\left[-C(t,0)\int_0^t g(x(u))du + \int_0^t C(t,s)g(x(s))ds\right].$$

Cancel terms, use the sign conditions, and use (3.3.6) in the last step of the process to unite the Liapunov functional and the equation obtaining

$$V'(t) = \int_0^t C_{st}(t,s)\left(\int_s^t g(x(u))du\right)^2 ds + C_t(t,0)\left(\int_0^t g(x(s))ds\right)^2$$
$$+ 2g(x)[a(t) - x(t)]$$
$$\leq 2g(x)a(t) - 2xg(x) \leq -xg(x) + \int_0^{2|a(t)|}\phi^{-1}(s)ds.$$

The lower bound given in (3.3.32) may be derived as in Theorem 2.1.10. The final conclusion is now immediate.

Exercise 3.3.7. There is something very interesting here. Work out the details of the following discussion. If g is bounded and if $\int_0^x g(s)ds$ is bounded, it does not seem possible that V could be mildly unbounded. But (3.3.32) indicates it is. Notice too that

$$V' \leq 2g(x)a(t) - 2xg(x)$$

says that V' is bounded above by some L on any fixed interval $[0,T]$. Thus, we can form

$$W = [V+1]e^{-Lt}.$$

Then W is mildly unbounded and $W' \leq 0$.

The applied mathematician correctly claims that our conditions of a' bounded or in L^p may be difficult to establish because of uncertainties and even stochastic forces. There is a simple way around that if $\frac{d}{dx}g(x) =: g^*(x)$ is bounded. For a given function $b(t)$, seek a function $a(t)$ which satisfies one of our boundedness theorems with $|a(t) - b(t)|$ bounded. Here is a sample theorem. We take a simple condition known to imply that $a \in \mathcal{BC}$ implies that the solution of (3.3.6) is in $\mathcal{BC}$ when $g(x) = x$. Many other conditions are known.

Theorem 3.3.6. *Suppose that $|g^*(x)| \leq 1$ and that there is an $\alpha < 1$ with $\int_0^t |C(t,s)|ds \leq \alpha$. If $x(t) = a(t) - \int_0^t C(t,s)g(x(s))ds$ and $y(t) = b(t) - \int_0^t C(t,s)g(y(s))ds$ with $a - b \in \mathcal{BC}$, so is $x - y$.*

3.3. EQUATIONS OF CHAPTER 2

Proof. Note that for fixed solutions x and y we have

$$x(t) - y(t) = a(t) - b(t) - \int_0^t C(t,s)[g(x(s)) - g(y(s))]ds$$
$$= a(t) - b(t) - \int_0^t C(t,s)g^*(\xi(s))[x(s) - y(s)]ds$$

by the mean value theorem for derivatives where $\xi(s)$ is between $x(s)$ and $y(s)$. The resulting integral equation has a bounded solution.

Note the transition from Theorem 3.3.1 to Theorem 3.3.2. By contriving a Liapunov functional with higher powers of $g(x)$ we are able to pass from the requirement of $a' \in L^1$ (which allows only bounded $a(t)$) to $a' \in L^{2n}$ which allows $a(t) = (t+1)^\beta$ for $0 < \beta < 1$. It is an absolutely enormous advance, especially in view of Theorem 3.3.6. Now look at Theorem 3.3.4 in which we allow $a' \in L^2$. This allows $a(t) = \ln(t+1)$ which is surprisingly large in view of classical results. But it would be a real *coup* to introduce a higher power of $g(x)$ in the Liapunov functional and allow $a' \in L^p$ for $0 < p < \infty$. Patience and imagination should achieve the result.

In Burton (2007b) for the linear case we obtain the counterpart of Theorem 3.3.3. But we also differentiate (3.3.6) and obtain the same result for $a' \in Y$; this includes $a(t) = t + \sin t$. It is a great surprise that the solution is in Y since this means that the function t is completely absorbed, yielding virtually no effect on the long-term behavior of the solution, while $\sin t$ exerts continued influence. Again, it would be a real *coup* to prove this for (3.3.9). The problem is that when $g(x) = x$ then (3.3.9) has $x' = -C(t,t)x + f(t,x(\cdot))$ which can be written as an integral equation, an effective mapping of $Y \to Y$. A clever map must be constructed for the nonlinear case.

We come now to a counterpart of the Adam and Eve theorem in which we need $|g(x)| \leq |x|$ and we ask that integration of the first coordinate of C be small, while the Adam and Eve result asked that integration of the second coordinate of C be small. The two results merge in the convolution case.

Theorem 3.3.7. *Let $a, g : \Re \to \Re$, $a(t) \in L^1[0, \infty)$, $|g(x)| \leq |x|$, with $a(t)$ and $g(x)$ continuous. Let $C : \Re \times \Re \to \Re$ be continuous, as is $\int_{t-s}^\infty |C(u+s,s)| du$. Suppose there is an $M < 1$ with $\int_0^\infty |C(u+t,t)| du \leq M$ and choose $k > 1$ with $Mk < 1$. For the equation*

$$x(t) = a(t) + \int_0^t C(t,s)g(x(s))ds \tag{3.3.33}$$

define

$$H(t) = k \int_0^t \int_{t-s}^\infty |C(u+s,s)| du |g(x(s))| ds.$$

Then there exists $\delta > 0$ such that

$$H'(t) \leq -\delta \left[|g(x)| + \int_0^t |C(t,s)g(x(s))| ds \right] + |a(t)| \quad (3.3.34)$$

so $|g(x(t))|$ and $|x(t)|$ are $L^1[0,\infty)$.

Proof. We have

$$|g(x)| \leq |x| \leq |a(t)| + \int_0^t |C(t,s)g(x(s))| ds$$

and

$$H'(t) = k \int_0^\infty |C(u+t,t)| du |g(x)| - k \int_0^t |C(t,s)g(x(s))| ds$$
$$\leq kM|g(x)| - (k-1) \int_0^t |C(t,s)g(x(s))| ds + |a(t)| - |g(x)|$$

from (3.3.33), so (3.3.34) holds. Moreover,

$$\delta |x| \leq \delta [|a(t)| + \int_0^t |C(t,s)g(x(s))| ds]$$

and so $H'(t) \leq -\delta |x| + (1+\delta)|a(t)|$ from which the conclusion follows.

We now give a corollary which relates this example to familiar aspects of Liapunov's direct method.

Corollary. *Let the assumptions of Theorem 3.2.7 be satisfied.*

(a) *If $a(t)$ and $C(t,s)$ satisfy a local Lipschitz condition in t (uniform in s when $0 \leq s \leq t$) and if $C(t,s)$ is bounded for $0 \leq s \leq t < \infty$, then $x(t) \to 0$ as $t \to \infty$.*

(b) *If there is a continuous function $\Phi : [0, \infty) \to [0, \infty)$ with $|C(t,s)| \leq \Phi(t-s)$ for $0 \leq s \leq t < \infty$ and if $a(t)$ and $\Phi(t) \to 0$ as $t \to \infty$, then $x(t) \to 0$ as $t \to \infty$.*

(c) *If there exists $\beta_1 > 0$ such that $|C(t,s)| \geq \beta_1 \int_{t-s}^\infty |C(u+s,s)| du$ for $t \geq 0$, then there exists $\gamma > 0$ with $H'(t) \leq -\gamma H(t) + |a(t)|$.*

(d) *If there exists $\beta_2 > 0$ such that $|C(t,s)| \leq \beta_2 \int_{t-s}^\infty |C(u+s,s)| du$ for $t \geq 0$, then along any solution $(k/\beta_2)[|x| - |a(t)|] \leq H(t)$.*

3.3. EQUATIONS OF CHAPTER 2

Proof. If the conditions in (a) hold and if $x(t) \not\to 0$, then there exists $\varepsilon > 0$ and $\{t_n\} \uparrow \infty$ with $|x(t_n)| \geq \varepsilon$. Let K be the Lipschitz constant for $a(t)$ and $C(t,s)$, $|C(t,s)| \leq B$, and consider a sequence $\{s_n\}$ with $t_n \leq s_n \leq t_n + L$ where L is fixed, but yet to be determined. Then

$$|x(t_n) - x(s_n)| \leq |a(t_n) - a(s_n)|$$
$$+ \Big| \int_0^{t_n} C(t_n, s) g(x(s)) ds - \int_0^{s_n} C(t_n, s) g(x(s)) ds \Big|$$
$$+ \Big| \int_0^{s_n} C(t_n, s) g(x(s)) ds - \int_0^{s_n} C(s_n, s) g(x(s)) ds \Big|$$
$$\leq K|t_n - s_n| + B \int_{t_n}^{s_n} |g(x(s))| ds + \int_0^{s_n} K|t_n - s_n| |g(x(s))| ds$$
$$\leq B \int_{t_n}^{s_n} |g(x(s))| ds + K|t_n - s_n| \Big(1 + \int_0^{s_n} |g(x(s))| ds\Big).$$

Now $g \in L^1$ implies that $\int_{t_n}^{s_n} |g(x(s))| ds =: \epsilon_n \to 0$ as $n \to \infty$. Also, there is a $D > 0$ such that $K(1 + \int_0^{s_n} |g(x(s))| ds) \leq D$. Thus,

$$|x(t_n) - x(s_n)| \leq B\varepsilon_n + D|t_n - s_n|.$$

Hence, there is an $N > 0$ and an $L > 0$ such that $n \geq N$ and $|t_n - s_n| \leq L$ imply that $|x(t_n) - x(s_n)| < \varepsilon/2$; thus, $|x(t)| \geq \varepsilon/2$ for $t_n \leq t \leq t_n + L$, a contradiction to $x \in L^1[0, \infty)$. This proves (a).

To prove (b), note that

$$|x(t)| \leq |a(t)| + \int_0^t \Phi(t-s)|x(s)| ds.$$

The integral is the convolution of an L^1-function ($|x(t)|$) with a function tending to zero; hence, the integral tends to zero.

To prove (c), we note from (3.3.34) that

$$H'(t) \leq -\delta\beta_1 \int_0^t \int_{t-s}^\infty |C(u+s,s)| du |g(x(s))| ds + |a(t)|$$
$$\leq -(\delta\beta_1/k)H(t) + |a(t)|,$$

as required.

We prove (d) by noting that

$$H(t) = k \int_0^t \int_{t-s}^{\infty} |C(u+s,s)| du |g(x(s))| ds$$
$$\geq (k/\beta_2) \int_0^t |C(t,s)g(x(s))| ds$$
$$\geq (k/\beta_2)[|x| - |a(t)|]$$

as required.

3.4 Resolvents and Nonlinearities

In Chapter 2 we studied scalar linear integral equations and the role of the resolvent in duplicating the forcing function so that large forcing functions frequently have little effect on the solution, while small forcing functions can exert enormous control over the behavior of solutions. Much of that work depends entirely on linearity. The purpose of this section is to see what can be proved in the nonlinear case and just how much techniques must be changed.

Much of Chapter 2 centered on the fact that $\int_0^t R(t,s)a(s)ds$ provided a copy of $a(t)$ and that $R(t,s)$ was derived from $C(t,s)$ alone. Thus, when we consider

$$x(t) = a(t) - \int_0^t C(t,s)g(x(s))ds$$

we realize that there is a resolvent and we are interested in knowing its properties and what significance it might have for the solution. A first step toward an understanding might be to consider

$$g(x) = x + r(x)$$

with $xr(x) \geq 0$ and the equations

$$x(t) = a(t) - \int_0^t C(t,s)[x(s) + r(x(s))]ds, \quad xr(x) \geq 0, \quad (3.4.1)$$

$$y(t) = a(t) - \int_0^t C(t,s)y(s)ds, \quad (3.4.2)$$

$$R(t,s) = C(t,s) - \int_s^t R(t,u)C(u,s)ds, \quad (3.4.3)$$

and

$$y(t) = a(t) - \int_0^t R(t,s)a(s)ds. \quad (3.4.4)$$

3.4. RESOLVENTS AND NONLINEARITIES

Theorem 3.4.1. *Suppose that the conditions of Theorem 3.3.1 hold so that $g(x) \in L^1[0, \infty)$ and let*

$$g(x) = x + r(x), \quad xr(x) \geq 0. \tag{3.4.5}$$

Then for

$$q(t) = \int_0^t C(t, s) r(x(s)) ds$$

we have

$$x(t) = a(t) - q(t) - \int_0^t R(t, s)[a(s) - q(s)] ds \in L^1[0, \infty) \tag{3.4.6}$$

and

$$q(t) - \int_0^t R(t, s) q(s) ds \in L^1[0, \infty).$$

Proof. As $g(x) \in L^1[0, \infty)$, then by (3.4.5) since x and $r(x)$ have the same sign it follows that x and $r(x)$ are both in $L^1[0, \infty)$. Thus, (3.4.6) is simply the statement that $x \in L^1$. Here are the details. Equation (3.4.1) has a fixed solution x. By Theorem 3.3.1 we have $g(x) = x + r(x) \in L^1$ and since x and $g(x)$ have the same sign then

$$\int_0^\infty |g(x(s))| ds = \int_0^\infty |x(s) + r(x(s))| ds = \int_0^\infty |x(s)| ds + \int_0^\infty |r(x(s))| ds.$$

Next, given C, then R is uniquely determined. For the fixed solution x, then the function $q(t)$ is a uniquely determined function of t. Moreover, we can write

$$x = [a(t) - \int_0^t C(t, s) r(x(s)) ds] - \int_0^t C(t, s) x(s) ds$$

or

$$x = a(t) - q(t) - \int_0^t C(t, s) x(s) ds$$

and the variation of parameters formula is

$$x(t) = a(t) - q(t) - \int_0^t R(t, s)[a(s) - q(s)] ds \in L^1.$$

Now (3.4.2) also satisfies the conditions of Theorem 3.3.1 so $y \in L^1[0, \infty)$ and by (3.4.4) we have

$$a(t) - \int_0^t R(t,s)a(s)ds \in L^1[0, \infty). \tag{3.4.8}$$

From (3.4.6) we have

$$x(t) - [a(t) - \int_0^t R(t,s)a(s)ds] = -[q(t) - \int_0^t R(t,s)q(s)ds]$$

where the left-hand-side is L^1 and so is the right-hand-side.

The resolvent is copying q as efficiently as it copies a and we have no apparent reason to believe that q or its derivative is integrable.

Theorem 3.4.2. *Let (3.4.5) and the conditions of Theorem 3.3.2 hold so that $a' \in L^{2n}$ implies that $g(x) \in L^{2n}$ for some fixed positive number n. If $q(t) = \int_0^t C(t,s)r(x(s))ds$ and if R is defined in (3.4.3) then for x and y the solutions of (3.4.1) and (3.4.2), respectively, we have the functions x, y, $q(t) - \int_0^t R(t,s)q(s)ds$, $a(t) - \int_0^t R(t,s)a(s)ds$, and $x - y$ all in $L^{2n}[0, \infty)$.*

Proof. In Theorem 3.3.2 we have $g(x) \in L^{2n}$ so $(x + r(x))^{2n} \geq x^{2n} \in L^{2n}$ and the same for $r(x)$. Theorem 3.3.2 also yields $y = a(t) - \int_0^t R(t,s)a(s)ds \in L^{2n}[0, \infty)$. Finally,

$$x - [a(t) - \int_0^t R(t,s)a(s)ds] = -[q(t) - \int_0^t R(t,s)q(s))]ds \in L^{2n}.$$

These two results suggest that much may be done by studying the resolvent for such nonlinear equations.

3.5 Notes

Theorem 3.1.4 is a modified form of one found in Burton (1994b). The material in Section 3.2 is taken from the standard classical literature and is found in almost any book on integral equations. Some of the material in Section 3.3 is taken from Burton (2007d) and was first published by the Tatra Mountains Mathematical Publications. Theorem 3.3.7 and its corollary are slight modifications of results from Burton (1996a) which was first published by Walter de Gruyter. That paper was actually the beginning of Liapunov functionals for integral equations. It shows beginnings not given here in which one starts with an integrodifferential equation, constructs a Liapunov functional of classical type and then shows how to extend it to integral equations. It is important to note that Kato (1994) studied that paper and then gave some general insights into how that could be done for a variety of problems. The paper of Kato is highly recommended for further study.

Chapter 4

Applications and Origins

4.0 Introduction

This chapter consists of three distinct parts.

In Section 4.1 we will look at many of the ways in which integral equations emerge from differential equations. We will see how high dimensional differential equations can be reduced to scalar integral equations either by variation of parameters or integration by parts. We will also see that this process has been one of the driving forces behind the development of fixed point theory. Finally, we will introduce a theorem which goes far in uniting fixed point theory and Liapunov's direct method.

In Section 4.2 we will give very brief sketches of some of the classical problems in integral equations which begin as integral equations and frequently are not nicely reducible to differential equations. It is our intent to only mention these problems and to give references where they can be studied more fully.

One of the greatest sources of integral equations is the area of feedback control. Such an example is given in Section 4.1, the Lurie problem, so it does double duty. Among the many choices left for what is to be covered we choose the problem of buckling rods since it yields a kernel totally contrary to our intuition. Finally, we also choose the problem of temperature in a semi-infinite rod because it shows a partial differential equation converted to an ordinary integral equation.

Both Sections 4.1 and 4.2 are rudimentary in nature. Mainly, we are offering the reader a bit of standard theory which can be enlarged upon by reading the existing literature. Some of those problems require very lengthy treatment for a reasonable understanding and we do not want to devote so much of the space here to things which are readily accessible elsewhere. All of the foregoing is classical and can be found in many places.

In Chapter 0 we presented an introductory lecture which offered something of an overview of the work which would appear in Chapters 1 and 2. In Section 4.3 we present a lecture in retrospect which delves into contrasting properties of kernels. It is intended to build some intuition about the character of integral equations.

4.1 Differential Equations

Integral equations appear in a wide variety of contexts and for so many different reasons. The most elementary encounter is seen in writing a differential equation as an integral equation so that various operator-type methods may be applied. In Chapter 3 we saw that

$$x' = f(t,x), \quad x(t_0) = x_0$$

can be written as

$$x(t) = x_0 + \int_{t_0}^{t} f(s, x(s))ds$$

so that we can apply mapping theory for existence of various types of solutions. In Chapter 1 we repeatedly used a variation of parameters formula to write a system of differential equations as a system of integral equations.

Taking that to its logical conclusion, Krasnoselskii (1958) (see also Smart (1980; p. 31)) studied a paper by Schauder on partial differential equations and deduced the following working hypothesis: "The inversion of a perturbed differential operator yields the sum of a contraction and a compact map." He then proved the following theorem.

Theorem 4.1.1 Krasnoselskii. *Let $\mathcal{M}$ be a closed convex non-empty subset of a Banach space $(\mathcal{S}, \|\cdot\|)$. Suppose that A and B map $\mathcal{M}$ into $\mathcal{S}$ such that*

(i) *$Ax + By \in \mathcal{M}(\forall x, y \in \mathcal{M})$,*

(ii) *A is continuous and $A\mathcal{M}$ is contained in a compact set,*

(iii) *B is a contraction with constant $\alpha < 1$.*

Then there is a $y \in \mathcal{M}$ with $Ay + By = y$.

History has shown Krasnoselskii to be so very right. Moreover, often when we do invert a problem we find that Krasnoselskii's theorem almost applies, but not quite. We find that we have a contraction, but it fails as we approach some point. We also find that the continuity of A fades in a certain region. And we find that we can not quite solve for $x = Ax + By$. This then generates a new theorem of the same type.

In a recent paper Sehie Park (2007) studied this set of extensions of Krasnoselskii's theorem and added one of his own which is designed to cover all of them. But for applications it still seems most productive to look at each of them separately to see just exactly what improvement is made. We will focus on one of those extensions throughout much of the remainder of this book. Recall that in Section 3.1 we proved an existence theorem by means of Schaefer's fixed point theorem in conjunction with a Liapunov functional. To invert the differential operator is to obtain an integral equation which will define a mapping. If we examine Krasnoselskii's theorem we see that the first term of the mapping is a contraction which usually will not smooth, as is required by Schaefer's theorem. The extension which we use is the one which combines Schaefer's theorem with Krasnoselskii's theorem, thereby continuing the marriage of the fixed point method with Liapunov's direct method.

We now list the references considered by Park and strongly recommend that the interested reader consider them when inverting a perturbed differential operator. One of these may provide the key to solving the problem at hand. Those references are: Avramescu and Vladimirescu (2003), Barroso (2003), Barroso and Teixeira (2005), Boyd and Wong (1969), Burton (1996b, 1998a), Cain and Nashed (1971), Calvert (1977), Dhage (2003a,b), Hoa and Schmitt (1994), Nashed and Wong (1969), Sehgal and Singh (1978), and Tan (1987). Add to this list Garcia-Falset (2008).

In Section 3.1 we looked at functions in the mapping set on a closed finite interval $[0, T]$ where T is arbitrary, but as large as we please. This gives us immediate access to the Ascoli-Arzela theorem in proofs of compactness. But in the problems often considered in this theorem the domain of the functions being mapped is an infinite interval and there are real problems with compactness. In Chapter 5 we will see the use of a weighted norm which yields compactness of closed equicontinuous sets which are bounded in the supremum norm. There are alternatives to this by using an Ascoli-Arzela theorem dealing with a set which shrinks as $t \to \infty$. One may see some detail in Burton and Furumochi (2001) or Burton and Zhang (2004) where the specific conditions of Krasnoselskii's theorem are considered in some depth.

Conditions for contraction give continuing difficulty. There is, of course, an almost countably infinite set of contraction mapping theorems and it is pointless to mention them here if they do not relate to the Krasnoselskii problem; in fact, those contraction conditions are numbered, not named by author. In the set of papers just mentioned by Park one finds a variety of contractions including nonlinear, Hoa-Schmitt, and large. Other contractions needed in conjunction with Krasnoselskii's theorem include a separate contraction defined by Liu and Li (2006) and (2007) and one associated also with the compactness problem defined by Burton and Zhang (2004). Both the aforementioned compactness questions and the contraction variations speak to the problem of changing Krasnoselskii's theorem just a bit in order to make it fit a new problem. There is a recent book by Vladimirescu and Avramescu (2006) which has some very informative material on Krasnoselskii's theorem and its applications.

4.1.1 The order of an equation

An nth order ordinary differential equation can often be written as a system of n first order equations. If we have

$$x^n = f(t, x, x', x'', ..., x^{n-1})$$

we write $x = x_1$, $x' = x_2$, $x'' = x_2' = x_3$,...,$x^n = x_n' = f(t, x_1, x_2, ..., x_n)$. Finally, we let X be the column vector having elements x_i and F the column vector function whose elements are the above right-hand-sides and we then have

$$X' = F(t, X),$$

a system of n first order differential equations. Next, we could write

$$X(t) = X(t_0) + \int_{t_0}^{t} F(s, X(s)) ds$$

and we would have a system of n integral equations. In virtually no sense have we made any progress. We have simply redefined matters. That system of integral equations does not represent any reduction of order.

In five steps we will show how very impressive an integral equation can be.

First, a linear first order ordinary differential equation is actually just a thinly disguised exercise in integration. Given

$$x' = a(t)x + b(t),$$

we see that to solve the equation is to integrate an unknown function. But there is a simple procedure of multiplying by an integrating factor so that the problem is reduced to the theoretically simple problem of integrating a known function.

Next, as we have mentioned several times before, the ideal theory of Ritt (1966) tells us that the solution of such a simple second order linear equation as

$$x'' + tx = 0$$

can not be expressed in terms of elementary functions, their composites, or integrals of composites. While existence theory applies and we can readily show that the solutions are bounded (but not the derivatives), the solution itself neatly escapes us. Moreover, the higher the order, the worse the situation becomes.

But every linear n^{th} order normalized ordinary differential equation can be expressed as a scalar integral equation. Some very important and highly nonlinear ordinary differential equations of order $n+1$ can also be reduced to a scalar integral equation. We start with a nonlinear equation which was extensively studied in the middle of the last century and which is now enjoying a great resurgence of interest, particularly when a delay is present.

4.1.2 Feedback Control Systems

Feedback control systems are found throughout engineering and virtually all integral equation monographs present them as examples of integral equations occurring directly without first going through a differential equation stage. Often they are linear in nature and give rise to linear integral equations of convolution type. Indeed, frequently even the integral equation is skipped and the investigator deals directly with transfer functions. The reader is referred to Corduneanu (1973) for a deep treatment. Brief introductions are found in Corduneanu (1991), Gripenberg, Londen, and Staffans (1990; pp. 8-9), and Miller (1971a; pp. 66-73). Many books on engineering are devoted entirely to such systems.

In view of so many good presentations, we prefer to use the space to discuss in some detail a feedback control problem with a long history and a recent rebirth which now presents a number of new problems for the investigator. It is a prime example of a high-dimension system of nonlinear differential equations, even with memory, which can be reduced to a very concise scalar integral equation.

The problem of Lurie concerns a control problem of $n+1$-dimension in the form

$$x' = Ax + bf(\sigma)$$
$$\sigma' = c^T x - rf(\sigma)$$

4.1. DIFFERENTIAL EQUATIONS

where x, c, and b are n-vectors, while σ, $f(\sigma)$, and r are scalars. Also A is an $n \times n$ constant matrix.

To explain what is being studied, we note that we begin with the linear constant coefficient system $y' = Ay$ and seek to improve the performance of the system by adding a control, $bf(\sigma)$. It is assumed that $\sigma f(\sigma) > 0$ for $\sigma \neq 0$. Now σ is determined from the feedback of the position $x(t)$ by the equation $\sigma' = c^T x(t) - rf(\sigma)$. That equation uses the information, $x(t)$, to manufacture a value $\sigma(t)$ which is then relayed back to $x' = Ax + bf(\sigma(t))$.

Modern versions of the problem insist that there are time delays. It takes a certain amount of time, T_1, to relay to σ the value of $x(t)$; thus, a better second equation is

$$\sigma' = c^T x(t - T_1) - rf(\sigma).$$

With this information we can obtain $\sigma(t)$; but it takes a certain amount of time, say T_2, to relay that to the first equation controlling x, yielding the improved first equation as

$$x' = Ax + bf(\sigma(t - T_2))$$

and, hence, the corrected system

$$x' = Ax(t) + bf(\sigma(t - T_2))$$
$$\sigma' = c^T x(t - T_1) - rf(\sigma(t)).$$

But, in either case we readily reduce the problem to a scalar integral equation. We will give the details for the original system and leave the delayed system for an exercise. The interested reader should be able to apply techniques of Chapter 2 and Theorem 3.3.5 to obtain stability results for this scalar equation.

Treating $bf(\sigma)$ as an inhomogeneous term, we use the variation of parameters formula to write

$$x(t) = e^{At} x(0) + \int_0^t e^{A(t-s)} bf(\sigma(s)) ds$$

so that we obtain the scalar equation

$$\sigma' = c^T \left[e^{At} x(0) + \int_0^t e^{A(t-s)} bf(\sigma(s)) ds \right] - rf(\sigma)$$

and then, upon integration and interchange of the order of integration, a scalar nonlinear integral equation is obtained of the form

$$\sigma = a(t) + \int_0^t H(t,s) f(\sigma(s)) ds.$$

This problem was introduced by Lurie (1951) and enjoyed much attention. The book by Lefschetz (1965) is devoted entirely to it, while significant material on it is also found in LaSalle and Lefschetz (1961). Cao, Li, and Ho (2005), Somolinos (1977), and Burton and Somolinos (2007) contain treatments with delays. During the last ten years there has been a great resurgence of interest. A check of the online Mathematical Reviews will yield almost one hundred papers on this problem, most of which are recent. Burton and Somolinos (2007) recently treated the problem with a variety of delays and with A having zero as a chracteristic root. The delayed equation is a good source of research problems.

4.1.3 Reduction of the general linear equation

The last subsection was presented for two reasons. First, it illustrated a fairly clear type of feedback system. But it also showed how large dimension systems can be collapsed into a simple scalar integral equation which can be studied by very elementary techniques. The work depended on the equation being in a particular form. The present subsection is intended to show that the same thing can be done for any linear normalized ordinary differential equation, regardless of the form.

Let f and $a_1(t), \ldots, a_n(t)$ be continuous on $[0, T)$ in

$$x^{(n)} + a_1(t) x^{(n-1)} + \cdots + a_n(t) x = f(t),$$

with $x(0), x'(0), \ldots, x^{(n-1)}(0)$ given initial conditions, and set $x^{(n)}(t) = z(t)$. Then

$$x^{(n-1)}(t) = x^{(n-1)}(0) + \int_0^t z(s)\,ds,$$

$$x^{(n-2)}(t) = x^{(n-2)}(0) + t x^{(n-1)}(0) + \int_0^t (t-s) z(s)\,ds,$$

$$x^{(n-3)}(t) = x^{(n-3)}(0) + t x^{(n-2)}(0) + \frac{t^2}{2!} x^{(n-1)}(0)$$
$$+ \int_0^t \frac{(t-s)^2}{2} z(s)\,ds,$$

$$\vdots$$

$$x(t) = x(0) + t x'(0) + \cdots + \frac{t^{n-1}}{(n-1)!} x^{(n-1)}(0)$$
$$+ \int_0^t \frac{(t-s)^{n-1}}{(n-1)!} z(s)\,ds.$$

If we replace these values of x and its derivatives in our differential equation we have a scalar integral equation for $z(t)$.

Not only is this a compact expression, but it is a sobering admonition that, while a scalar linear first order differential equation is a thinly disguised exercise in elementary integration, a scalar linear integral equation commands our full attention and respect.

4.1.4 Inverting operators I

Equations of the form

$$V(t,x) = S\left(t, \int_0^t H(t,s,x(s))\,ds\right) \tag{4.1.1.1}$$

arise in a natural way. It is supposed that there is an $\alpha > 0$ such that V, $S : [-\alpha, \alpha] \times [-\alpha, \alpha] \to R$ and $H : [-\alpha, \alpha] \times [-\alpha, \alpha] \times [-\alpha, \alpha] \to R$ are continuous, $S(0,0) = V(t,0) = 0$. The problem is to find a $\beta > 0$ and a function $\varphi : [-\beta, \beta] \to R$, $\varphi(0) = 0$, and $x = \varphi(t)$ satisfies (4.1.1.1). By changes of variable, many equations will fit this form.

In the present problem V will define a contraction $\tilde{V}$ on the complete metric space of continuous $\psi : [-\beta, \beta] \to R$ with the supremum metric, while S will define a compact mapping, $\tilde{S}$. Thus, we will seek a fixed point of the mapping $P\varphi = \tilde{S}\varphi + (I - \tilde{V})\varphi$. We are using the terminology of Smart (1980; p. 25) to say that a mapping is compact if it maps a set M in a topological space X into a compact subset of X.

The simplest example concerns the standard implicit function theorem. Given the scalar equation

$$f(t,x) = 0 \text{ with } f(0,0) = 0, \tag{4.1.1.2}$$

the classical problem is to find a $\beta > 0$ and a continuous function $\varphi : [-\beta, \beta] \to R$, $\varphi(0) = 0$, and $f(t, \varphi(t)) \equiv 0$; in other words, can we solve $f(t,x) = 0$ for $x = \varphi(t)$?

There are three classical attacks on the problem: one can use techniques of advanced calculus (cf. Taylor and Mann (1983; pp. 225–232)), fixed point theory (cf. Smart (1980; p. 6)), and differential equations (cf. Hartman (1984; pp. 5, 11–12)), under the common assumption that

$$\frac{\partial f}{\partial x}(0,0) \neq 0. \tag{4.1.1.3}$$

The intuitive reason for (4.1.1.3) is clear. If f is differentiable, then it can be approximated arbitrarily well by a linear function near $(0,0)$ so that $z = f(t,x)$ can be approximated by a plane intersecting the plane

$z = 0$. The precise statement is given as follows from Smart (1980; p. 6); an n-dimensional analog with parameters can be found in Hartman (1964; p. 5).

Theorem (Implicit function). *Let N be a neighborhood of $(0,0)$ in which $f : N \to R$ is continuous, $\partial f / \partial x$ exists in N, is continuous at $(0,0)$, $\partial f(0,0)/\partial x \neq 0$, and $f(0,0) = 0$. Then there is a unique continuous function φ with $f(t, \varphi(t)) = 0$.*

Condition (4.1.1.3) allows one to construct a contraction mapping on a complete metric space with a fixed point φ. In the same way, when (4.1.1.3) holds we can reverse the following steps to obtain φ. We have from (4.1.1.2) that

$$\frac{\partial f}{\partial t}(t, x) + \frac{\partial f(t, x)}{\partial x} \frac{dx}{dt} = 0 \tag{4.1.1.4}$$

and for (t, x) near $(0, 0)$, then

$$\frac{dx}{dt} = -(\partial f(t,x)/\partial t)/(\partial f(t,x)/\partial x) =: G(t, x) \tag{4.1.1.5}$$

so with $x(0) = 0$ we obtain

$$x(t) = \int_0^t G(s, x(s)) \, ds. \tag{4.1.1.6}$$

If G is continuous, then (4.1.1.6) has a solution φ by the Peano existence theorem (cf. Hartman (1964; p. 10) or Smart (1980; p. 44)).

On the other hand, if (4.1.1.3) fails then we write (4.1.1.4) as

$$P(t,x)x' = F(t,x), \quad P(0,0) = 0, \tag{4.1.1.7}$$

where we emphasize that P is not necessarily $\partial f / \partial x$. We can invert that differential operator in (4.1.1.7) by writing

$$\int_0^x P_t(t,s) \, ds + P(t,x)x' = F(t,x) + \int_0^x P_t(t,s) \, ds$$

or

$$\frac{d}{dt} \int_0^x P(t,s) \, ds = F(t,x) + \int_0^x P_t(t,s) \, ds$$

and so an integration and use of $x(0) = 0$ yields

$$\int_0^x P(t,s) \, ds = \int_0^t \left[F(s, x(s)) + \int_0^{x(s)} P_t(s, v) \, dv \right] ds \tag{4.1.1.8}$$

which is a form of (4.1.1.1) with $V(t, 0) = 0$.

4.1. DIFFERENTIAL EQUATIONS

If we can prove that (4.1.1.8) has a solution and reverse the steps from (4.1.1.8) back to (4.1.1.2), then the problem is solved.

At first this seems to be the perfect form for Krasnoselskii's theorem; however, when we set up the mapping we see that we must employ one of the many linearization tricks and that destroys the contraction. From (4.1.1.8) we write the mapping equation

$$(H\phi)(t) = \phi(t) - \int_0^{\phi(t)} P(t,s)ds + \int_0^t \left[F(s,\phi(s)) + \int_0^{\phi(s)} P_t(s,v)dv \right] ds$$

on a certain space of functions. The last term smooths so it would be Krasnoselskii's choice for $A\phi$. It is the very condition $\frac{\partial f(0,0)}{\partial x} = 0$ which frequently prevents

$$\phi(t) - \int_0^{\phi(t)} P(t,s)s$$

from being a contraction. The x derivative of

$$x - \int_0^x P(t,s)ds$$

is

$$L(t,x) = 1 - P(t,x)$$

(which is our guide for a Lipschitz constant) and we remember from (4.1.1.7) that $P(0,0) = 0$. Our contraction is lost at $(0,0)$.

This is typical of the kind of small things which go wrong in trying to implement Krasnoselskii's theorem and such difficulties have generated every one of the extensions enumerated by Park (2007) mentioned earlier in this section.

One solution here was given in Burton (1996b) which advanced the idea of a "large contraction" which is a contraction outside a neighborhood of a given point. That definition was substituted into Krasnoselskii's theorem and the argument was saved. In fact, it turned out to be rather fundamental, as may be seen in three papers of Corduneanu (1997), (2000), and (2001), for example.

Definition 4.1.1. *Let (M,ρ) be a metric space and $B : M \to M$. B is said to be a* large contraction *if $\varphi, \psi \in M$, with $\varphi \neq \psi$ then $\rho(B\varphi, B\psi) < \rho(\varphi, \psi)$ and if $\forall \varepsilon > 0 \ \exists \delta < 1$ such that $[\varphi, \psi \in M, \rho(\varphi, \psi) \geq \varepsilon] \Rightarrow \rho(B\varphi, B\psi) \leq \delta\rho(\varphi, \psi)$.*

In virtually every piece of nonlinear analysis we perform, in one way or another we use a linearization trick. Given a nice local contraction

$$Bx = -x^3$$

we want to use a variation of parameters argument to define a mapping. Thus, we write

$$Bx = -x + (x - x^3).$$

Our contraction, $-x^3$, has become $x - x^3$ which loses its contraction constant as $x \to 0$. What we show is that this is no problem. It is still a large contraction and arguments can proceed. Krasnoselskii was "right on target."

It turns out that almost any kind of contraction condition in Krasnoselskii's theorem will work. For example, Liu and Li (2006), (2007) define a separate contraction and show that Krasnoselskii's theorem is still true with that kind of contraction.

Definition 4.1.2. *Let (X,d) be a metric space and $f : X \to X$. The mapping f is said to be a separate contraction if there exist two functions $\phi, \psi : R^+ \to R^+$ satisfying*
(i) $\psi(0) = 0$, ψ is strictly increasing,
(ii) $d(f(x), f(y)) \leq \phi(d(x,y))$,
and
(iii) $\psi(r) \leq r - \phi(r)$ for $r = d(x,y) > 0$.

They also show that if f is a large contraction, then it is a separate contraction.

Most of the material from this subsection was taken from Burton (1996b), as indicated in the bibliography, and was first published by the American Mathematical Society.

4.1.5 Inverting Operators II

While the last example showed natural aspects of Krasnoselskii's result, it did little for our present program of using Liapunov's direct method on integral equations. Recall that in Section 3.1 we combined Schaefer's fixed point theorem with Liapunov functionals. Schaefer's theorem required that the mapping send bounded sets into compact sets; in short, the mapping needed to smooth. Krasnoselskii's mapping A does smooth, but B does not. Moreover, it can be very tedious to construct the set M of Krasnoselskii's theorem, whereas Schaefer's theorem does not require any such details. Here is the basic question. Could we combine the best aspects of both

4.1. DIFFERENTIAL EQUATIONS

theorems and use Liapunov functions to obtain the boundedness required in Schaefer's theorem? That is exactly what is done in Burton and Kirk (1998). Here is the motivation and details.

Our next problem has a delay in both x and x'. Because of the delay in x it is a functional differential equation, while the delay in x' makes it a neutral functional differential equation. Let $|\alpha| < 1$, $a > 0$, $h > 0$, and let q be continuous. Invert

$$x' = \alpha x'(t - h) + ax - q(t, x, x(t - h)) \qquad (4.1.5.1)$$

and find a solution for a given continuous initial function φ.

Write (4.1.5.1) as

$$(x - \alpha x(t - h))' = a(x - \alpha x(t - h)) + a\alpha x(t - h) - q(t, x, x(t - h)),$$

multiply by e^{-at}, and group terms as

$$\left[(x - \alpha x(t - h))e^{-at}\right]' = \left[a\alpha x(t - h) - q(t, x, x(t - h))\right]e^{-at}.$$

We search for a solution having the property that

$$(x(t) - \alpha x(t - h))e^{-at} \to 0 \text{ as } t \to \infty$$

so that an integration from t to infinity yields

$$-(x(t) - \alpha x(t - h))e^{-at} = \int_t^\infty \left[a\alpha x(s - h) - q(s, x(s), x(s - h))\right]e^{-as}ds$$

and, finally

$$x(t) = \alpha x(t - h) + \int_t^\infty \left[q(s, x(s), x(s - h)) - a\alpha x(s - h)\right]e^{a(t - s)}ds. \quad (4.1.5.2)$$

A general form for such equations is

$$x(t) = f(x(t - h)) + \int_t^\infty Q(s, x(s), x(s - h))C(t - s)ds + p(t). \quad (4.1.5.3)$$

It is exactly the type of inversion which Krasnoselskii described. The integral may smooth, but $f(x(t - h))$ does not.

We call this a neutral delay integral equation of advanced type and it is a very interesting equation. It may have a solution on all of $\Re$ or it may have a solution on $[0, \infty)$ generated by an initial function φ on $[-h, 0]$. In the latter case, notice that we can not obtain a local solution: we must get the full solution on $[0, \infty)$. Thus, we will need to employ a fixed point theorem to get existence and that means that we will get a fixed point in the solution space; hence, we must know in advance the form of the solution space. We study this problem in Chapter 6 and find that the solution will have discontinuities at $t = nh$, but the jumps will tend to zero as $n \to \infty$. (This is parallel to solutions of functional differential equations smoothing (cf. El'sgol'ts (1966)). Equally important is the need to know in advance the growth of the solution so that functions in the solution space will have a weighted norm allowing such growth.

4.1.6 Inverting operators III

This problem begins the same as the last one, but now we want a periodic solution. Let $|\alpha| < 1$, $a > 0$, $h > 0$, $p(t+T) = p(t)$, $g(t+T, x, y) = g(t, x, y)$. Invert

$$x' = \alpha x'(t-h) + ax - g(t, x, x(t-h)) \tag{4.1.6.1}$$

and find a T-periodic solution. We wrote (4.1.5.1) as

$$[(x - \alpha x(t-h))e^{-at}]' = [a\alpha x(t-h) - q(t, x, x(t-h))]e^{-at}.$$

If $x \in \mathcal{P}_T$ solves that equation, integrate from $t - T$ to t and use $x(t+T) = x(t)$ to obtain

$$(x(t) - \alpha x(t-h))e^{-at} - (x(t) - \alpha x(t-h))e^{-a(t-T)}$$
$$= \int_{t-T}^{t} [a\alpha x(s-h) - q(s, x(s), x(s-h)]e^{-as} ds$$

so that

$$x(t) = \alpha x(t-h)$$
$$+ \frac{1}{1 - e^{aT}} \int_{t-T}^{t} [a\alpha x(s-h) - q(s, x(s), x(s-h))]e^{a(t-s)} ds. \tag{4.1.6.2}$$

Compare this with (6.1.1.3) which was obtained from an ordinary differential equation.

Our mapping for Krasnoselskii's theorem is

$$(P\phi)(t) = \alpha \phi(t-h)$$
$$+ \frac{1}{1 - e^{aT}} \int_{t-T}^{t} [a\alpha \phi(s-h) - q(s, \phi(s), \phi(s-h))]e^{a(t-s)} ds.$$

4.1.7 Union of Krasnoselskii's and Schaefer's theorems

In both of the last two subsections we obtained mappings which fit Krasnoselskii's theorem. But our goal is to handle such problems by means of Liapunov functions as we did in Section 3.1 using Schaefer's theorem. But Schaefer's theorem will not apply since the mapping generated directly from (4.1.5.3) will not map bounded sets into compact sets. The following proposition and theorem found in Burton and Kirk (1998) will allow us to do just what we wish in this regard.

Proposition 4.1.1. *If $(\mathcal{B}, \|\cdot\|)$ is a normed space, if $0 < \lambda < 1$, and if $B : \mathcal{B} \to \mathcal{B}$ is a contraction mapping with contraction constant α, then $\lambda B \frac{1}{\lambda} : \mathcal{B} \to \mathcal{B}$ is also a contraction mapping with contraction constant α, independent of λ; in particular*

$$\|\lambda B(x/\lambda)\| \leq \alpha \|x\| + \|B0\|.$$

With this proposition we were able to prove the following result.

Theorem 4.1.2. *Let $(\mathcal{B}, \|\cdot\|)$ be a Banach space, $A, B : \mathcal{B} \to \mathcal{B}$, B a contraction with contraction constant $\alpha < 1$, and A continuous with A mapping bounded sets into compact sets. Either*

(i) $x = \lambda B(x/\lambda) + \lambda Ax$ *has a solution in $\mathcal{B}$ for $\lambda = 1$, or*

(ii) *the set of all such solutions, $0 < \lambda < 1$, is unbounded.*

The result has generated much interest. It has been used by several investigators to prove existence of solutions of integral equations. Others have extended it to Frechet spaces. Much of the work is in terms of semi-norms. When B is linear then the contraction can be changed so that A^p is a nonlinear contraction for some positive integer p; in that case it can be shown that A itself is a nonlinear contraction by using a different norm. We can also substitute separate contraction for the contraction. In the same spirit that Krasnoselskii's theorem is changed so many times to accommodate some difficulty, any of these changes may be exactly what is needed in a given problem. For details we refer the reader to Avramescu (2003), Avramescu and Vladimirescu (2003), (2004a,b)(2005), Dhage (2002), Dhage and Ntouyas (2002), Purnaras (2006), Liu and Li (2006), (2007).

The problems of the last two subsections are perfectly suited to this result and they will be treated in Chapter 6.

4.2 Classical Problems

4.2.1 Buckling of Rods

Reynolds (1984) considers an integral equation which is associated with the buckling of viscoelastic rods. It is a scalar equation of the form

$$x(t) = f(t) + \int_0^t k(t,s)x(s)h(s)ds$$

which is singular at 0 in the sense that

$$\int_0^1 |h(s)|ds = \infty,$$

but, for all $t \in (0,1]$, then

$$\int_t^1 |h(s)|ds < \infty.$$

Under a certain set of conditions he is able to show that the equation has uncountably many solutions. It is interesting for so many reasons. First, it comes to us as an integral equation and we can not possibly differentiate it and turn it into an integrodifferential equation. Moreover, one can see that none of the techniques developed in Chapter 2 can possibly apply to it. Finally, if we think about problems with memory, it has the most remarkable kind of memory. That kernel assigns an infinite weight to the solution at $t = 0$. We have previously discussed memory problems and have pointed out that common problems put the greatest weight on the part of $x(s)$ near $s = t$, while the weight reduces as $t - s$ increases; this is the whole idea of a convex kernel. Here we have the opposite situation. The solution can never forget what has happened to it at $t = 0$. While it is a situation so different from those of previous discussions, one can hardly avoid the thought that the genetic code at our conception plays a central role in our lives from the first outraged wail to the last labored gasp.

4.2.2 Temperature in a semi-infinite Rod

Komarath Padmavally (1958) studies the temperature $U(x,t)$ of the point x of a rod at time t. The rod is assumed to be initially at 0 throughout and its end is heated by a known variable temperature $\phi(t)$. Others had

4.2. CLASSICAL PROBLEMS

studied the problem earlier with ϕ constant. The temperature function U satisfies the partial differential system

$$U_{xx}(x,t) = U_t(x,t), \qquad x > 0, \qquad t > 0,$$

$$U(x,0) = 0,$$

and

$$-U_x(0,t) = g[U(0,t), \phi(t)]$$

where $g(u,v)$ denotes the rate of flow of heat from the source at temperature v to the end of the rod at temperature u.

Under a set of conditions it can be shown that any function satisfying this partial differential system also satisfies

$$U(x,t) = \int_0^t \frac{g[U(0,s), \phi(s)]}{\sqrt{\pi(t-s)}} e^{-(x^2/4(t-s))} ds$$

and if $U(x,t)$ is continuous in $x \geq 0, t \geq 0$, then

$$U(0,t) = \int_0^t \frac{g[U(0,s), \phi(s)]}{\sqrt{\pi(t-s)}} ds.$$

Again, we could hardly hope to convert this to an ordinary differential equation by differentiation. However, while it is singular, the weight occurs at t, rather than at 0 so the memory is consistent with our experiences. We should not give up on the idea of using Liapunov functionals for problems with mildly singular kernels. For some of these problems we can construct a Liapunov functional of the form

$$V(t) = \int_0^t C_s(t,s) \left(\int_s^t g(x(s)) ds \right)^2 + C_t(t,0) \left(\int_0^t g(x(s)) ds \right)^2$$

for $t > 0$. While Leibnitz's rule can not be used owing to the singularity, in some such problems one can show under reasonable conditions that V does have a continuous derivative.

The heat problem has been studied by many investigators. Gripenberg et al. (1991; p. 649) gives many references and devotes two of the preceeding sections to the development. Many good problems are described by Corduneanu (1991), especially pp. 22-23.

For more recent contributions of this general nature in which heat equations are converted to integral equations where blow-up and quenching occurs, as well as periodic solutions, see Kirk and Roberts (2002), Kirk and Olmstead (2000), (2002), and (2005), as well as extensive references contained therein. These authors convert partial differential equations to integral equations using Green's functions. The kernels almost always have mild singularities which are at least locally integrable so that many of the techniques used here will apply. A notable example is found in Kirk and Olmstead (2002; p. 134, Equation (39)) with a globally L^1-kernel so fixed point mappings are readily defined, as are limiting equation arguments of the type found in Miller (1971a; p. 172).

On the other hand there are famous problems related to integral equations which can be found in the most unlikely places. One of these problems concerning a body sliding down a cycloid is found in Chapter XCVI of the classic whaling tale told by Herman Melville (1950; p. 419) in his book Moby Dick. Correct details are found in the Mathematics Dictionary by James and James (1959; p. 37) in the second edition.

4.3 The Nature of the Kernel

In the work here one can detect the use of eight different Liapunov functions and functionals. They are of three basic and very different types.

In the convolution case with an L^1 kernel and L^p forcing function we can often determine boundedness from the Laplace transform and the roots of a transcendental equation. None of that will be considered in this book. When these conditions fail, we are in a world without order and we search for conditions to ensure boundedness. In Chapter 2 and the latter part of Chapter 3 we presented a large number of results concerning boundedness of solutions with little motivation or relation to a given problem. In this section we hope to impart some ideas for the intuition.

An integral equation has a memory and we would so like to understand what kind of memory promotes bounded solutions and what kind does not. We have seen a vast number of results of this type. Sometimes we find that for a small kernel the solution is bounded, but sometimes we have a large kernel with derivatives of certain signs which promote boundedness. We wonder if the solution will become unbounded if we change the signs of the derivative.

It turns out that this question arises in the literature in a very prominent place by a leading investigator and it is glossed over without comment. In this section we will present that problem and study the possibilities.

The enticing idea emerges that small kernels promote boundedness if $a(t)$ is small, while boundedness can also be promoted with large $a(t)$ pro-

4.3. THE NATURE OF THE KERNEL

vided that the kernel is quite smooth. Let us see how it unfolds and where it can lead us.

We study integral equations of the form $x(t) = a(t) - \int_0^t C(t,s)x(s)ds$ with sharply contrasting kernels typified by $C^*(t,s) = \ln(e + (t-s))$ and $D^*(t,s) = [1 + (t-s)]^{-1}$. The kernel assigns a weight to $x(s)$ and these kernels have exactly opposite effects of weighting. Each type is well represented in the literature. Our first project is to show that for $a \in L^2[0, \infty)$, then solutions are largely indistinguishable regardless of which kernel is used. This is a surprise and it leads us to study the essential differences. In fact, those differences become large as the magnitude of $a(t)$ increases.

The form of the kernel alone projects necessary conditions concerning the magnitude of $a(t)$ which could result in bounded solutions. Thus, the next project is to determine how close we can come to proving that the necessary conditions are also sufficient. In fact, the necessary condition is that for a modified C^* then $a(t)$ should not exceed $(t+1)^2$ for bounded solutions, while we can obtain sufficient conditions for $a(t) = (t+1)^p$ where $0 < p < 3/2$, an unexpectedly large function.

The third project is to show that solutions will be bounded for given conditions on C regardless of whether a is chosen large or small; this is important in real-world problems since we would like to have $a(t)$ as the sum of a bounded, but badly behaved function, and a large well behaved function.

The work here is, in every respect, of nonconvolution type. But it will be easier to explain the direction we will take by discussing some simple convolution kernels.

Let

$$C^*(t,s) = \ln(e + (t-s)) \qquad (4.3.1)$$

and

$$D^*(t,s) = \frac{1}{(t-s)+1}, \qquad (4.3.2)$$

noting that

$$C^*(t,t) = 1, \qquad D^*(t,t) = 1.$$

We use the symbol $*$ here because these functions will later be generalized and denoted by C and D in (4.3.11) and (4.3.12). In particular, positive constants can be added to and multiplied by these functions without changing our basic assumptions.

Note also that $C^*(t,0) = \ln(e+t)$ is increasing to infinity, while $D^*(t,0)$ is decreasing to zero.

Here is the first question we study. If $a \in L^2[0, \infty)$ what are the essential qualitative differences between solutions of

$$x(t) = a(t) - \int_0^t C^*(t,s)x(s)ds \qquad (4.3.3^*)$$

and

$$z(t) = a(t) - \int_0^t D^*(t,s)z(s)ds \qquad (4.3.4^*)$$

and do investigators have reason to care?

For later reference, Equations (4.3.3) and (4.3.4) will follow (4.3.12).

There would be no story to tell unless it were the case that there is little qualitative difference and that many investigators have studied such problems since 1928. We find that under general conditions patterned from (4.3.1) and (4.3.2) that $z \in L^2[0, \infty)$, while for $y = \int_0^t x(s)ds$ we have $y \in L^2[0, \infty)$ and $\int_0^t x(s)ds \to 0$ as $t \to \infty$. More can be said. It is a surprise because C^* and D^* have fairly opposite properties and $(4.3.4^*)$ is known to have nice qualitative properties so we would suspect that $(4.3.3^*)$ does not.

Kernel (4.3.1) is the prototype and is, so to speak, the middle of the road. We will devote the next subsection to showing how closely the solution of $(4.3.3^*)$ is to the solution of $(4.3.4^*)$ when $a \in L^2$ and then show that equations in the class of $(4.3.3^*)$ are so very powerful when $a(t)$ is extremly large; the solution is kept small in spite of enormous perturbations. In the process we will vary C^* considering also the kernels

$$r(t-s) + \ln(e + (t-s)), \qquad r \geq 0,$$

and

$$1 + \text{Arctan}(t-s)$$

as members of the same class of kernels as in (4.3.1), but being above and below (4.3.1), respectively.

Let us interpret C^* and D^*. For fixed t the integrals involve history of the solution on the interval $[0, t]$. For a given s in that interval we are multiplying $x(s)$ by a weight and then the integral is adding up all of those products to determine, along with $a(t)$, the value of the solution. Notice that at $s = 0$ then C^* has the weight $\ln(e + t)$, which is large for t large, while for $s = t$ then C^* has the weight $C^*(t,t) = 1$; the value of $x(t)$ is being overwhelmed by the early values of x, while recent values, by comparison, are practically ignored. It is customary to refer to (4.3.2)

4.3. THE NATURE OF THE KERNEL

as an example of a fading memory kernel. Accordingly, we could refer to (4.3.1) as a growing memory kernel, typically seen in problems driven by genetics where individual characteristics become more pronounced as an individual ages.

It is not at all unusual to see problems which have a growing memory. We mentioned earlier the work of Reynolds (1984) on the buckling of viscoelastic rods. His problem is singular and the solution is continually driven by an infinite weight at zero. On the other hand, the literature is replete with fading memory problems, many of which are singular, but with a locally integrable singularity. We have mentioned work on superfluidity by Levinson (1960), or heat transfer by Padmavally (1958), for example.

In an attempt to convey to the reader an image of what is happening we suggest the following. Think of (4.3.1) as "genetically" driven while (4.3.2) is "environmentally" driven. In the first case an individual's characteristics are continually magnified as a result of genetics; the infant comes to resemble the parent more and more as time goes on. In the second case, the individual's characteristics are changing because of diet, exercise, and general environment; sadly, good habits of youth translate into far too little benefit if not practiced in our old age, a typical example of fading memory.

The problem addressed here originates in a paper of Levin (1965) which contains an ambiguous statement. Levin reviews an equation

$$x'(t) = -\int_0^t a(t-s)g(x(s))ds \qquad (4.3.5)$$

with

$$a(t) \in C[0,\infty), \quad (-1)^k a^{(k)}(t) \geq 0$$
$$\text{for} \quad (0 < t < \infty; k = 0,1,2,3) \qquad (4.3.6)$$

and $g(x)$ continuous on $(-\infty, \infty)$ with

$$xg(x) > 0 \quad \text{for} \quad x \neq 0. \qquad (4.3.7)$$

Equation (4.3.5) has application in many areas beginning with mathematical biology, reactor dynamics, and viscoelasticity. Volterra (1928) proposed a related form for a problem in mathematical biology, suggesting that one might construct a Liapunov functional. That functional was constructed by Levin (1963), yielding strong qualitative results for (4.3.5).

But in Levin's paper of 1965 he is interested in

$$x(t) = f(t) - \int_0^t b(t-s)g(x(s))ds \qquad (4.3.8)$$

under the assumptions

$$b(t) \in C^1[0,\infty), \quad (-1)^k b^{(k)}(t) \geq 0 \quad \text{for} \quad (0 \leq t < \infty; k = 0, 1) \quad (4.3.9)$$

and

$$f(t) \in C^1[0,\infty), \quad \int_0^\infty |f'(t)|dt < \infty, \tag{4.3.10}$$

yielding $|x(t)|$ bounded and other qualitative results.

The work begins when Levin remarks that (4.3.5) can be converted to (4.3.8) by integration and that $b(t) = \int_0^t a(s)ds$ and $f(t) = x(0)$ for that conversion. While it is true that (4.3.5) can be written as

$$x(t) = x(0) - \int_0^t \int_0^{t-u} a(s)ds g(x(u))du,$$

for $b(t) = \int_0^t a(s)ds$ and $a(t) \geq 0$, then we do have $b(t) \geq 0$, as required in (4.3.9), but $b'(t) = a(t) \geq 0$ violating (4.3.9) in the nontrivial case.

The question arises: Can we violate $b'(t) \leq 0$ and still retain nice qualitative properties? Here, we construct Liapunov functionals proving this under more general conditions on the kernel. We arrive at the stunning conclusion that it makes little difference whether the kernel increases or decreases, when other important conditions hold. Indeed, if Levin had continued with the case which violated (4.3.9) he would have been dealing with an equation whose solutions were little changed for $a \in L^2$, but infinitely better able to withstand large perturbations without letting the solution become large.

4.3.1 Necessary conditions for boundedness

In both (4.3.5) and (4.3.8) Levin is concerned with showing that the solution remains small. One of the goals of the investigator is to identify kernels which will promote stability. It is very simple to show that the kernel (4.3.1) is potentially far more stable than (4.3.2).

We ask the question: How large can $a(t)$ be and still have $x(t)$ or $z(t)$ bounded? We are dealing with very large kernels and it is going to require strong methods to prove boundedness. It is time to test just how strong the Liapunov methods are.

Theorem 4.3.1. *If r is a positive constant and $C(t,s) = r(t-s) + C^*(t,s)$, then $x(t) = a(t) - \int_0^t C(t,s)x(s)ds$ has a bounded solution only if there is a constant $M > 0$ with*

$$|a(t)| \leq M\left(1 + (t+1)^2\right).$$

4.3. THE NATURE OF THE KERNEL

Proof. If $|x(t)| \leq K$ then we have

$$|a(t)| \leq |x(t)| + \int_0^t [r(t-s) + \ln(e + (t-s))]|x(s)|ds$$
$$\leq K + K\int_0^t [rs + \ln(e+s)]ds$$
$$\leq M + M(t+1)^2$$

for some $M > 0$.

Is this a sharp estimate? Are our techniques good enough that if $a(t) = (t+1)^2$, can we expect to prove that $x(t)$ is bounded using that kernel? So far, we prove that for this $C(t,s)$ then we can take $a(t) = (t+1)^p$ where $0 < p < 3/2$ in order to obtain $x(t)$ bounded.

Lest we become too disappointed with that result, let us realize that what we have proved is that $a(t) = (t+1)^p$ for $p < 3/2$ is a harmless perturbation. And that should give us some pause when we consider the feeble perturbations which motivated this study.

Next, if we are to have $z(t)$ bounded, how large can we choose $a(t)$?

Theorem 4.3.2. *If $D(t,s) = D^*(t,s)$, then $z(t) = a(t) - \int_0^t D(t,s)z(s)ds$ has a bounded solution only if there is an $M > 0$ with*

$$|a(t)| \leq M[1 + \ln(t+1)].$$

Proof. If $|z(t)| \leq K$ then

$$|a(t)| \leq |z(t)| + \int_0^t D^*(t,s)|z(s)|ds$$
$$\leq K + K\int_0^t (u+1)^{-1}du$$
$$\leq M + M\ln(t+1)$$

for some $M > 0$.

4.3.2 Small perturbations

In this section we want to show that if $a \in L^2$ then (4.3.1) and (4.3.2) result in similar behavior of the solutions. In the context of genetics and environment, we interpret this to mean that if $a(t)$ represents challenges and opportunities, then for very small $a(t)$ the genetically driven individual and the environmentally driven individual do about the same. In the next section we will show that when the challenges and opportunities are great, then the genetically driven individual will perform far better than the environmentally driven individual.

We look to (4.3.1) and (4.3.2) for guidance in our assumptions by defining new functions C and D with the following derivatives being at least continuous:

$$C(t,t) \geq \alpha > 0, \quad C_s(t,s) \leq 0, \quad C_{ss}(t,s) \leq 0,$$
$$C_{st}(t,s) \geq 0, \quad C_{sst}(t,s) \geq 0, \tag{4.3.11}$$

and

$$D_s(t,s) \geq 0, \quad D(t,0) \geq 0,$$
$$D_t(t,0) \leq 0, \quad D_{st}(t,s) \leq 0. \tag{4.3.12}$$

These are large kernels and it should not be thought that an element, $C(t,s)$, satisfying (4.3.11) is necessarily larger or smaller than an element, $D(t,s)$, satisfying (4.5.12). For example,

$$D(t,s) = M + D^*(t,s), \quad M \geq 0,$$

satisfies (4.5.12) and, if M is large, then it lies entirely above

$$C(t,s) = 1 + \text{Arctan}(t-s),$$

satisfying (4.5.11). So often in the theory of integral equations methods call for kernels to be of convolution type and $L^1[0, \infty)$, those requirements will never hold for (4.3.11) and they need not hold for (4.3.12). Hence, different methods are needed. Liapunov functionals supply the need in a very simple way.

With these assumptions we now define the more general equations as

$$x(t) = a(t) - \int_0^t C(t,s)x(s)ds \tag{4.3.3}$$

and

$$z(t) = a(t) - \int_0^t D(t,s)z(s)ds. \tag{4.3.4}$$

Under these conditions (4.3.4) becomes exactly our old (2.1.1) of Theorem 2.1.10 and we repeat that result now.

4.3. THE NATURE OF THE KERNEL

Theorem 4.3.3. *If $a : [0, \infty) \to R$ is continuous, while (4.3.12) holds for (4.3.4) then along the solution of (4.3.4) the functional*

$$V(t) = \int_0^t D_s(t,s) \left(\int_s^t z(u)du \right)^2 ds + D(t,0) \left(\int_0^t z(s)ds \right)^2$$

satisfies

$$V'(t) \leq -z^2(t) + a^2(t).$$

(i) If $a \in L^2[0, \infty)$, so is z; moreover, $V(t)$ is bounded.
(ii) If there are constants B and K with

$$\sup_{t \geq 0} \int_0^t D_s(t,s)ds = B < \infty \quad \text{and} \quad \sup_{t \geq 0} D(t,0) = K < \infty$$

then along the solution of (4.3.4) we have

$$(a(t) - z(t))^2 \leq 2(B+K)V(t)$$

where (4.3.4) does not require $a \in L^2$. However, if $a \in L^2$ and bounded then both $V(t)$ and z are bounded.

If we are keeping track of the number of Liapunov functionals used, we might keep in mind that this V actually originated with an infinite delay equation and had the form

$$V(t) = \int_{-\infty}^t D_s(t,s) \left(\int_s^t z(u)du \right)^2 ds$$

so the present one is second in line.

Notice that to this point we conclude that (4.3.4) with (4.3.12) is quite straightforward with $a \in L^2$ implying that $z \in L^2$ and we consider that result sufficient. But matters are more difficult for (4.3.3). However, with more work it does turn out that for $a(t)$ small then (4.3.12) and (4.3.13) yield surprisingly similar behavior.

Next, we return to the methods of Section 2.3 and refer the reader again to Theorems 2.2.8 and 2.3.6. Integrate (4.3.3) by parts and write

$$x(t) = a(t) - C(t,t) \int_0^t x(s)ds + \int_0^t C_s(t,s) \int_0^s x(u)duds$$

so that by taking

$$y(t) = \int_0^t x(s)ds$$

we have

$$y'(t) = a(t) - C(t,t)y(t) + \int_0^t C_s(t,s)y(s)ds. \tag{4.3.13}$$

The reader may verify that it is possible to find C and D satisfying (4.3.11) and (4.3.12), respectively, whose sum will satisfy the conditions here. Thus, one may consider equations driven both genetically and environmentally.

In reading Theorems 4.3.5 and 4.3.6 it may help to think of them in terms of (i) and (ii) of Theorem 4.3.3 holding.

Theorem 4.3.4. *Suppose that (4.3.11) holds. If y is a solution of (4.3.13) and if we define V by*

$$V(t) = y^2(t) - \int_0^t C_{ss}(t,s)\left(\int_s^t y(u)du\right)^2 ds - C_s(t,0)\left(\int_0^t y(u)du\right)^2$$

then the derivative of V satisfies

$$V'(t) \leq (1/\alpha)a^2(t) - \alpha y^2(t).$$

Thus,

$$y^2(t) + \alpha \int_0^t y^2(s)ds \leq V(0) + (1/\alpha)\int_0^t a^2(s)ds.$$

If, in addition, $a \in L^2[0,\infty)$ and if both $C(t,t)$ and $\int_0^t |C_s^2(t,s)|ds$ are bounded, then

$$|x(t) - a(t)|$$

is bounded.

Proof. We have

$$V'(t) = 2ya(t) - 2C(t,t)y^2 + 2y\int_0^t C_s(t,s)y(s)ds$$

$$- \int_0^t C_{sst}(t,s)\left(\int_s^t y(u)du\right)^2 ds - C_{st}(t,0)\left(\int_0^t y(u)du\right)^2$$

$$- 2yC_s(t,0)\int_0^t y(u)du - 2y\int_0^t C_{ss}(t,s)\int_s^t y(u)duds.$$

4.3. THE NATURE OF THE KERNEL

Integration of the last term by parts yields

$$-2y\left[C_s(t,s)\int_s^t y(u)du\Big|_0^t + \int_0^t C_s(t,s)y(s)ds\right]$$
$$= -2y\left[-C_s(t,0)\int_0^t y(u)du + \int_0^t C_s(t,s)y(s)ds\right]$$

so that by collecting terms and taking into account sign conditions we now arrive at

$$V'(t) \leq 2ya(t) - 2\alpha y^2 \leq (1/\alpha)a^2(t) - \alpha y^2(t).$$

We may write

$$y^2(t) \leq V(t) \leq V(0) + (1/\alpha)\int_0^t a^2(s)ds - \alpha \int_0^t y^2(s)ds$$

so that

$$y^2(t) + \alpha \int_0^t y^2(s)ds \leq V(0) + (1/\alpha)\int_0^t a^2(s)ds.$$

With $a \in L^2[0,\infty)$ we have $y^2(t) + \alpha \int_0^t y^2(s)ds$ bounded. Now, from (4.3.13) we have $x = y'$ so

$$|x(t) - a(t)| \leq |C(t,t)||y(t)| + \sqrt{\int_0^t C_s^2(t,s)ds \int_0^t y^2(s)ds}$$

which is bounded.

Theorem 4.3.4 shows that $y^2(t) = \left(\int_0^t x(s)ds\right)^2 \in L^1[0,\infty)$ which, of course, says that there is a sequence $\{t_n\} \uparrow \infty$ along which that integrand tends to zero. But since that integrand is an integral, the integrand actually converges to zero. Here are the details.

In understanding this result, recall that Theorem 4.3.4 gave conditions ensuring that $|x(t) - a(t)|$ is bounded so if a is bounded, then that yields x bounded, as required below.

4. APPLICATIONS AND ORIGINS

Theorem 4.3.5. If

$$\int_0^t y^2(s)ds = \int_0^t \left(\int_0^s x(u)du \right)^2 ds$$

is bounded and if x is bounded, then

$$y(t) = \int_0^t x(u)du \to 0$$

as $t \to \infty$. Moreover, for each $L > 0$, it is true that $\int_{t-L}^t x(u)du \to 0$ as $t \to \infty$. In particular, if $\{[s_n, t_n]\}$ is a sequence of intervals on which $x(t)$ is of one sign, where $s_n \to \infty$, then $\int_{s_n}^{t_n} x(s)ds \to 0$ as $n \to \infty$.

Proof. If the theorem is false then there is an $\epsilon > 0$ and a sequence $\{t_n\} \uparrow \infty$ with

$$\left(\int_0^{t_n} x(u)du \right)^2 \geq \epsilon.$$

Since the integral of y^2 converges, for each n there is a $t > t_n$ with $(\int_0^t x(u)du)^2 < \epsilon/2$, so there is a sequence $\{\lambda_n\}$ of positive numbers with

$$\left(\int_{t_n}^{t_n + \lambda_n} x(u)du \right)^2 = \epsilon/2$$

and the equality is false for a smaller λ. Clearly, $\lambda_n \to 0$ as $n \to \infty$, otherwise we would have

$$\left(\int_0^{t_n + s} x(u)du \right)^2 \geq \epsilon/2$$

for all $s \in [0, \lambda_n]$, contradicting the convergence. As $x(t)$ is bounded and $\lambda_n \to 0$ we have a contradiction. Notice that $\int_0^t x(u)du$ and $\int_0^{t+L} x(u)du \to 0$ as $t \to \infty$ so the same is true for their difference.

Notice also that in (4.3.2) we have $D^*(t, s) \to 0$ as $t - s \to \infty$, while in (4.3.1) we have $C^*(t, s) \to \infty$ as $t - s \to \infty$. If we equalize these and let $C(t, s) \to L < \infty$ as $t - s \to \infty$, then we can obtain a much stronger result.

4.3. THE NATURE OF THE KERNEL

Theorem 4.3.6. Let x solve (4.3.3), let (4.3.11) hold, let y solve (4.3.13), and let $y(t) = \int_0^t x(u)du \to 0$ as $t \to \infty$. If for all large fixed T we have

$$C(t,T) - C(t,0) \to 0 \quad \text{as} \quad t \to \infty$$

and both

$$C(t,t) \quad \text{and} \quad C(t,T)$$

are bounded independently of t and T, then

$$|x(t) - a(t)| \to 0$$

as $t \to \infty$.

Proof. In (4.3.13) we see that $C(t,t)y(t) \to 0$ as $t \to \infty$. Next, for T large, for $t > T$, and for $\|\cdot\|$ denoting the supremum norm, we have

$$\int_0^t |C_s(t,s)y(s)|ds = \int_0^T |C_s(t,s)y(s)|ds + \int_T^t |C_s(t,s)y(s)|ds$$

$$\leq \|y\| \int_0^T -C_s(t,s)ds + \left(\sup_{s \geq T}|y(s)|\right)\int_T^t -C_s(t,s)ds$$

$$= \|y\|[C(t,0) - C(t,T)] + \sup_{s \geq T}|y(s)|[-C(t,t) + C(t,T)].$$

Consider the last line. Let $\epsilon > 0$ be given. For the last term, since $C(t,t)$ and $C(t,T)$ are bounded, take T so large that the last term is bounded by ϵ. With that T fixed, let t be so large that the first term is also bounded by ϵ.

Now, we will have good reason (discussed at the beginning of the last subsection) for wanting to show that solutions of (4.3.3) are bounded when we only ask that $a(t)$ is bounded. One such result will now be given.

Theorem 4.3.7. Suppose that (4.3.11) holds, that $a(t)$ is bounded, and that there is an $M > 0$ with

$$-\int_0^t C_{ss}(t,s)(t-s)\int_s^t a^2(u)duds + t|C_s(t,0)|\int_0^t a^2(u)du \leq M.$$

Then for V defined in Theorem 4.3.4 we have both V and $y^2(t) = \left(\int_0^t x(s)ds\right)^2$ bounded.

Proof. If V is not bounded then there is a sequence $\{t_n\} \uparrow \infty$ with $V(t_n) \geq V(s)$ for $0 \leq s \leq t_n$ and there is a $\gamma > 0$ with $y^2(t_n) \leq \gamma$, as may be seen from the derivative of V. Taking $t = t_n$ we then have from $V'(t) \leq (1/\alpha)a^2(t) - \alpha y^2(t)$ that

$$0 \leq V(t) - V(s) \leq (1/\alpha) \int_s^t a^2(u)du - \alpha \int_s^t y^2(u)du$$

or that

$$\int_s^t y^2(u)du \leq (1/\alpha^2) \int_s^t a^2(u)du.$$

Use this and the Schwarz inequality in V to obtain

$$V(t) \leq y^2(t) - \int_0^t C_{ss}(t,s)(t-s) \int_s^t (1/\alpha^2)a^2(u)duds$$
$$+ t|C_s(t,0)| \int_0^t (1/\alpha^2)a^2(u)du$$
$$\leq \gamma + (1/\alpha^2)M.$$

The result follows from this.

We readily find $a(t)$ for which conditions hold for $C(t,s) = 1 + \text{Arctan}(t-s)$.

4.3.3 Sufficient conditions for boundedness

We now turn to the necessary conditions which we have derived and derive conditions on D and C so that we obtain sufficient conditions for boundedness. These boundedness results are based on $a(t)$ being differentiable. In the last section we will seek boundedness without differentiating $a(t)$.

The next result is just our old Theorem 2.2.6 which we restate for reference.

Theorem 4.3.8. *Let* $a' \in L^2[0,\infty)$, $D(t,t) \geq \alpha > 0$, *and*

$$-2\alpha + \int_0^\infty |D_1(u+t,t)|du + \int_0^t |D_t(t,s)|ds \leq -\beta$$

for some $\beta > 0$. Then $z \in L^2[0,\infty)$ and is bounded where z is a solution of (4.3.4).

4.3. THE NATURE OF THE KERNEL

Our D^* will not quite satisfy the conditions of Theorem 4.3.8 and its necessary condition of Theorem 4.3.2. Thus, we turn to C and ask if we can simply differentiate (4.3.3) and conquer $a(t) = \int_0^t \ln[e+s]ds$ using $C(t,s) = \ln[e+(t-s)]$; in fact, we will have to differentiate twice.

Theorem 4.3.9. Let $C(t,s)$ satisfy

$$C(t,t) \geq \alpha > 0,$$
$$C_t(t,s) \geq 0, \quad C_{tt}(t,s) \leq 0, \quad C_{ts}(t,s) \geq 0, \quad C_{tts}(t,s) \leq 0.$$

Then $a' \in L^2$ implies that the solution x of (4.3.3) is also in L^2; moreover, $x(t)$ is bounded.

Proof. The derivative of (4.3.3) is

$$x' = a'(t) - C(t,t)x - \int_0^t C_t(t,s)x(s)ds.$$

Define

$$V(t) = x^2 + \int_0^t C_{ts}(t,s)\left(\int_s^t x(u)du\right)^2 ds + C_t(t,0)\left(\int_0^t x(u)du\right)^2;$$

Since $x^2 \leq V$, if V is bounded so is x. Differentiate V along a solution of that derivative of (4.3.3) and obtain

$$V'(t) = 2xa'(t) - 2C(t,t)x^2$$
$$- 2x\int_0^t C_t(t,s)x(s)ds + C_{tt}(t,0)\left(\int_0^t x(u)du\right)^2$$
$$+ 2xC_t(t,0)\int_0^t x(u)du + \int_0^t C_{tst}(t,s)\left(\int_s^t x(u)du\right)^2 ds$$
$$+ 2x\int_0^t C_{ts}(t,s)\int_s^t x(u)duds.$$

If we integrate the last term by parts we obtain

$$2x\left[C_t(t,s)\int_s^t x(u)du \Big|_0^t + \int_0^t C_t(t,s)x(s)ds\right]$$
$$= 2x\left[-C_t(t,0)\int_0^t x(u)du + \int_0^t C_t(t,s)x(s)ds\right].$$

This results in

$$V'(t) \leq 2xa'(t) - 2\alpha x^2$$

A standard inequality, followed by integration, now finishes the proof.

4. APPLICATIONS AND ORIGINS

If we hope to conquer $a(t) = \int_0^t \ln[e+s]ds$ with $C^*(t,s)$ it is clear that we must take another derivative of (4.3.3).

We displayed three "genetic" type kernels:

$r(t-s) + C^*(t,s), \quad r > 0,$

$C^*(t,s)$

$1 + \text{Arctan}(t-s).$

For $a \in L^2$ we found that the smallest one generated behavior of x more closely approximating that of $D^*(t,s)$. In this subsection we will show that the largest one will yield $x(t)$ bounded when $a'' \in L^2$, allowing $a(t) = (t+1)^p$, where $p < 3/2$, for example.

We are now going to continue the process and obtain second order equations. First, return to (4.3.3) and differentiate twice to obtain

$$x''(t) = a''(t) - [(C(t,t))' + C_t(t,t)]x - C(t,t)x' - \int_0^t C_{tt}(t,s)x(s)ds. \quad (4.3.14)$$

We come now to the critical part. Given (4.3.14), can we find an appropriate Liapunov functional? We can, and with fascinating ease by the simple device of integration by parts; we have learned to use integrals of the form given here in many contexts for construction of Liapunov functionals and adding them together.

From (4.3.14) we have

$$x' = q$$

$$q' = a'' - [(C(t,t))' + C_t(t,t)]x - C(t,t)q - \int_0^s C_{tt}(t,u)dux(s)\Big|_0^t$$

$$+ \int_0^t \int_0^s C_{tt}(t,u)duq(s)ds$$

or

$$x' = q$$

$$q' = a'' - [(C(t,t))' + C_t(t,t) + \int_0^t C_{tt}(t,u)du]x(t)$$

$$+ \int_0^t \int_0^s C_{tt}(t,u)duq(s)ds - C(t,t)q.$$

In the theorem below the reader will reasonably ask where these conditions come from and we want to show that they are natural.

4.3. THE NATURE OF THE KERNEL

Consider the prototype
$$C(t,s) = r(t-s) + \ln[e + (t-s)], \quad r > 0.$$
Then
$$C(t,t) = 1 =: \alpha$$
$$C_t(t,s) = r + [e + (t-s)]^{-1},$$
$$C_{tt}(t,s) = -[e + (t-s)]^{-2},$$
$$C_{ttt}(t,s) = 2[e + (t-s)]^{-3},$$
$$(C(t,t))' = 0.$$

Notice that as long as $C(t,s)$ is of convolution type then
$$\int_0^t C_{tt}(t,u)du = -C_t(t,u)\Big|_0^t = -C_t(t,t) + C_t(t,0).$$
Thus, in the theorem below we would have
$$K(t) = 0 + C_t(t,t) - C_t(t,t) + C_t(t,0) = r + [e+t]^{-1} \geq r > 0$$
and
$$K'(t) = -[e+t]^{-2}.$$

Theorem 4.3.10. *Suppose there is a positive constant α with*
$$C_{tt}(t,u) \leq 0,$$
$$K(t) =: (C(t,t))' + C_t(t,t) + \int_0^t C_{tt}(t,u)du \geq 0,$$
$$K'(t) \leq 0,$$
$$C_{ttt}(t,s) \geq 0,$$
$$C(t,t) \geq \alpha > 0.$$

Then for V defined by
$$V(t) = K(t)x^2 + q^2 - \int_0^t C_{tt}(t,s)\left(\int_s^t q(u)du\right)^2 ds$$
the derivative of V along a solution of (4.3.14) satisfies
$$V'(t) \leq -\alpha q^2(t) + \frac{1}{\alpha}(a''(t))^2.$$
In particular, if $a'' \in L^2[0,\infty)$ then $q = x' \in L^2[0,\infty)$.

Proof. We have

$$V'(t) = 2qxK(t) + K'(t)x^2 + 2qa'' - K(t)2qx - 2q^2C(t,t)$$
$$+ 2q\int_0^t\int_0^s C_{tt}(t,u)du\, q(s)ds - \int_0^t C_{ttt}(t,s)\left(\int_s^t q(u)du\right)^2 ds$$
$$- 2q\int_0^t C_{tt}(t,s)\int_s^t q(u)du\, ds.$$

If we integrate the last term by parts we have

$$-2q\left[\int_0^s C_{tt}(t,u)du \int_s^t q(u)du \Big|_0^t + \int_0^t\int_0^s C_{tt}(t,u)du\, q(s)ds\right]$$

and notice that the first term is zero, while the last term cancels with another term in the derivative of V. Collecting terms and taking into account the sign conditions in the theorem we see that

$$V'(t) \leq -2\alpha q^2(t) + 2q(t)a''(t) \leq -\alpha q^2(t) + \frac{1}{\alpha}(a''(t))^2,$$

as required. This will give $q \in L^2$, V bounded.

We see that if we use $C^*(t,s)$ then $K(t) \to 0$ and we do not have $x(t)$ bounded on the basis of this theorem. But if we use $r(t-s) + C^*(t,s)$ then $K(t) \geq r > 0$. For $a(t) = (t+1)^p$ we would need $0 < p < 3/2$ to have $a'' \in L^2$.

Corollary. *Let the conditions of the last theorem hold and suppose there is a $\beta > 0$ with*

$$K(t) \geq \beta.$$

Then $x(t)$ is bounded.

Proof. We have $\beta x^2(t) \leq V(t)$ and so V bounded implies x bounded.

4.3.4 A balanced case

In the last subsection we saw x and z bounded when $a' \in L^2$ and that is a strong condition; real-world problems frequently have uncertainties which would make that so hard to ascertain. We would like to say that if $a_1 - a_2$ is bounded, then the difference between the solutions of $x = a_i(t) - \int_0^t C(t,s)x(s)ds$, $i = 1,2$, is bounded. And that difference is a solution of $x(t) = a_1(t) - a_2(t) - \int_0^t C(t,s)x(s)ds$. The application is that

4.3. THE NATURE OF THE KERNEL

$a_1(t)$ might satisfy the differentiablity conditions, while a_2 does not. So we would get boundedness of solutions with a_2 by studying a_1. In conclusion, then, we want to show that the solution of $x(t) = a(t) - \int_0^t C(t,s)x(s)ds$ is bounded when $a(t)$ is bounded.

We now pattern $C(t,s)$ after $k + \text{Arctan}(t-s)$ by asking that there is a $k > 0$ with

$$C(t,t) \geq k \tag{4.3.15}$$

and that there is an $\alpha > 0$ with

$$\int_0^t |C_s(t,s)|ds \leq k - \alpha \tag{4.3.16}$$

and

$$\int_0^t |C_t(t,s)|ds \leq k - \alpha. \tag{4.3.17}$$

Notice that a necessary condition for the solution of

$$x(t) = a(t) - \int_0^t [k + \text{Arctan}(t-s)]x(s)ds$$

to be bounded, say $|x(t)| \leq M$, is that

$$|a(t)| \leq M + M\int_0^t [k + \text{Arctan}(t-s)]ds \leq J(1+t)$$

for some $J > 0$.

Theorem 4.3.11. *Let (4.3.15) and (4.3.17) hold and let $a(t) = J(1+t)^p$, $0 \leq p \leq 1$. Then the solution of (4.3.3) is bounded.*

Proof. Notice that

$$x'(t) = Jp(1+t)^{p-1} - C(t,t)x - \int_0^t C_t(t,s)x(s)ds, \quad x(0) = J$$

and for fixed p there is an L with

$$|Jp(1+t)^{p-1}| \leq L.$$

Find $x_1 > 0$ such that $x_1 > |J|$ and $L - kx_1 + (k-\alpha)x_1 < 0$. There is an interval $[0, t_1)$ with $|x(t)| < x_1$. Suppose that for such an interval $|x(t_1)| = x_1$. The Razumikhin function

$$V(t) = |x(t)|$$

satisfies

$$V'(t) \leq |Jp(1+t)^{p-1}| - k|x| + \int_0^t |C_t(t,s)||x(s)|ds$$

so that for $t \leq t_1$ we have

$$V'(t) \leq L - k|x(t)| + |x(t_1)|(k-\alpha)$$

and it is negative at $t = t_1$, a contradiction to V increasing at t_1.

We now want to show that the solution of (4.3.3) is still bounded if $a(t)$ differs from $J(t+1)^p$ by at most a bounded function. In particular we do not require $a(t)$ to be differentiable. Thus, we consider (4.3.3) and write

$$y' = a(t) - C(t,t)y + \int_0^t C_s(t,s)y(s)ds$$

where $y(t) = \int_0^t x(s)ds$.

Theorem 4.3.12. *Let $|a(t)| \leq M$ for some $M > 0$, let $C(t,t)$ be bounded, and let (4.3.15) and (4.3.16) hold. Then the solution of (4.3.3) is bounded.*

Proof. Exactly the same argument as in the last theorem shows that y is bounded. But

$$|x(t)| = |y'(t)| \leq |a(t)| + |C(t,t)||y(t)| + \|y\| \int_0^t |C_s(t,s)|ds.$$

This proves the result.

We now pattern $D(t,s)$ after $k + [t-s+1]^{-1}$ so that $D(t,0) \downarrow k > 0$, whereas $C(t,0) = k + \arctan t \uparrow k + (\pi/2)$ and inquire if the behavior is the same as we just saw for C. We ask that there exist positive constants k, α with

$$D(t,t) \geq k, \tag{4.3.18}$$

$$\int_0^t |D_s(t,s)|ds \leq k - \alpha, \tag{4.3.19}$$

and

$$\int_0^t |D_t(t,s)|ds \leq k - \alpha. \tag{4.3.20}$$

4.3. THE NATURE OF THE KERNEL

A necessary condition for the solution of

$$z(t) = a(t) - \int_0^t [k + [t - s + 1]^{-1} z(s) ds$$

to be bounded, say $|z(t)| \leq M$, is for

$$|a(t)| \leq M + M \int_0^t \left[k + [t-s+1]^{-1} \right] ds \leq J(1+t)$$

for some $J > 0$.

Theorem 4.3.13. *Let (4.3.18) and (4.3.20) hold and let $a(t) = J(1+t)^p$, $0 \leq p \leq 1$. Then the solution of (4.3.4) is bounded.*

The proof is identical to that of Theorem 4.3.11.

Theorem 4.3.14. *Let $|a(t)| \leq M$ for some $M > 0$, let $C(t,t)$ be bounded, and let (4.3.18) and (4.3.19) hold. Then the solution of (4.3.4) is bounded.*

The proof is identical to that of Theorem 4.3.12.

We began by mentioning that we would see eight Liapunov functionals. Look ahead to Chapter 6. We have

$$x(t) = f(x(t-h)) + \int_t^\infty [g(x(s)) + r(x(s-h))] C(t,s) ds + p(t) \quad (6.4.18)$$

with Liapunov functional

$$V(t) = \int_t^\infty \lambda^2 C_s(t,s) \left(\int_s^t [g(x(u)) + r(x(u-h))] du \right)^2 ds. \quad (6.4.23)$$

Finally, we note that we may use the Levin-type Liapunov functional for the resolvent equation for $D(t,s)$ and the derived resolvent equation for $C(t,s)$ to obtain close relations between the resolvents for those two functions.

This section is taken from Burton (2008a), as detailed in the bibliography, which is published by the University of Szeged, Szeged, Hungary.

Chapter 5

Infinite Delay: Schaefer's Theorem

5.1 Introduction

This chapter is best described as a series of lectures on nonlinear integral equations with infinite delay which are never converted to a differential equation. Section 5.1.4 is an introductory lecture laced with conjectures and a definite effort is made to present simple and intuitive results. The nonlinearity is always either x^n or $x^{1/n}$ so that assumptions on the behavior of the nonlinearity need not be mentioned, resulting in a drastic reduction of hypotheses. In the same vein the last section, 5.4, concerns a lecture on an intriguing property; again, it is largely free of complications, but less so than Section 5.1.4. But the sections in between consist of fairly difficult and complicated problems involving periodic solutions and asymptotic behavior. The nonlinearity is $g(t,x)$ and it takes considerable space to describe the essential properties of g. In Chapter 3 we suggested exercises in which the reader was asked to combine two types of Liapunov functionals and obtain general theorems. That type of work is displayed in Section 5.3. The basic theme of the chapter concerns a convex kernel. Section 5.3 shows one way to deviate from the convexity and still retain the conclusions obtained from convexity. This is important because in real-world problems something as controlled as convexity can only be achieved by assumption and never by observation and measurement. Each section is mainly independent, but all utilize the same basic Liapunov functional which we have seen several times in earlier chapters.

Tricomi was one of the early investigators and expositors of Volterra integral equations and a close friend and confidant of Volterra. His book

5.1. INTRODUCTION

(1985) contains a wealth of information about early investigations and history. Volterra studied the Volterra equation of the *first kind* in the form

$$\int_0^t C(t,s)x(s)ds = a(t)$$

where x is the unknown function and the Volterra equation of the *second kind* in the form

$$x(t) = a(t) - \lambda \int_0^t C(t,s)x(s)ds$$

which is the type mainly studied in this book. We usually take x and a as real n-vector functions and C to be an $n \times n$ matrix of functions, while λ is a parameter which we usually take to be 1, but it plays a crucial role in other presentations. See, for example, Yosida (1991; p. 145) where he obtains a formal power series solution in powers of λ and a sequence ϕ_n obtained recursively.

Tricomi (1985; p. 40) notes that these equations are considered to be singular if the kernel is not integrable or if the lower limit is $-\infty$. We focus here on

$$x(t) = a(t) - \int_{-\infty}^t C(t,s)x(s)ds \tag{*}$$

for a variety of reasons. The first that we will mention is that for fixed points there is the natural mapping defined by

$$(P\phi)(t) = a(t) - \int_{-\infty}^t C(t,s)\phi(s)ds.$$

If we ask that a and C be continuous and that there exist a $T > 0$ with

$$a(t+T) = a(t) \text{ and } C(t+T, s+T) = C(t,s)$$

then a translation argument will show that if $(\mathcal{P}_T, \|\cdot\|)$ is the Banach space of T-periodic functions $\phi : \Re \to \Re^n$ with the supremum norm, then $P : \mathcal{P}_T \to \mathcal{P}_T$. We will then be poised to find a fixed point, a periodic solution of (*). Generally, that will be achieved using some variant of Schaefer's theorem (Theorem 3.1.2) which requires an *a priori* bound. That will be obtained using a Liapunov functional. Moreover, in some cases the Liapunov functional will not be required to be positive. The bound will be obtained from the derivative alone.

We will also deviate from Tricomi's terminology by saying informally that an equation $x(t) = a(t) - \int_{-\infty}^t D(t,s,x(s))ds$ has a mild singularity if

$D(t, s, x)$ is infinite at some point, but for each function ϕ which is bounded and continuous on any interval $(-\infty, t_1]$ we have $\int_{-\infty}^{t} D(t, s, \phi(s))ds$ continuous on $(-\infty, t_1)$. Finally, we frequently form Liapunov functionals of the form

$$\int_{-\infty}^{t}\int_{t-s}^{\infty}|C(u+s,s)|du|x(s)|ds$$

and we will always assume that this can be defined and differentiated by Leibnitz's rule.

5.1.1 Initial Conditions

Generally, existence and uniqueness theorems are concerned with initial value or boundary value problems. To specify a solution of the scalar equation

$$x(t) = b(t) - \int_{-\infty}^{t} C(t,s)g(s,x(s))ds$$

we may require a continuous initial function $\phi : (-\infty, 0] \to \Re$ such that

$$a(t) := b(t) - \int_{-\infty}^{0} C(t,s)g(s,\phi(s))ds$$

is continuous and we then seek a solution of

$$x(t) = a(t) - \int_{0}^{t} C(t,s)g(s,x(s))ds, \quad x(0) = a(0),$$

for $t > 0$. In effect, the initial function turned the singular Volterra equation into a non-singular Volterra equation of the second kind. We note too that there is a discontinuity between the initial function ϕ and the solution x unless $\phi(0) = a(0)$. See the discussion with Figure 3.1. There are times when such a discontinuity is undesirable. It is proved in Chapter 7, Proposition 7.2.1, under general conditions that we can redefine ϕ on an arbitrarily short interval $[t_0, 0]$ so that the new ϕ satisfies $\phi(0) = a(0)$. See the discussion with Figure 3.1. The point is that we seek a function x on the whole interval $\Re$ which satisfies the integral equation for $t > 0$ and agrees with the initial function on the initial interval. That solution will certainly be continuous for $t > 0$ and for $t < 0$; in some cases we are interested in the two pieces matching at $t = 0$. Obviously, only in rare cases will the initial function itself satisfy the differential equation.

5.1.2 Memory and Convexity

Consider a convolution equation

$$x(t) = a(t) - \int_{-\infty}^{t} C(t-s)x(s)ds$$

and fix $t = t_1$. The value of $x(t_1)$ depends on $x(s)$ for $-\infty < s \leq t_1$. Moreover, every such value of $x(s)$ is weighted by a factor $C(t_1 - s)$. As s takes on all such values, we are looking at $C(u)$ for $0 \leq u < \infty$, $u = t_1 - s$. Here is a heuristic rationale. Our experiences would lead us to expect $C(u) \to 0$ as $u \to \infty$; we remember the present most clearly and our memory fades with time. It would certainly seem reasonable to ask that $C'(t) \leq 0$. But $C(t)$ should be positive so we can reasonably ask that $C''(t) \geq 0$. Finally, we might want the sum of the weights to be finite, so we ask that $\int_0^\infty C(u)du < \infty$. But if we think of the nonconvolution case, then we must take into account the experiment changing with time. Perhaps there is a day-night fluctuation of temperature affecting the experiment and so we consider $C(t, s)$ and ask that $C_t(t, s) \leq 0$, while $C_{st}(t, s) \geq 0$.

There are more direct ways of setting the conditions. Various physical processes can be abstractly defined in ways that assume the convexity. Certain one-dimensional viscoelastic problems are formulated as

$$x'(t) = -\int_{t-r}^{t} a(t-s)g(x(s))ds$$

where

$$a(r) = 0, \quad a(t) \geq 0, \quad a'(t) \leq 0, \quad a''(t) \geq 0, \quad 0 \leq t \leq r.$$

Here, x is the strain and a is the relaxation function.

There is little or no chance of such intricate conditions being actually verified in a real-world problem. In an earlier book (Burton (2006b; p. 4)) we objected to such assumptions and showed that some problems of this class can be solved using contraction mappings which average functions instead of asking such exact properties of the functions involved. Here, we use a variety of assumptions. Frequently, we ask conditions on C which are clearly convexity assumptions and those are easily seen.

5.1.3 Much less than Convexity

We divide the discussion and emphasize alternatives to convexity. We may use conditions which ask nothing about the derivative of C, and this can introduce another of the four remarkable situations which we observe in this book. For the same equation we can offer a set of conditions which rely on the derivatives of C to make the Liapunov functional produce an *a priori* bound yielding periodic solutions, or we can offer a set of conditions which rely on the integrals of C to make a different Liapunov functional produce the bounds. As those conditions are totally independent, we are led to believe that both are extraneous and that the theorem could be proved without either assumption. It is an attractive problem.

Particularly in cases of finite delay, we will ask less than convexity and simply require $C_{st} \leq 0$. In one section we ask that the kernel be a sum of functions, one of which is convex and the other is not.

5.1.4 An Introductory Lecture: Periodicity

Much has been written about periodic solutions of differential equations. There are two simple and most pleasing general results. The first is the Poincaré-Bendixson theorem (cf. Burton (2005c; p. 210)) concerning a pair of first order equations

$$x' = P(x, y)$$
$$y' = Q(x, y)$$

with P and Q continuous and locally Lipschitz.

Theorem. *If there is a solution bounded in the future then there is an equilibrium point or a periodic solution in its ω−limit set.*

Without terminology, the result says that if there is a bounded solution it either gets close infinitely often to a point where P and Q both vanish or it approaches a periodic solution.

Very late in the theory of differential equations Massera (1950) (cf. Burton (2005c; p. 217)) proved that for a first order equation

$$x' = f(t, x)$$

with f continuous and locally Lipschitz in x and periodic in t, if there is a bounded solution then there is a periodic solution. He proved a partial extension to second order systems.

There are also isolated and special results. Epstein (1962) proved that a linear periodic system $x' = A(t)x$ with $A(t)$ odd has all periodic solutions. But these are the rare exceptions. Most of the periodic results are long and detailed with restrictive sign conditions.

5.1. INTRODUCTION

We come upon a scalar equation

$$x(t) = a(t) + \int_{-\infty}^{t} D(t, s, x(s))ds$$

in a variety of ways, hoping to prove that there is a periodic solution. Sometimes these equations arise as limiting forms of $x(t) = a(t) + \int_0^t D(t, s, x(s))ds$ which might be generated from a partial differential equation, or it may arise directly from a physical problem. See Miller (1971a; p. 172) or Burton (2005c; p. 105) for information on limiting equations.

To be more explicit, in a recent paper Kirk and Olmstead (2002) use a Green's function to convert a partial differential equation to an integral equation on the t-interval $[0, \infty)$. It has a mild singularity, but in their equation (39) on p. 153 they are considering an L^1-kernel which is well suited to the Miller (1971a; p. 172) transformation to an infinite delay problem of the type we consider here. In their case they do not have a periodicity property, but there are other heat problems with that property as may be seen, for example, in Padmavally (1958).

We have seen that

$$x(t) = a(t) + \int_0^t C(t, s)x(s)ds$$

may have asymptotically periodic solutions whose domains are $[0, \infty)$. If we can transform it into an infinite delay problem then that solution becomes periodic with compact domain, an enormous simplification.

Such scalar equations can represent high order ordinary differential equations so it would seem folly to hope for results as simple as that of Massera. Yet, there are indications that it might be true and we explore that possibility here. This offers a simple introduction to the subject and a possible research problem.

In any case one is led to the search for a periodic solution of the scalar equation

$$x(t) = a(t) + \int_{-\infty}^{t} D(t, s, x(s))ds \tag{5.1.1}$$

with $T > 0$ so that

$$a(t + T) = a(t), \quad D(t + T, s + T, x) = D(t, s, x), \tag{5.1.2}$$

with a continuous. All of the results will be easily illustrated from the single function

$$D(t, s, x) = m[(t - s) + 1]^{-k} g(x)$$

and the algebra is simple. We have studied this problem in Burton (1993, 1994a), Burton and Furumochi (1995, 1996), and Burton and Makay (2002) under considerably stronger assumptions.

Our conjecture here is that if $\int_{-\infty}^{t} D(t,s,x(s))ds$ converges for any continuous and periodic function x and if D is reasonably smooth, then there always is a periodic solution. This is suggested in an old result, which we offer as Theorem 5.1.1. But it is offered more strongly in Theorem 5.1.4 in which we do ask that the equation be of sublinear type; however, once the periodic solution is established, then all the action is taking place in a strip of $|x| \leq K$ for some $K > 0$. It is then totally immaterial what the behavior of D is with respect to x outside that strip. We study this problem when D is globally Lipschitz, locally Lipschitz, and non-Lipschitz in x.

The present problem is not unlike that considered in Section 4.3 where we showed that kernels with rather opposite behavior produced similar solutions. Smoothness of the kernel seemed to be the important property.

Let $(\mathcal{P}_T, \|\cdot\|)$ denote the Banach space of continuous scalar T-periodic functions with the supremum norm and assume that for $\phi \in \mathcal{P}_T$ then

$$\int_{-\infty}^{t} D(t,s,\phi(s))ds \in \mathcal{P}_T. \tag{5.1.3}$$

This will allow problems with mild singularities. The following simple result is well-known. Indeed, essentially the same is found in Chapter 2 for the linear case.

Theorem 5.1.1. *Let (5.1.2) and (5.1.3) hold. Suppose there is a function $B(t,s)$ with*

$$|D(t,s,x) - D(t,s,y)| \leq B(t,s)|x-y| \tag{5.1.4}$$

for $-\infty < s \leq t < \infty$, $x, y \in \Re$ and $\alpha < 1$ with $\int_{-\infty}^{t} B(t,s)ds$ defined and

$$\int_{-\infty}^{t} B(t,s)ds \leq \alpha. \tag{5.1.5}$$

Then (5.1.1) has a solution in $\mathcal{P}_T$.

Proof. Define a mapping $P: \mathcal{P}_T \to \mathcal{P}_T$ by $\phi \in \mathcal{P}_T$ implies that

$$(P\phi)(t) = a(t) + \int_{-\infty}^{t} D(t,s,\phi(s))ds$$

and notice that by (5.1.3) it is well-defined, while by (5.1.4) if $\phi, \eta \in \mathcal{P}_T$ then

$$|(P\phi)(t) - (P\eta)(t)| \leq \int_{-\infty}^{t} B(t,s)|\phi(s) - \eta(s)|ds \leq \alpha \|\phi - \eta\|$$

5.1. INTRODUCTION

by (5.1.5). Thus, P is a contraction and there is a unique fixed point in $\mathcal{P}_T$.

Notice that there is no sign condition; everything depends on a global Lipschitz condition, (5.1.4), and smallness condition, (5.1.5). The idea is to write $D(t,s,x) = C(t,s)g(x)$, drop the global Lipschitz condition, let $\int_{-\infty}^{t} |C(t,s)|ds$ be large, and show that by either making C smooth or g small we can still conclude that there is a periodic solution.

We now investigate whether loss of the global Lipschitz condition can affect the result. Consider

$$x(t) = a(t) - \int_{-\infty}^{t} C(t,s)x^n(s)ds \tag{5.1.6}$$

where n is an odd positive integer and let

$$a(t+T) = a(t), \quad C(t+T, s+T) = C(t,s). \tag{5.1.7}$$

Assume that for $\phi \in \mathcal{P}_T$ then

$$\int_{-\infty}^{t} |C(t,s)\phi^n(s)|ds \quad \text{is continuous.} \tag{5.1.8}$$

Our work will be based on the Liapunov functional working together with the fixed point theorem of Schaefer which we list again for ready reference.

Schaefer's Theorem. *Let $(X, \|\cdot\|)$ be a normed space, P a continuous mapping of X into X which is compact on each bounded subset of X. Then either*

(i) *the equation $x = \lambda P x$ has a solution in X for $\lambda = 1$, or*

(ii) *the set of all such solutions x, for $0 < \lambda < 1$, is unbounded.*

We have repeatedly used the following Liapunov functional and we remind the reader how it is constructed.

$$x(t) = a(t) - \int_{-\infty}^{t} C(t,s)g(s,x(s))ds$$

write

$$(x(t) - a(t))^2 = \left(-\int_{-\infty}^{t} C(t,s)g(s,x(s))ds\right)^2.$$

For the moment, assume $C_s(t,s) \geq 0$, that there is an $M > 0$ with $\int_{-\infty}^{t} C_s(t,s)ds \leq M$, and that $C(t,s)(t-s) \to 0$ as $t \to \infty$.

244   5. INFINITE DELAY: SCHAEFER'S THEOREM

If we integrate the right-hand-side by parts and use the Schwarz inequality then we have

$$(x(t) - a(t))^2 = \left(C(t,s) \int_s^t g(u, x(u)) du \right) \Big|_{-\infty}^t$$

$$- \int_{-\infty}^t C_s(t,s) \left(\int_s^t g(u, x(u)) du \right)^2$$

$$\leq \int_{-\infty}^t C_s(t,s) ds \int_{-\infty}^t C_s(t,s) \left(\int_s^t g(u, x(u)) du \right)^2 ds$$

$$\leq M \int_{-\infty}^t C_s(t,s) \left(\int_s^t g(u, x(u)) du \right)^2 ds.$$

We have arrived at the Liapunov functional

$$V(t) = \int_{-\infty}^t C_s(t,s) \left(\int_s^t g(u, x(u)) du \right)^2 ds.$$

Notice that there is something of a "wedge" under V in the form $(1/M)(x(t) - a(t))^2$. Moreover, if $C_s > 0$, then for $a(t) = 0$ the functional V becomes positive definite in the classical sense.

We specialize the above work and use the Liapunov functional

$$V(t) = \lambda \int_{-\infty}^t C_s(t,s) \left(\int_s^t x^n(u) du \right)^2 ds. \qquad (5.1.9)$$

to prove that there is an *a priori* $L^p[0,T]$ bound on the norm of the solution of

$$x(t) = \lambda \left[a(t) - \int_{-\infty}^t C(t,s) x^n(s) ds \right] \qquad (5.1.10)$$

and then parlay that into a supremum norm bound. For our mapping, $X = \mathcal{P}_T$ and we define $P : \mathcal{P}_T \to \mathcal{P}_T$ by $\phi \in \mathcal{P}_T$ implies that

$$(P\phi)(t) = a(t) - \int_{-\infty}^t C(t,s) \phi^n(s) ds. \qquad (5.1.11)$$

We will see many derivatives of C here, but that can be misleading as examples will show. If we take $C(t,s) = C(t-s)$ with $C(t) = t(t-1)$ for $0 \leq t \leq 1$ and $C(t) = 0$ for $t \geq 1$, then the limits on the integral of the Liapunov functional will change and problems with derivatives will vanish. In this case, we would see C_s change sign and that is a property in which

5.1. INTRODUCTION

we will always be interested. To leave open a number of possibilities of the type just mentioned we will refrain from placing strict conditions on C, but ask that V can always be differentiated by Leibnitz's rule. Later, differentiability is reduced and even eliminated.

In the next theorem one may note that (5.1.12) and (5.1.13) allow us to define and differentiate (5.1.9) for bounded functions x, while (5.1.14) allows us to perform a certain integration by parts. Thus, it is (5.1.15) alone which seems to be pertinent for the result itself.

Theorem 5.1.2. Let (5.1.7) and (5.1.8) hold. Assume that

$$\int_{-\infty}^{t} C_s(t,s)(t-s)^2 ds \quad \text{is continuous,} \tag{5.1.12}$$

as is

$$\int_{-\infty}^{t} C_{st}(t,s)(t-s)^2 ds, \tag{5.1.13}$$

and that

$$C(t,s)(t-s) \to 0 \quad \text{as} \quad s \to -\infty. \tag{5.1.14}$$

If, in addition,

$$C_{st}(t,s) \leq 0, \tag{5.1.15}$$

then for any fixed solution x of (5.1.10) in $\mathcal{P}_T$ we have

$$\int_0^T x^{n+1}(s) ds \leq \int_0^T a^{n+1}(s) ds. \tag{5.1.16}$$

Proof. Use the solution x of (5.1.10) in $\mathcal{P}_T$ and define $V(t)$ in (5.1.9) so that

$$V'(t) = \lambda \int_{-\infty}^{t} C_{st}(t,s) \left(\int_s^t x^n(u) du \right)^2 ds$$

$$+ 2\lambda x^n(t) \int_{-\infty}^{t} C_s(t,s) \int_s^t x^n(u) du\, ds.$$

We integrate the last term by parts obtaining

$$2\lambda x^n(t) \left[C(t,s) \int_s^t x^n(u) du \Big|_{-\infty}^{t} + \int_{-\infty}^{t} C(t,s) x^n(s) ds \right]$$

so that by (5.1.14) the first term is zero and we have by (5.1.10) and (5.1.15)

$$V'(t) \leq 2\lambda x^n(t)\left[\int_{-\infty}^t C(t,s)x^n(s)ds\right]$$
$$= 2x^n(t)[\lambda a(t) - x(t)]$$
$$\leq \frac{2}{n+1}[a^{n+1}(t) - x^{n+1}(t)].$$

Now it is readily verified that since $x \in \mathcal{P}_T$, so is V. Thus, $V(T) = V(0)$ and

$$0 = V(T) - V(0) \leq \frac{2}{n+1}\left(\int_0^T a^{n+1}(s)ds - \int_0^T x^{n+1}(s)ds\right)$$

or

$$\int_0^T x^{n+1}(s)ds \leq \int_0^T a^{n+1}(s)ds,$$

as required.

Next, we use the L^p bound and (5.1.10), the integral equation, to obtain a supremum norm bound.

Theorem 5.1.3. *Let (5.1.7), (5.1.8), and (5.1.16) hold. Suppose there is a constant Q such that*

$$\int_{-\infty}^{t_1}|C(t_1,s) - C(t_2,s)|ds \leq Q|t_1 - t_2| \quad \text{if} \quad 0 \leq t_1 \leq t_2 \leq T \quad (5.1.17)$$

and also that

$$\sup_{0 \leq t \leq T}\sum_{j=0}^{\infty}\left(\int_{t-T}^t C^{n+1}(t+jT,s)ds\right)^{\frac{1}{n+1}} < \infty. \tag{5.1.18}$$

Then (5.1.6) has a solution in $\mathcal{P}_T$.

Proof. In (5.1.11) we defined $P : \mathcal{P}_T \to \mathcal{P}_T$. We will find a number L such that if $x \in \mathcal{P}_T$ is a solution of (5.1.10) in $\mathcal{P}_T$ then $\|x\| < L$, where $\|\cdot\|$ is the supremum norm.

Condition (5.1.17) readily shows that P maps bounded sets into equicontinuous sets. Let the bounded set be fixed and let ϕ be contained

5.1. INTRODUCTION

in that set. There are then positive constants J and Y with $\|\phi\| \leq J$ and $\|\phi^n\| \leq Y$. Thus, if $0 \leq t_1 \leq t_2 \leq T$, then

$$|(P\phi)(t_2)-(P\phi)(t_1)| \leq |a(t_2) - a(t_1)|$$
$$+ \left|\int_{-\infty}^{t_1} [C(t_2,s) - C(t_1,s)]\phi^n(s)ds\right|$$
$$+ \left|\int_{t_1}^{t_2} C(t_2,s)\phi^n(s)ds\right|$$
$$\leq |a(t_2) - a(t_1)| + (YQ + YE)|t_1 - t_2|$$

where $E = \sup_{0 \leq s \leq T, 0 \leq t_2 \leq T} |C(t_2,s)|$.

Continuity of P follows from (5.1.8) and the uniform continuity of ϕ^n when $\phi_n \in \mathcal{P}_T$ and $\phi_n \to \phi$.

For $x \in \mathcal{P}_T$ a solution of (5.1.10) we have

$$|x(t) - \lambda a(t)| \leq \left|\int_{-\infty}^t C(t,s)x^n(s)ds\right|$$
$$= \left|\sum_{j=0}^\infty \int_{t-(j+1)T}^{t-jT} C(t,s)x^n(s)ds\right|$$
$$= \left|\sum_{j=0}^\infty \int_{t-T}^t C(t, s-jT)x^n(s)ds\right|$$
$$= \left|\sum_{j=0}^\infty \int_{t-T}^t C(t+jT, s)x^n(s)ds\right|$$
$$\leq \sum_{j=0}^\infty \left(\int_{t-T}^t C^{n+1}(t+jT,s)ds\right)^{\frac{1}{n+1}} \left(\int_{t-T}^t a^{n+1}(s)ds\right)^{\frac{n}{n+1}},$$

which has a finite bound independent of λ. Referring now to Schaefer's theorem we see that the second alternative is ruled out and the conclusion holds.

If we replace x^n in (5.1.6) by $x^{1/m}$ where $m > 1$ is odd, then the same Liapunov functional would work in that case. But we opt for a different method of proof and move from the requirement of two derivatives on C to the contrasting condition that C be "twice" integrable on the whole line.

Consider the equation

$$x(t) = a(t) - \int_{-\infty}^{t} C(t,s)x^{1/m}(s)ds \tag{5.1.19}$$

where (5.1.7), periodicity, holds and $m > 1$ is odd. We will first suppose that $C_s(t,s)$ exists and later suppose that $C(t,s)$ is bounded by a function, $E(t,s)$, for which $E_s(t,s)$ exists. A far stronger result will be proved in the first case.

Consider the conditions of the following theorem. First, in order for (5.1.19) to be a well-defined problem we would expect

$$\int_{-\infty}^{t} |C(t,s)|ds < \infty; \tag{*}$$

thus, we would expect $C(t,s) \to 0$ as $s \to -\infty$ a bit faster than $1/s$. Hence, (5.1.20) is expected.

Moreover, in the convolution case, (5.1.21) would first say that $\int_0^\infty |C(u)|du < \infty$, which is just (*) again, and is less than Theorem 5.1.1 required because it need not be less than one. In addition, there is no Lipschitz condition. However, the second part of (5.1.21) now asks that $\int_t^\infty |C(u)|du \in L^1[0,\infty)$ which is more than Theorem 5.1.1 required. That last condition is needed only to define the Liapunov functional and, hence, may be extraneous.

Finally, with (5.1.20) holding we have

$$\int_{-\infty}^{t} C_s(t,s)(t-s)ds = C(t,s)(t-s)\Big|_{-\infty}^{t} + \int_{-\infty}^{t} C(t,s)ds$$
$$= \int_{-\infty}^{t} C(t,s)ds$$

which is periodic and, hence, bounded if continuous.

For smooth kernels, this next result says essentially that if (5.1.19) is well-defined and (5.1.21) holds, then it has a periodic solution. That periodic solution lies in a strip $|x| \leq K$, for some $K > 0$, so the fact that (5.1.19) is sublinear is a device in the proof, but not apparently in the essential nature of the problem. However, the bound K does depend on the magnitude of $a(t)$.

Theorem 5.1.4. *Suppose that (5.1.7) and (5.1.17) hold, while (5.1.8) holds for $n = 1/m$. In addition, suppose that*

$$C(t,s)(t-s) \to 0 \quad \text{as} \quad s \to -\infty \quad \text{for fixed } t, \tag{5.1.20}$$

5.1. INTRODUCTION

that there is an $\alpha < \infty$ with

$$\int_0^\infty |C(u+t,t)|du \leq \alpha \quad \text{and} \quad \int_{-\infty}^t \int_{t-s}^\infty |C(u+s,s)|du\,ds \text{ exists,} \quad (5.1.21)$$

and that

$$\int_{-\infty}^t |C_s(t,s)|[(t-s)+1]ds \text{ is bounded.} \quad (5.1.22)$$

Then (5.1.19) has a solution in $\mathcal{P}_T$.

Proof. With a view to using Schaefer's theorem we start with

$$x(t) = \lambda\left[a(t) - \int_{-\infty}^t C(t,s)x^{1/m}(s)ds\right], \quad (5.1.19_\lambda)$$

define the corresponding mapping P from it, and for $x \in \mathcal{P}_T$ a solution of $(5.1.19_\lambda)$, define the new Liapunov functional

$$V(t) = \lambda \int_{-\infty}^t \int_{t-s}^\infty |C(u+s,s)|du\,|x^{1/m}(s)|ds.$$

The derivative along this solution $x \in \mathcal{P}_T$ of the integral equation $(5.1.19_\lambda)$ satisfies

$$V'(t) = \lambda \int_0^\infty |C(u+t,t)|du\,|x^{1/m}(t)| - \lambda \int_{-\infty}^t |C(t,s)x^{1/m}(s)|ds$$

$$\leq \alpha|x^{1/m}(t)| - \lambda \int_{-\infty}^t |C(t,s)x^{1/m}(s)|ds$$

$$\leq \alpha|x^{1/m}| + |a(t)| - |x(t)|$$

$$\leq -\beta|x^{1/m}(t)| + (\gamma + |a(t)|)$$

for some positive constants β and γ.

As x is supposed to be a solution in $\mathcal{P}_T$ we see that $V \in \mathcal{P}_T$. Thus,

$$0 = V(T) - V(0) \leq -\beta \int_0^T |x^{1/m}(s)|ds + \int_0^T (\gamma + |a(t)|)dt$$

so

$$\int_0^T |x^{1/m}(s)|ds \leq (1/\beta)\int_0^T (\gamma + |a(t)|)dt =: J.$$

5. INFINITE DELAY: SCHAEFER'S THEOREM

Moreover, if t is chosen so that $V(t)$ is the maximum of that periodic function, V, then for $s < t$ we have

$$0 \leq V(t) - V(s) \leq \int_s^t (\gamma + |a(u)|)du - \beta \int_s^t |x^{1/m}(u)|du$$

or for any t and for $s < t$ it follows that

$$\int_s^t |x^{1/m}(u)|du \leq (1/\beta)\int_s^t (\gamma + |a(u)|)du + J.$$

Thus, there is a constant $K > 0$ with

$$\int_s^t |x^{1/m}(u)|du \leq J + (t-s)K.$$

An integration by parts in (5.1.19$_\lambda$) yields

$$x(t) = \lambda\bigg[a(t) + C(t,s)\int_s^t x^{1/m}(u)du\bigg|_{-\infty}^t$$
$$- \int_{-\infty}^t C_s(t,s)\int_s^t x^{1/m}(u)duds\bigg]$$

or by (5.1.20)

$$|x(t)| \leq |a(t)| + \int_{-\infty}^t |C_s(t,s)|\int_s^t |x^{1/m}(u)|duds$$
$$\leq |a(t)| + \int_{-\infty}^t |C_s(t,s)|[K(t-s) + J]ds.$$

By (5.1.22) this is bounded. The compactness follows from (5.1.17) and the continuity follows from (5.1.8). (See Lemma 6.3.2.) By Schaefer's theorem the mapping has a fixed point.

We now suppose that $C_s(t,s)$ fails to exist and that there is a continuous function $E(t,s)$ and a positive constant M with

$$|C(t,s)| \leq E(t,s), \int_{-\infty}^t E(t,s)ds \leq M,$$

$$\int_{-\infty}^t E_s(t,s)[1 + (t-s)]^2 ds \quad \text{is bounded, and}$$

$$E(t,s)(t-s) \to 0 \quad \text{as} \quad s \to -\infty \quad \text{for fixed} \quad t. \quad (5.1.23)$$

Notice that this will still not allow a mild singularity in C.

5.1. INTRODUCTION

Theorem 5.1.5. Let (5.1.23) hold. If $x(t)$ is a bounded solution of (5.1.19$_\lambda$) then

$$(x(t) - \lambda a(t))^2 \leq M \int_{-\infty}^{t} E_s(t,s) \left(\int_s^t |x^{1/m}(u)| du \right)^2 ds. \quad (5.1.24)$$

Proof. We have

$$(x(t) - \lambda a(t))^2 = \left| \int_{-\infty}^{t} C(t,s) x^{1/m}(s) ds \right|^2$$

$$\leq \left| \int_{-\infty}^{t} E(t,s) |x^{1/m}(s)| ds \right|^2$$

$$= \left(-E(t,s) \int_s^t |x^{1/m}(u)| du \Big|_{-\infty}^{t} \right.$$

$$\left. + \int_{-\infty}^{t} E_s(t,s) \int_s^t |x^{1/m}(u)| du\, ds \right)^2$$

$$\leq \int_{-\infty}^{t} E_s(t,s) ds \int_{-\infty}^{t} E_s(t,s) \left(\int_s^t |x^{1/m}(u)| du \right)^2 ds$$

$$\leq M \int_{-\infty}^{t} E_s(t,s) \left(\int_s^t |x^{1/m}(u)| du \right)^2 ds$$

proving the result.

Theorem 5.1.6. For $n = 1/m$, let (5.1.7), (5.1.8), (5.1.17), and (5.1.21) hold.
(i) If $x \in \mathcal{P}_T$ solves (5.1.19$_\lambda$) then there is a $\beta > 0$, a $\gamma > 0$, and a $t \in [0, T]$ such that for $-\infty < s < t$ we have

$$\int_s^t |x^{1/m}(u)| du \leq (1/\beta) \int_s^t (\gamma + |a(u)|) du. \quad (5.1.25)$$

(ii) If, in addition, (5.1.23) holds so that (5.1.24) is satisfied then (5.1.19) has a solution in $\mathcal{P}_T$.

Proof. Following the proof of Theorem 5.1.4 we again start with (5.1.19$_\lambda$) and define

$$V(t) = \lambda \int_{-\infty}^{t} \int_{t-s}^{\infty} |C(u+s,s)| du |x^{1/m}(s)| ds$$

whose derivative was shown to satisfy

$$V'(t) \leq -\beta |x^{1/m}(t)| + (\gamma + |a(t)|).$$

As x is supposed to be a solution in $\mathcal{P}_T$ we see that $V \in \mathcal{P}_T$. Again, for $V(t)$ the maximum of that periodic function and for $s < t$ we have

$$0 \leq V(t) - V(s) \leq \int_s^t (\gamma + |a(u)|)du - \beta \int_s^t |x^{1/m}(u)|du$$

as in the proof of Theorem 5.1.4 so that

$$\int_s^t |x^{1/m}(u)|du \leq (1/\beta) \int_s^t (\gamma + |a(u)|)du$$

and then

$$\int_s^t |x^{1/m}(u)|du \leq J + (t-s)K$$

for any $s < t$. Using this in (5.1.24) yields

$$(x(t) - \lambda a(t))^2 \leq M \int_{-\infty}^t E_s(t,s)[J + (t-s)K]^2 ds,$$

and the right-hand-side is bounded by (5.1.23) and the boundedness of $a(t)$, yielding a suitable *a priori* bound on x. The equicontinuity and continuity of P are exactly as before.

We are working here with Liapunv functionals. In several cases one may use Razumikhin techniques to some advantage.

One of our objectives is to consider problems originating as partial differential equations (See Miller (1971a; p. 60, p. 172, and p. 208)) which were then parlayed into integral equations and then into infinite delay problems by means of limiting processes. In some such problems we find mild singularities and, at least in the convolution case for the limiting process, C is to be an L^1-function. We now show how Theorem 5.1.4 can be changed to cover just such problems.

In case of mild singularities, (5.1.17) would not hold and the proof of equicontinuity given in the proof of Theorem 5.1.3 would not work. However, there are alternative methods as one readily sees in the case of

$$C(t,s) = e^{-(t-s)}(t-s)^{-1/2}$$

when we work out the left-hand-side of (5.1.17). Thus, in our result below we simply ask for the compactness.

5.1. INTRODUCTION

Theorem 5.1.7. *Suppose that (5.1.7) and (5.1.8) hold for $n = 1/m$ and that the mapping $P : \mathcal{P}_T \to \mathcal{P}_T$ defined by*

$$(P\phi)(t) = a(t) - \int_{-\infty}^{t} C(t,s)\phi^{1/m}(s)ds$$

maps bounded subsets of $\mathcal{P}_T$ into compact subsets and that P is continuous. Let (5.1.21) hold and suppose that

$$\sup_{0 \leq t \leq T} \sum_{n=0}^{\infty} \left(\int_{t-T}^{t} C^2(t+nT,s)ds \right)^2 < \infty. \tag{5.1.26}$$

Then (5.1.19) has a solution in $\mathcal{P}_T$.

Proof. We follow the proof of Theorem 5.1.4 with $(5.1.19_\lambda)$ again and define the Liapunov functional

$$V(t) = \lambda \int_{-\infty}^{t} \int_{t-s}^{\infty} |C(u+s,s)| du |x^{1/m}(s)| ds.$$

Then we notice in the differentiation of V the last lines may be changed and we can have

$$V'(t) = \lambda \int_{0}^{\infty} |C(u+t,t)| du |x^{1/m}(t)| - \lambda \int_{-\infty}^{t} |C(t,s)x^{1/m}(s)| ds$$

$$\leq \alpha |x^{1/m}(t)| - \lambda \int_{-\infty}^{t} |C(t,s)x^{1/m}(s)| ds$$

$$\leq \alpha |x^{1/m}| + |a(t)| - |x(t)|$$

$$\leq -\beta |x^{2/m}(t)| + (\gamma + |a(t)|)$$

for some positive constants β and γ. That gives the change

$$0 = V(T) - V(0) \leq -\beta \int_{0}^{T} |x^{2/m}(s)| ds + \int_{0}^{T} (\gamma + |a(t)|) dt$$

so

$$\int_{0}^{T} |x^{2/m}(s)| ds \leq (1/\beta) \int_{0}^{T} (\gamma + |a(t)|) dt =: H.$$

Now from (5.1.19$_\lambda$) we have

$$|x(t)| - |a(t)| \leq \sum_{n=0}^{\infty} \int_{t-(n+1)T}^{t-nT} |C(t,s)||x^{1/m}(s)|ds$$

$$= \sum_{n=0}^{\infty} \int_{t-T}^{t} |C(t,s-nT)||x^{1/m}(s)|ds$$

$$\leq \sum_{n=0}^{\infty} \left(\int_{t-T}^{t} C^2(t+nT,s)ds \right)^{1/2} \left(\int_{t-T}^{t} x^{2/m}(s)ds \right)^{1/2}$$

$$\leq H^{1/2} \left(\int_{t-T}^{\infty} C^2(t+nT,s)ds \right)^{1/2}.$$

By (5.1.26) this yields the required bound and the conclusion follows from Schaefer's theorem.

We will now produce a differential inequality from a Liapunov functional and the properties of C_{st}. Consider the equation

$$x(t) = a(t) - \int_{-\infty}^{t} C(t,s)g(x(s))ds \tag{5.1.27}$$

where a, C, g are all continuous, a is bounded, $xg(x) > 0$ for $x \neq 0$, and (5.1.7) holds. First, if we examine common kernels such as $(t-s+1)^{-2}$ or $e^{-(t-s)}$ then we notice that frequently there is a continuous function $\gamma : [0, \infty) \to (0, \infty)$ with

$$C_{ts}(t,s) \leq -\gamma(t)C_s(t,s). \tag{5.1.28}$$

If we then define

$$V(t) = \lambda \int_{\infty}^{t} C_s(t,s) \left(\int_s^t g(x(u))du \right)^2 ds$$

we find that the derivative of V along the solution of the equation

$$x(t) = \lambda \left[a(t) - \int_{-\infty}^{t} C(t,s)g(x(s))ds \right] \tag{5.1.27$_\lambda$}$$

will satisfy

$$V'(t) \leq -\gamma(t)V(t) + 2g(x)[\lambda a(t) - x(t)]. \tag{5.1.29}$$

When $|x| \leq 2\|a\|$ then $2|g(x)a(t)| \leq 2\|a\|g^*$ for $g^* = \sup_{0 \leq |x| \leq 2\|a\|} |g(x)|$. Then for $|x| \geq 2\|a\|$ we have $2|g(x)a(t)| \leq |g(x)||x|$. We therefore see that

$$V'(t) \leq -\gamma(t)V(t) + 2\|a\|g^*. \tag{5.1.30}$$

5.1. INTRODUCTION

Thus, Liapunov functionals for integral equations can satisfy the same kinds of differential inequalities widely seen for differential equations. In some cases that differential inequality will yield a bound suitable for Schaefer's theorem.

In our work to this point we have always taken two steps. First, we obtain an integral bound and then parlay it into a supremum bound. The following will show how the first part can be accomplished in very general cases. Suppose that we have the form (5.1.27) where

$$C_{st}(t,s) \leq 0$$

and

$$g(x) =: xF(x)$$

where $F(x) > 0$ and continuous. We define V as above and obtain $V'(t) \leq 2g(x)[\lambda a(t) - x(t)]$. As $a \in \mathcal{P}_T$ we can write

$$|a(t)| \leq \|a\|.$$

Thus, when $|x| \geq 2\|a\|$ we have

$$2|g(x)a(t)| \leq 2\|a\| |xF(x)| \leq x^2 F(x)$$

so that

$$V'(t) \leq -x^2 F(x).$$

When $|x| \leq 2\|a\|$ then for

$$F^* := \sup_{|x| \leq 2\|a\|} F(x)$$

we have

$$2|a(t)xF(x)| \leq (2\|a\|)^2 F^*.$$

In any case, we will have

$$V'(t) \leq -x^2 F(x) + (2\|a\|)^2 F^*.$$

This will yield

$$\int_0^T x^2(s) F(x(s)) ds \leq (2\|a\|)^2 F^* T.$$

The second step of our problem is to parlay this into a supremum norm bound. This is a broad unsolved problem.

We are now going to combine Theorems 5.1.2, 5.1.3, and 5.1.7 in a way which gives us choices of hypotheses and those choices seem totally independent.

Theorem 5.1.8. *In Equation (5.1.19) take $m = 3$ and suppose that (5.1.7), (5.1.8), (5.1.17), and (5.1.26) hold. If either (5.1.21) or {(5.1.12), (5.1.13), (5.1.14), and (5.1.15)} hold then (19) has a solution in $\mathcal{P}_T$.*

Proof. First, because (5.1.17) holds we can prove that the mapping P in Theorem 5.1.7 maps bounded subsets of $\mathcal{P}_T$ into compact subsets, as we did in the proof of Theorem 5.1.3. Thus, if (5.1.21) holds then our first choice becomes exactly Theorem 5.1.7 so there is a periodic solution.

Next, suppose that (5.1.12)–(5.1.15) hold. According to the proof of Theorem 5.1.2 the functional V in (5.1.9) will be defined for $n = 1/3$ and we will again have

$$V'(t) \leq \frac{2}{n+1}\left[a^{n+1}(t) - x^{n+1}(t)\right],$$

together with (5.1.16) which now reads

$$\int_0^T x^{4/3}(s)ds \leq \int_0^T a^{4/3}(s)ds =: K.$$

But $x^{4/3}(s) + 1 \geq x^{2/3}(s)$ and so we have

$$\int_0^T x^{2/3}(s)ds \leq K + T =: H$$

where H will now be used again as the constant in the proof of Theorem 5.1.7. Thus, the proof of Theorem 5.1.7 can be completed with that H and the conclusion holds.

We are left with an intriguing problem. Perhaps there is a great array of independent conditions such as the two sets illustrated in the theorem. On the other hand, it may be that (5.1.7), (5.1.8), (5.1.17), and a convergence condition are all that is needed and the two hypotheses offered here are totally extraneous. As mentioned earlier, it is interesting that we need n to be odd, but there is no sign condition on $C(t, s)$. We feel that this is a reducible condition.

Finally, compare Theorems 5.1.6 and 5.1.3. Theorem 5.1.6 contains nothing about derivatives of C. Except for (5.1.14), everything in Theorem 5.1.3 rests on derivatives of C. For smooth functions, (5.1.14) is closely related to (5.1.8).

The material in this section is taken from Burton (2008b), as detailed in the bibliography, and was first published by the House of the Book of Science.

5.2 Qualitative Theory

We view this section as the most straightforward of all the infinite delay work. It is based on a simple Liapunov functional, the supremum norm, and an unmodified form of Schaefer's theorem. The equation is

$$x(t) = a(t) - \int_{-\infty}^{t} D(t,s)g(s,x(s))ds \qquad (5.2.1)$$

where $a: \Re \to \Re$, $D: \Re \times \Re \to \Re$, $g: \Re \times \Re \to \Re$ are all continuous, g is bounded for x bounded, there is an $M > 0$ such that if $0 \leq \lambda \leq 1$ then

$$2g(t,x)[\lambda a(t) - x] \leq -|xg(t,x)| + M \qquad (5.2.2)$$

(this condition can be weakened using (5.2.20)), there is a $K > 0$ with

$$|g(t,x)| \leq K + |xg(t,x)|, \qquad (5.2.3)$$

(this is notation, rather than an assumption)

$$D_s(t,s) \geq 0, D_{st}(t,s) \leq 0, \text{ both continuous,} \qquad (5.2.4)$$

and D satisfies conditions to be stated later.

To specify a solution of (5.2.1) we require a bounded continuous function $\varphi: (-\infty, 0] \to \Re$ and ask that

$$U(t,\varphi) := \int_{-\infty}^{0} D(t,s)g(s,\varphi(s))ds$$

be continuous for $t \geq 0$. Then U is treated as part of $a(t)$ so that (5.2.1) becomes

$$x(t) = a(t) - U(t,\varphi) - \int_{0}^{t} D(t,s)g(s,x(s))ds. \qquad (5.2.1^*)$$

This equation has a solution $x(t,\varphi)$ which agrees with φ on $(-\infty, 0]$ and satisfies (5.2.1) on an interval $[0,\alpha)$, provided in (5.2.1) that $\varphi(0) = a(0) - U(0,\varphi)$; if the solutions remain bounded then $\alpha = \infty$. Convergence conditions on D are required. See Chapter 3 for details on existence. Theorem 5.2.1 treats only solutions which are continuous in that they match

5. INFINITE DELAY: SCHAEFER'S THEOREM

up with their initial functions. A continuity argument shows that such solutions do exist, while Theorem 5.2.2 explicitly shows such existence. We have already mentioned that in Proposition 7.2.1 it is shown that for each given continuous ϕ we can find another function which differs arbitrarily little from ϕ for which the continuity condition does hold.

5.2.1 Boundedness

We require that

$$\int_{-\infty}^{t} \left[|D(t,s)| + D_s(t,s)(t-s)^2 + |D_{st}(t,s)|(t-s)^2\right] ds \qquad (5.2.5)$$

be continuous,

$$\lim_{s \to -\infty} (t-s)D(t,s) = 0 \text{ for fixed } t, \qquad (5.2.6)$$

and that there are constants A and B with $\qquad (5.2.7)$

(i) $\int_{-\infty}^{t} D_s(t,s)(t-s)^2 ds \leq A$,

(ii) $\int_{-\infty}^{t} D_s(t,s) ds \leq B$.

To review Section 5.2, in the next theorem x is continuous on all of $\Re$ and satisfies (5.2.1) on $[0,\infty)$. The initial function, ϕ, is bounded and continuous.

Theorem 5.2.1. *If (5.2.2) – (5.2.7) hold then $(a(t) - x(t))^2$ is bounded for any continuous solution $x(t)$ of (5.2.1) on $[0,\infty)$.*

Proof. Define the Liapunov functional

$$V(t) = \int_{-\infty}^{t} D_s(t,s) \left(\int_s^t g(v,x(v))dv\right)^2 ds.$$

If $x(t)$ is any continuous solution of (5.2.1) on $[0,\infty)$ with bounded initial function then V is defined and

$$V'(t) = \int_{-\infty}^{t} D_{st}(t,s) \left(\int_s^t g(v,x(v))dv\right)^2 ds$$

$$+ 2g(t,x(t)) \int_{-\infty}^{t} D_s(t,s) \int_s^t g(v,x(v))dv\, ds.$$

5.2. QUALITATIVE THEORY

If we integrate the last term by parts we have

$$2g(t,x(t))\left[D(t,s)\int_s^t g(v,x(v))dv\Big|_{-\infty}^t + \int_{-\infty}^t D(t,s)g(s,x(s))ds\right].$$

The first term vanishes at both limits by (5.2.6); the first term of V' is not positive; and if we use (5.2.1) on our last term then for $t \geq 0$ we obtain

$$V'(t) \leq 2g(t,x(t))[a(t) - x(t)]$$
$$\leq -|xg(t,x)| + M$$

by (5.2.2). The next details here have been seen in Chapters 2 and 3 for the finite memory case. Notice that

$$x(t) - a(t) = -\int_{-\infty}^t D(t,s)g(s,x(s))ds$$

$$= D(t,s)\int_s^t g(u,x(u))du\Big|_{-\infty}^t$$

$$-\int_{-\infty}^t D_s(t,s)\int_s^t g(u,x(u))du\,ds$$

so that by the Schwarz inequality

$$\left(\int_{-\infty}^t D_s(t,s)\int_s^t g(v,x(v))dv\,ds\right)^2 \leq$$

$$\int_{-\infty}^t D_s(t,s)ds \int_{-\infty}^t D_s(t,s)\left(\int_s^t g(v,x(v))dv\right)^2 ds$$

$$\leq BV(t)$$

by (5.2.7) (ii). We have just integrated the left side by parts, obtaining

$$\left(\int_{-\infty}^t D(t,s)g(s,x(s))ds\right)^2$$

so that by (5.2.1) we now have $(a(t) - x(t))^2 \leq BV(t)$. We will complete the proof by showing that V is bounded.

If V is not bounded, then there is a sequence

$$\{t_n\} \uparrow \infty \text{ with } V(t_n) \geq V(s) \text{ for } 0 \leq s \leq t_n.$$

Thus,

$$0 \leq V(t_n) - V(s) \leq -\int_s^{t_n} |x(v)g(v,x(v))|dv + M(t_n - s)$$

or

$$\int_s^{t_n} |x(v)g(v,x(v))|dv \le M(t_n - s) \qquad (*)$$

and so

$$V(t_n) = \int_{-\infty}^0 D_s(t_n, s)\left(\int_s^{t_n} g(v, x(v))dv\right)^2 ds$$
$$+ \int_0^{t_n} D_s(t_n, s)\left(\int_s^{t_n} g(v, x(v))dv\right)^2 ds$$
$$\le 2\int_{-\infty}^0 D_s(t_n, s)\left[\left(\int_s^0 g(v, x(v))dv\right)^2\right.$$
$$+ \left.\left(\int_0^{t_n} g(v, x(v))dv\right)^2\right] ds$$
$$+ \int_0^{t_n} D_s(t_n, s)\left[\int_s^{t_n} \Big(K + |x(v)g(v, x(v))|\Big)dv\right]^2 ds$$

by (5.2.3). If we let φ be the initial function for x and denote by $\|g(.,\varphi)\|$ the supremum of $|g(s,\varphi(s))|$ for $-\infty < s \le 0$, then we have by (*) that

$$V(t_n) \le 2\|g(.,\varphi)\|^2 \int_{-\infty}^0 D_s(t_n, s)s^2 ds$$
$$+ 2\int_{-\infty}^0 D_s(t_n, s)\big[Kt_n + Mt_n\big]^2 ds$$
$$+ \int_0^{t_n} D_s(t_n, s)\big[K(t_n - s) + M(t_n - s)\big]^2 ds$$
$$\le 2\|g(.,\varphi)\|^2 \int_{-\infty}^0 D_s(t_n, s)s^2 ds$$
$$+ 2(K+M)^2 \int_{-\infty}^0 D_s(t_n, s)t_n^2 ds$$
$$+ [K+M]^2 \int_0^{t_n} D_s(t_n, s)(t_n - s)^2 ds$$
$$\le 2\|g(.,\varphi)\|^2 \int_{-\infty}^0 D_s(t_n, s)s^2 ds$$
$$+ 2(K+M)^2 A + (K+M)^2 A$$

by (5.2.7)(i) since $\int_{-\infty}^0 D_s(t_n,s)t_n^2 ds \leq \int_{-\infty}^0 D_s(t_n,s)(t_n-s)^2 ds$ and $\int_0^{t_n} D_s(t_n,s)(t_n-s)^2 ds \leq \int_{-\infty}^{t_n} D_s(t_n,s)(t_n-s)^2 ds$. We obtain

$$V(t_n) \leq 3A(K+M)^2 + 2A\|g(\cdot,\varphi)\|.$$

5.2.2 Periodicity I

Here, we will get our supremum bound for Schaefer's theorem from the positive properties of the Liapunov function, rather than from an L^p bound and the Schwarz inequality, as in Theorem 5.1.3.

Suppose that there is a $T > 0$ such that

$$D(t+T, s+T) = D(t,s) \tag{5.2.8}$$

and that

$$a(t+T) = a(t), g(t+T, x) = g(t,x). \tag{5.2.9}$$

We also suppose that there is a $Q > 0$ such that

$$\int_{-\infty}^{t_1} |D(t_1,s) - D(t_2,s)| ds \leq Q|t_1 - t_2|, \quad \text{if} \quad 0 \leq t_1 \leq t_2 \leq T. \tag{5.2.10}$$

It will follow from (5.2.5) and (5.2.8) that (5.2.7) holds because those integrals are periodic and continuous. Also, for each fixed $p > 0$ then

$$\Phi(t) := \int_{-\infty}^p D_s(t,s)(t-s)^2 ds \to 0 \text{ as } t \to \infty, \tag{5.2.11}$$

as may be seen from the substitutions $\Phi(t+nT)$ and $s = u + nT$.

Theorem 5.2.2. *If (5.2.2) – (5.2.10) hold, then (5.2.1) has a T-periodic solution.*

Proof. Let $0 \leq \lambda \leq 1$ and write

$$x(t) = \lambda a(t) - \int_{-\infty}^t D(t,s)\lambda g(s, x(s)) ds. \tag{5.2.1}_\lambda$$

Lemma 5.2.1. *There is a $J > 0$ such that any T-periodic solution x of $(5.2.1_\lambda)$ satisfies $\|x\| \leq J$ where $\|\cdot\|$ is the supremum norm.*

Proof. Suppose that x is a T-periodic solution of (5.2.1$_\lambda$) with $\|x\| = X$. Since x is T-periodic, so is $V(t)$ where V is now defined by

$$V(t) = \lambda^2 \int_{-\infty}^{t} D_s(t,s) \left(\int_{s}^{t} g(v,x(v))dv \right)^2 ds.$$

That is, in the proof of Theorem 5.2.1 we now identify g as λg and $a(t)$ as $\lambda a(t)$ so that we will have by (5.2.2) that

$$V'(t) \leq -\lambda g(t,x)[x - \lambda a(t)] \leq -\lambda|xg(t,x)| + \lambda M.$$

Since V is T-periodic there is a sequence $\{t_n\} \uparrow \infty$ with $V(t_n) \geq V(s)$ for $s \leq t_n$. Thus,

$$0 \leq V(t_n) - V(s) \leq -\lambda \int_{s}^{t_n} |x(v)g(v,x(v))|dv + \lambda M(t_n - s).$$

Therefore,

$$V(t_n) \leq \lambda^2 \int_{-\infty}^{t_n} D_s(t_n,s) \left(\int_{s}^{t_n} [K + |x(v)g(v,x(v))|]dv \right)^2 ds$$

$$\leq \lambda^2 \int_{-\infty}^{t_n} D_s(t_n,s)(K(t_n-s) + M(t_n-s))^2 ds$$

$$\leq \lambda^2 (K+M)^2 A.$$

Since $(\lambda a(t) - x(t))^2 \leq BV(t)$ there is a $J > 0$ with $\|x\| \leq J$ and Lemma 5.2.1 is true.

Now, let $(\mathcal{P}, \|\cdot\|)$ be the Banach space of continuous T-periodic functions with the supremum norm and define H by $\varphi \in \mathcal{P}$ and $0 \leq \lambda \leq 1$ implies that

$$(H\varphi)(t) = a(t) - \int_{-\infty}^{t} D(t,s)g(s,\varphi(s))ds. \tag{5.2.12}$$

Lemma 5.2.2. *$H : \mathcal{P} \to \mathcal{P}$ and H maps bounded sets into compact sets.*

Proof. A change of variable shows that if $\varphi \in \mathcal{P}$ then $(H\varphi)(t+T) = (H\varphi)(t)$. We proceed as in the proof of Theorem 5.1.3. Since $\varphi \in \mathcal{P}$, there

5.2. QUALITATIVE THEORY

is a J with $\|\varphi\| = J$ and a Y with $\|g(t,\varphi)\| = Y$. Thus, if $0 \leq t_1 \leq t_2 \leq T$, then

$$|(H\varphi)(t_1) - (H\varphi)(t_2)| \leq |a(t_1) - a(t_2)|$$
$$+ \left| \int_{-\infty}^{t_1} \Big[D(t_1,s) - D(t_2,s) \Big] g(s,\varphi(s)) ds \right|$$
$$+ \left| \int_{t_1}^{t_2} D(t_2,s) g(s,\varphi(s)) ds \right|$$
$$\leq |a(t_1) - a(t_2)| + (YQ + YE)|t_1 - t_2|$$

where $E = \sup_{0 \leq s \leq T, \, 0 \leq t_2 \leq T} |D(t_2,s)|$. Thus, H is equi-continuous, while boundedness is clear. So, by the Ascoli theorem, it lies in a compact set.

Lemma 5.2.3. *H is continuous in φ.*

Proof. Let $\varphi_1, \varphi_2 \in \mathcal{P}$ so that $\|\varphi_i\| < J$ for some $J > 0$. Then by the uniform continuity of $g(t,x)$ and by (5.2.5) and (5.2.8) we can make

$$|(H\varphi_1)(t) - (H\varphi_2)(t)|$$
$$= \left| \int_{-\infty}^{t} D(t,s) \big[g(s,\varphi_1(s)) - g(s,\varphi_2(s)) \big] ds \right|$$

as small as we please.

Theorem 5.2.2 now follows from Schaefer's result (Theorem 3.1.2) since a solution of $\varphi = \lambda H\varphi$ solves (5.2.1$_\lambda$) and those solutions are bounded by Lemma 5.2.1.

5.2.3 Asymptotic Behavior

We now suppose that there is a constant A with

$$\int_0^t D_s(t,s)[(t-s) + (t-s)^2] ds \leq A, \quad a(t) \to 0 \text{ as } t \to \infty, \quad (5.2.13)$$

that there is a function $M : [0,\infty) \to (0,\infty)$ with

$$2g(t,x)[a(t) - x] \leq -|xg(t,x)| + M(t), \quad M(t) \downarrow 0 \text{ as } t \to \infty, \quad (5.2.14)$$

and that for each $J > 0$ and $\delta > 0$ there is an $S > 0$ such that

$$|x| \leq J \text{ implies } |g(t,x)| \leq \delta + S|x|. \quad (5.2.15)$$

Theorem 5.2.3. *Let (5.2.4) – (5.2.7), (5.2.11), and (5.2.13) – (5.2.15) hold. Then every continuous solution of (5.2.1) with bounded initial function tends to zero as $t \to \infty$.*

Proof. Since (5.2.14) implies (5.2.2) by Theorem 5.2.1 all solutions are bounded. Let $x(t)$ be a fixed solution so that $|x(t)| \leq J$ and $|g(t, x(t))| \leq Y$ for J and Y positive numbers. From the proof of Theorem 5.2.1, if $x(t) \not\to 0$ then $V(t) \not\to 0$ since $a(t) \to 0$ and $(a(t) - x(t))^2 \leq BV(t)$. Thus, we let $\limsup_{t \to \infty} V(t) = P > 0$. Then for each $\varepsilon > 0$ there is a $K > 0$ and a sequence $\{t_n\} \uparrow \infty$ with $V(t_n) \geq V(s) - \epsilon$ for $K \leq s \leq t_n$. Thus

$$-\varepsilon \leq V(t_n) - V(s) \leq M(K)(t_n - s)$$
$$- \int_s^{t_n} |x(v)g(v, x(v))| dv$$

or

$$\int_s^{t_n} |x(v)g(v, x(v))| dv \leq \varepsilon + M(K)(t_n - s). \tag{5.2.16}$$

Let δ and ε be small positive numbers and find the aforementioned K. Then

$$V(t_n) \leq \int_{-\infty}^{K} D_s(t_n, s)(t_n - s)^2 ds Y^2$$
$$+ \int_K^{t_n} D_s(t_n, s)(t_n - s) \int_s^{t_n} g^2(v, x(v)) dv ds.$$

Denote the first term on the right by $L(n)$ and have

$$V(t_n) \leq L(n)$$
$$+ \int_K^{t_n} D_s(t_n, s)(t_n - s) \int_s^{t_n} |g(v, x(v))| \big[\delta + S|x(v)|\big] dv\, ds$$
$$\leq L(n) + \int_K^{t_n} D_s(t_n, s)(t_n - s) \bigg[Y\delta(t_n - s)$$
$$+ \int_s^{t_n} S|x(v)g(v, x(v))| dv \bigg] ds$$
$$\leq L(n) + \delta Y \int_K^{t_n} D_s(t_n, s)(t_n - s)^2 ds$$
$$+ S \int_K^{t_n} D_s(t_n, s)(t_n - s) \big[\varepsilon + M(K)(t_n - s)\big] ds$$

$$= L(n) + \delta Y \int_K^{t_n} D_s(t_n, s)(t_n - s)^2 ds$$

$$+ \varepsilon S \int_K^{t_n} D_s(t_n, s)(t_n - s) ds$$

$$+ SM(K) \int_K^{t_n} D_s(t_n, s)(t_n - s)^2 ds$$

$$\leq L(n) + \delta Y A + \varepsilon S A + SM(K)A.$$

Now, for a given $\delta > 0$, find S in (5.2.15). Then choose ε so small that $\varepsilon S A < \delta$. Next, choose K so large that $SM(K)A < \delta$. Now $L(n) \to 0$ as $t_n \to \infty$ by (5.2.11), so that as $\delta \to 0$, we see that $V(t_n) \to 0$ as $t_n \to \infty$. Hence, $x(t) \to 0$.

5.2.4 Periodicity II

We turn away from V positive and obtain a periodic solution under much weaker condition than (5.2.4). To have the conditions clearly at hand we will repeat some of them now. We are concerned again with the integral equation

$$x(t) = a(t) - \int_{-\infty}^{t} D(t,s) g(s, x(s)) ds \tag{5.2.1}$$

where $a : \Re \to \Re$, $D : \Re \times \Re \to \Re$, $g : \Re \times \Re \to \Re$ are all continuous, there is an $M > 0$ such that if $0 \leq \lambda \leq 1$ then

$$2g(t,x)[\lambda a(t) - x] \leq -|xg(t,x)| + M, \tag{5.2.2}$$

there is a $K > 0$ with

$$|g(t,x)| \leq K + |xg(t,x)|, \tag{5.2.3}$$

$$D_{st}(t,s) \leq 0, \tag{5.2.4*}$$

and that there is a $T > 0$ such that

$$\int_{-\infty}^{t} |D(t,s)| ds \tag{5.2.5*}$$

converges for $0 \leq t \leq T$. We also assume that

$$D(t+T, s+T) = D(t,s) \tag{5.2.8}$$

and that

$$a(t+T) = a(t), \quad g(t+T, x) = g(t, x). \tag{5.2.9}$$

Finally, suppose that there is a $Q > 0$ such that

$$\int_{-\infty}^{t} |D(t_1, s) - D(t_2, s)| ds \leq Q|t_1 - t_2|,$$

$$\text{if } 0 \leq t_1 \leq t_2 \leq T. \tag{5.2.10}$$

The problem is to show that, under certain additional convergence conditions then (5.2.1) has a T-periodic solution.

Let $(\mathcal{P}_T, \|\cdot\|)$ be the Banach space of continuous functions $\phi : \Re \to \Re$ where $\phi(t+T) = \phi(t)$ with the supremum norm. Our work depends on the following condition and it will be written in another form, as well.

Let $M > 0$ be given and let $\mathcal{J}_{MT} \subset \mathcal{P}_T$ be the set of ϕ with

$$\int_0^T |\phi(s)g(s, \phi(s))| ds \leq MT. \tag{5.2.17}$$

Our basic assumption is that for each $M > 0$ there is an $L = L(M) > 0$ such that

$$\phi \in \mathcal{J}_{MT} \implies \int_{-\infty}^t |D(t,s)g(s, \phi(s))| ds \leq L(M) \tag{5.2.18}$$

for $0 \leq t \leq T$.

Now (5.2.18) can be written much more clearly as follows. If $x \in \mathcal{P}_T$ and if (5.2.18) holds then the integral I, below, converges and

$$I := \int_{-\infty}^t |D(t,s)g(s, x(s))| ds$$

$$= \lim_{R \to -\infty} \int_{-R}^t |D(t,s)g(s, x(s))| ds$$

$$= \lim_{n \to \infty} \int_{-nT}^t |D(t,s)g(s, x(s))| ds$$

$$= \lim_{k \to \infty} \sum_{n=0}^k \int_{t-(n+1)T}^{t-nT} |D(t,s)g(s, x(s))| ds$$

$$= \lim_{k \to \infty} \sum_{n=0}^k \int_{t-T}^t |D(t, s-nT)g(s-nT, x(s-nT))| ds$$

$$= \lim_{k \to \infty} \sum_{n=0}^k \int_{t-T}^t |D(t+nT, s)g(s, x(s))| ds.$$

Thus, (5.2.18) becomes

$$\phi \in \mathcal{J}_{MT} \implies \sum_{n=0}^{\infty} \int_{t-T}^{t} |D(t+nT,s)g(s,x(s))|ds \leq L(M). \quad (5.2.19)$$

In a given problem we would prove (5.2.19), prove that the steps could be reversed, and get (5.2.18).

We would expect to use Young's inequality to get (5.2.19), but as examples, take successively $g(t,x) = x$, $g(t,x) = x^3$, and $g(t,x) = x^{1/3}$, use the Schwarz inequality, and find that (5.2.19) becomes a very reasonable assumption when (5.2.17) is applied.

Theorem 5.2.3. *Let (5.2.2), (5.2.3) (5.2.4*), (5.2.5*), (5.2.8)–(5.2.10), and (5.2.18) hold. Then (5.2.1) has a T-periodic solution.*

Proof. Our proof parallels that of Theorem 5.2.2. For V defined in the proof of Lemma 5.2.1 we have

$$V'(t) \leq -\lambda|xg(t,x)| + \lambda M.$$

If $x \in \mathcal{P}_T$, so is V and thus

$$0 = V(T) - V(0) \leq \lambda[-\int_0^T |x(s)g(s,x(s))|ds + MT]$$

so that for $\lambda > 0$, we have

$$\int_0^T |x(s)g(s,x(s))|ds \leq MT.$$

If $\lambda = 0$, then $x(t) = 0$. We now have by (5.2.19) that $|x(t)| \leq |a(t)| + L$, an *a priori* bound.

The proof that H in (5.2.12) is continuous and maps bounded sets into compact sets is just as in the proof of Theorem 5.2.2.

Most of the material in this section was taken from Burton (1993).

5.3 A Refined Liapunov Functional

In this section we refine the Liapunov functional and give many results on qualitative properties derived from this one Liapunov functional. We attempt to relate the arguments to classical techniques in differential equations. Some of the results contain long and detailed conditions, but the real hope is that they will stimulate ideas for simplifications and extensions. The reader will find this section to be in marked contrast to any other of this text.

Let $D: \Re \times \Re \to \Re$ with both D and $\int_{-\infty}^{t} |D(t,s)|ds$ being continuous, let $a: \Re \to \Re$ be continuous, and let $g: \Re \times \Re \to \Re$ and $g_i: \Re \to \Re$ all be continuous with $xg(t,x) > 0$ if $x \neq 0$, $|g_1(x)| \leq |g(t,x)| \leq |g_2(x)|$, $xg_1(x) > 0$ if $x \neq 0$. Consider the equation

$$x(t) = a(t) - \int_{-\infty}^{t} D(t,s)g(s,x(s))ds. \tag{5.3.1}$$

If $\phi : (-\infty, t_0] \to \Re$ is a given bounded and continuous initial function, then there is a continuous solution $x(t, t_0, \phi)$ defined on an interval $[t_0, \alpha)$ and satisfying (5.3.1) on that interval, while agreeing with ϕ on $(-\infty, t_0]$, provided that ϕ is chosen so that (5.3.1) is an identity at $t = t_0$ (see, Section 3.1). If the solution remains bounded then it can be continued for all future time. It is always assumed that ϕ is chosen so that the solution is continuous. As mentioned in the last section, for a given bounded continuous ϕ there is a ψ arbitrarily close to ϕ for which the continuity holds.

In this section we reduce the convexity conditions on D from those in Section 5.2, but increase the conditions on $a(t)$ and $g(t,x)$.

We suppose that there are continuous functions $B, Q: \Re \times \Re \to \Re$ with

$$B(t,s) = D(t,s) + Q(t,s), \tag{5.3.2}$$

$$B_s(t,s) \geq 0, \quad B_{st}(t,s) \leq 0, \tag{5.3.3}$$

$$\int_{-\infty}^{t} [|B(t,s)| + B_s(t,s)(t-s)^2 \\ + |B_{st}(t,s)| + |Q(t,s)|]ds \tag{5.3.4}$$

is continuous,

$$\lim_{s \to -\infty} (t-s)B(t,s) = 0 \text{ for fixed } t, \tag{5.3.5}$$

$$\int_{0}^{\infty} |Q(u+t,t)|du + \int_{-\infty}^{t} \int_{t-s}^{\infty} |Q(u+s,s)|du\,ds \tag{5.3.6}$$

exists for $t \geq 0$.

Much can be deduced from the following result. We shall give a few possibilities.

5.3. A REFINED LIAPUNOV FUNCTIONAL

THEOREM 5.3.1. *If $x(t)$ is a solution of (5.3.1) on $[t_0, \alpha)$, then for $B(t,t) > 0$ and $k \geq 0$ the functional*

$$V(t) = \int_{-\infty}^{t} B_s(t,s) \left(\int_{s}^{t} g(v, x(v)) dv \right)^2 ds$$

$$+ k \int_{-\infty}^{t} \int_{t-s}^{\infty} |Q(u+s,s)| du \, g^2(s, x(s)) ds \qquad (5.3.7)$$

satisfies

$$\left[a(t) - x(t) + \int_{-\infty}^{t} Q(t,s) g(s, x(s)) ds \right]^2 \leq V(t) B(t,t) \qquad (5.3.8)$$

and

$$V'_{(5.3.1)}(t) \leq 2g(t, x(t))[a(t) - x(t)]$$

$$- (k-1) \int_{-\infty}^{t} |Q(t,s)| g^2(s, x(s)) ds$$

$$+ \left[\int_{-\infty}^{t} |Q(t,s)| ds + k \int_{0}^{\infty} |Q(u+t,t)| du \right] g^2(t,x). \qquad (5.3.9)$$

Proof. First, we apply Schwarz's inequality and use (5.3.7) to obtain

$$V(t) \geq \left(\int_{-\infty}^{t} B_s(t,s) \int_{s}^{t} g(v, x(v)) dv \, ds \right)^2 / B(t,t).$$

Integrate by parts and use (5.3.5), together with the fact that there is a bounded initial function to obtain

$$V(t) B(t,t)$$

$$\geq \left[B(t,s) \int_{s}^{t} g(v, x(v)) dv \bigg|_{s=-\infty}^{s=t} + \int_{-\infty}^{t} B(t,s) g(v, x(v)) dv \right]^2.$$

The first term on the right is zero. When B is separated as in (5.3.2) and (5.3.1) is used, we have

$$V(t) B(t,t) \geq \left[a(t) - x(t) + \int_{-\infty}^{t} Q(t,s) g(s, x(s)) ds \right]^2$$

so that (5.3.8) holds.

Denote the last term in V by $Z(t)$ and compute

$$V'(t) = \int_{-\infty}^{t} B_{st}(t,s) \left(\int_s^t g(v,x(v))dv \right)^2 ds$$

$$+ \int_{-\infty}^{t} B_s(t,s) 2 \int_s^t g(v,x(v))dv\, ds\, g(t,x(t)) + Z'(t)$$

$$\leq 2g(t,x(t)) \left[B(t,s) \int_s^t g(v,x(v))dv \bigg|_{s=-\infty}^{s=t} \right.$$

$$\left. + \int_{-\infty}^{t} B(t,s) g(s,x(s)) ds \right] + Z'(t)$$

$$= 2g(t,x(t)) \left[\int_{-\infty}^{t} D(t,s) g(s,x(s)) ds \right.$$

$$\left. + \int_{-\infty}^{t} Q(t,s) g(s,x(s)) ds \right] + Z'(t)$$

$$\leq 2g(t,x(t)) \big[a(t) - x(t) \big] + g^2(t,x(t)) \int_{-\infty}^{t} |Q(t,s)| ds$$

$$+ \int_{-\infty}^{t} |Q(t,s)| g^2(s,x(s)) ds$$

$$+ k \int_0^{\infty} |Q(u+t,t)| du\, g^2(t,x(t))$$

$$- k \int_{-\infty}^{t} |Q(t,s)| g^2(s,x(s)) ds$$

$$= 2g(t,x(t)) \big[a(t) - x(t) \big]$$

$$+ g^2(t,x(t)) \left[\int_{-\infty}^{t} |Q(t,s)| ds + k \int_0^{\infty} |Q(u+t,t)| du \right]$$

$$- (k-1) \int_{-\infty}^{t} |Q(t,s)| g^2(s,x(s)) ds,$$

as required.

Many interesting consequences can be derived from (5.3.8) and (5.3.9). We begin with two extreme cases.

5.3. A REFINED LIAPUNOV FUNCTIONAL

Corollary 5.3.1. *If $a(t) = Q(t,s) = 0$ and if $\int_{-\infty}^{t} D_s(t,s)ds \leq 1/M$ for some M, then along any solution $x(t)$ we have*

$$Mx^2(t) \leq V(t)$$

and

$$V'(t) \leq -2g(t, x(t))x(t) \leq -2g_1(x(t))x(t).$$

Thus, $x(t)$ is bounded, $x = 0$ is stable, and

$$\int^{\infty} g_1(x(t))x(t)dt < \infty.$$

We later give three kinds of conditions to ensure that $x(t) \to 0$ as $t \to \infty$ when $a(t) \to 0$.

Remark. Notice that Corollary 5.3.1 has no growth condition on g, but there will be in Corollary 5.3.2 when $Q(t,s) \neq 0$. In effect, Q is a "perturbation term" and the bounds on Q offer a measure of how far D can deviate from the conditions on B. In addition, Condition (5.3.10) will itself be a growth condition on g when $Q \neq 0$. To start the completion of Corollary 5.3.1, examine Corollary 5.3.3 and note that when $Q = 0$, then no growth condition on g is required to conclude that $x(t) \to 0$. Corollary 5.3.4 asks that g satisfy a local Lipschitz condition in order to conclude that $x(t) \to 0$.

There are many possible variants of the next lemma. It is a natural extension of the statement that the convolution of an L^1-function with a function tending to zero, itself tends to zero.

Lemma 5.3.1. *Let $h : [0, \infty) \to [0, \infty)$ with $\int_0^\infty h(s)ds < \infty$ and let $C : \Re \times \Re \to \Re$ be continuous with $|C(t,s)| \leq K$ if $0 \leq s \leq t$ for some $K > 0$. Suppose also that for each $P > 0$ we have $\limsup_{t \to \infty, 0 \leq s \leq P} |C(t,s)| = 0$. Then $\int_0^t C(t,s)h(s)ds \to 0$ as $t \to \infty$.*

Proof. Let $\epsilon > 0$ be given and choose $P > 0$ so that $\int_P^\infty K h(t)dt < \epsilon/2$. Then

$$\int_0^t |C(t,s)|h(s)ds \leq \int_0^P |C(t,s)|h(s)ds + K\int_P^\infty h(s)ds$$

$$\leq \sup_{0 \leq s \leq P} |C(t,s)| \int_0^P h(s)ds + \epsilon/2.$$

5. INFINITE DELAY: SCHAEFER'S THEOREM

The next lemma will be used repeatedly.

Lemma 5.3.2. *Let $xg(t,x) > 0$ if $x \neq 0$, g be continuous and bounded for x bounded, $|a(t)| \leq A/2$ for some $A > 0$ and all t, and let $c_1 > 0$. Then there is an $M > 0$ with*

$$-2c_1 xg(t,x) + 2|g(t,x)|\,|a(t)| \leq -c_1 xg(t,x) + M|a(t)|.$$

Proof. We have

$$K(t,x) := -2c_1 xg(t,x) + 2|g(t,x)a(t)|$$
$$\leq -c_1 xg(t,x) + |g(t,x)|[-c_1|x| + 2|a(t)|].$$

If $|x| \geq 2A/c_1$, then $-c_1|x| + 2|a(t)| \leq -2A + A < 0$. If $|x| \leq 2A/c_1$, then $2|g(t,x)| \leq M$ for some $M > 0$, and the proof is complete.

Corollary 5.3.2. *Suppose there is a $k \geq 1$ and a $\beta < 2$ such that*

$$\beta xg(t,x) \geq \left[\int_{-\infty}^{t} |Q(t,s)|ds + k\int_{0}^{\infty}|Q(u+t,t)|du\right]g^2(t,x) \quad (5.3.10)$$

and that $a(t)$ is both bounded and $L^1[0,\infty)$. Then for any solution $x(t)$ of (5.3.1) on $[t_0,\infty)$ we have $\int_{t_0}^{\infty} x(t)g(t,x(t))dt < \infty$. If, in addition, $\int_{-\infty}^{t} D_s(t,s)ds$ is bounded, $|g(t,x)| \leq J|x|$ for some $J > 0$, if $\int_{-\infty}^{t_0}|D(t,s)|ds \to 0$ as $t \to \infty$, if $\int_{t_0}^{t}|D(t,s)|ds$ is bounded, and if for each $P > 0$ we have $\limsup\limits_{t\to\infty\ 0\leq s\leq P}|D(t,s)| = 0$, then $x(t) \to a(t)$ as $t \to \infty$.

Proof. By (5.3.9), (5.3.10), and Lemma 5.3.2 we have $V'(t) \leq -c_1 x(t)g(t,x(t)) + M|a(t)|$. Since $V \geq 0$, the first conclusion holds. Next, if ϕ is the bounded initial function on $(-\infty, t_0]$ with $g^* \geq |g(t,\phi(t))|$ on $(-\infty, t_0]$, then

$$\left|\int_{-\infty}^{t} D(t,s)g(s,x(s))ds\right|$$
$$\leq \left[\int_{t_0}^{t}|D(t,s)|ds \int_{t_0}^{t}|D(t,s)|g^2(s,x(s))ds\right]^{1/2}$$
$$+ \int_{-\infty}^{t_0}|D(t,s)|g^* ds. \qquad (*)$$

Since $|g(t,x)| \leq J|x|$, it follows that $\int_{t_0}^{\infty} g^2(s,x(s))ds < \infty$ because $\int_{t_0}^{\infty} x(s)g(s,x(s))ds < \infty$. Thus, by Lemma 5.3.1, $(*)$ tends to zero as $t \to \infty$ and the conclusion follows from (5.3.1).

5.3. A REFINED LIAPUNOV FUNCTIONAL

A classical result from Yoshizawa (1966; p. 191) for a finite delay equation $x' = F(t, x_t)$ states that if there is a $V(t, \phi)$ and increasing functions W_i with

(i) $W_1(|\phi(0)|) \leq V(t, \phi) \leq W_2(\|\phi\|)$,
(ii) $V'(t, x_t) \leq -W_3(|x(t)|)$, and
(iii) $|F(t, \phi)|$ is bounded for ϕ bounded,

then $x = 0$ is uniformly asymptotically stable. Condition (iii) assures us that a bounded solution is Lipschitz; hence, $\int^\infty W_3(|x(t)|)dt < \infty$ implies that $x(t)$ must tend to zero. The following result is a counterpart for integral equations and it leads us to *a priori* bounds for periodic solutions.

Corollary 5.3.3. *Let $a(t) \in L^1[0, \infty)$, $a(t) \to 0$ as $t \to \infty$, and either $Q(t, s) = 0$ or $|g(t, x)| \leq J|x|$ for some $J > 0$. Also, for each $t_0 \in R$ and each $P > 0$ let both*

$$\int_{-\infty}^{t_0} |Q(t,s)|ds \to 0 \text{ as } t \to \infty \text{ and } \limsup_{t \to \infty, \ 0 \leq s \leq P} Q(t, s) = 0.$$

Finally, suppose there are $k \geq 1$ and $\beta < 2$ such that (5.3.10) holds and an $M > 0$ such that $|a(t_1) - a(t_2)| \leq M|t_1 - t_2|$, $\int_{-\infty}^t |B_s(t, s)|ds \leq M$, $\int_{t_0}^t |Q(t, s)|ds \leq M$, and

$$\int_{-\infty}^{t_1} |D(t_1, s) - D(t_2, s)|ds \leq M|t_1 - t_2| \tag{5.3.11}$$

for $0 \leq t_1 \leq t_2 < \infty$ and $|t_1 - t_2|$ small. Then every solution $x(t)$ is defined on $[t_0, \infty)$ and $x(t) \to 0$ as $t \to \infty$.

Proof. By the proof of Corollary 5.3.2 we have V bounded and $\int_{t_0}^\infty x(t)g(t, x(t))dt < \infty$. By assumption $|g_1(x)| \leq |g(t, x)| \leq |g_2(x)|$ where $xg_1(x) > 0$ if $x \neq 0$. If $x(t) \not\to 0$, then there is an $\epsilon > 0$ and a sequence $\{t_n\} \uparrow \infty$ with $|x(t_n)| \geq \epsilon$. Since V is bounded, if $Q = 0$, then from (5.3.8) and $a(t) \to 0$, we have $x(t)$ bounded. If $Q \neq 0$, then $|g(t, x)| \leq J|x|$ so $\int^\infty x(t)g(t, x(t))dt < \infty$ yields $\int^\infty g^2(t, x(t))dt < \infty$; and this implies that

$$\left| \int_{-\infty}^t Q(t,s)g(s,x(s))ds \right|$$
$$\leq \left| \int_{-\infty}^{t_0} Q(t,s)g(s,x(s))ds \right|$$
$$+ \left[\int_{t_0}^t |Q(t,s)|ds \int_{t_0}^t |Q(t,s)|g^2(s,x(s))ds \right]^{1/2}$$

and this tends to zero. By (5.3.8), again, $x(t)$ is bounded and so, in any case, $|g(t,x(t))| \leq L$ for some $L > 0$.

From the boundedness of $x(t)$ and $\int^\infty x g_1(x) dt < \infty$, we can suppose that $|x(t_n)| = \epsilon$ and choose another sequence $\{s_n\} \uparrow \infty$ with $|x(s_n)| = \epsilon/2$ and $\epsilon \geq |x(t)| \geq \epsilon/2$ for $t_n \leq t \leq s_n$.

Thus,

$$\epsilon/2 \leq |x(t_n) - x(s_n)| \leq |a(t_n) - a(s_n)|$$
$$+ \left| \int_{-\infty}^{t_n} D(t_n, s) g(s, x(s)) ds - \int_{-\infty}^{t_n} D(s_n, s) g(s, x(s)) ds \right|$$
$$+ \left| \int_{-\infty}^{t_n} D(s_n, s) g(s, x(s)) ds - \int_{-\infty}^{s_n} D(s_n, s) g(s, x(s)) ds \right|$$
$$\leq M|t_n - s_n| + L \int_{-\infty}^{t_n} |D(t_n, s) - D(s_n, s)| ds$$
$$+ \left| \int_{s_n}^{t_n} D(s_n, s) g(s, x(s)) ds \right| \leq (M + LM + LB)|t_n - s_n|$$

where $\int_{t_n}^{s_n} |D(s_n, s)| ds \leq B$. This yields $|t_n - s_n| \geq \delta$ for some $\delta > 0$, contradicting $\int^\infty x(t) g_1(x(t)) dt < \infty$ while $|x(t)| \geq \epsilon/2$ on $[t_n, s_n]$. This completes the proof.

There is a third way to drive $x(t)$ to zero.

Corollary 5.3.4. *Let $Q(t,s) = 0$, $a(t) \to 0$ as $t \to \infty$, $a(t) \in L^1[0, \infty)$, and $\int_{-\infty}^t D_s(t,s) ds$ be bounded. Suppose also that for each $P > 0$ we have $\int_{-\infty}^P D_s(t,s)(t-s)^2 ds \to 0$ as $t \to \infty$ and that there is an M independent of P with $\int_P^t D_s(t,s)(t-s) ds \leq M$. If, in addition, for each $K > 0$ there is a $J > 0$ such that $|x| \leq K$ implies $|g(t,x)| \leq J|x|$, then $x(t) \to 0$ as $t \to \infty$.*

Proof. From (5.3.8), if $V(t) \to 0$, so does $x(t)$. Since $x(t)$ is bounded by (5.3.8) and the fact that V is bounded, we have $\int^\infty g^2(t, x(t)) dt < \infty$. By Schwarz's inequality we obtain

$$V(t) \leq \int_{-\infty}^t D_s(t,s)(t-s) \int_s^t g^2(v, x(v)) dv\, ds$$
$$\leq \int_{-\infty}^P D_s(t,s)(t-s) \int_s^t g^2(v, x(v)) dv\, ds$$
$$+ \int_P^t D_s(t,s)(t-s) \int_P^\infty g^2(v, x(v)) dv\, ds$$

5.3. A REFINED LIAPUNOV FUNCTIONAL

$$\leq J^2 \int_{-\infty}^{P} D_s(t,s)(t-s)^2 ds$$
$$+ \left(\int_{P}^{\infty} g^2(v,x(v))dv\right) \int_{P}^{t} D_s(t,s)(t-s)ds.$$

The last integral is bounded by M, while its coefficient tends to zero as $P \to \infty$. This completes the proof.

In the next result, notice that the *a priori* bound does not require V to be positive. The *a priori* bound comes from V' alone.

Corollary 5.3.5. *Let (5.3.10) hold and suppose there is a $T > 0$ with $a(t+T) = a(t)$, $g(t+T,x) = g(t,x)$, $D(t+T, s+T) = D(t,s)$, and $B_s(t+T, s+T) = B_s(t,s)$. Suppose, in addition, that $xg_1(x) \to \infty$ as $|x| \to \infty$, and that there is an $M > 0$ with $|g(t,x)| \leq M|x|$, $\sup_{0 \leq t \leq T} \int_{-\infty}^{t} |D(t,s)|ds \leq M$, $|a(t_1) - a(t_2)| \leq M|t_1 - t_2|$, $|g(t,x_1) - g(t,x_2)| \leq M|x_1 - x_2|$, and that $\int_{-\infty}^{t_1} |D(t_1,s) - D(t_2,s)|ds \leq M|t_1 - t_2|$ for $0 \leq t_1 \leq t_2 \leq T$. Then there is a $K > 0$ such that if $x(t)$ is any T-periodic solution of (5.3.1), then $\sup_{0 \leq t \leq T} |x(t)| =: \|x\| \leq K$ and there is a T-periodic solution.*

Proof. Let $0 \leq \lambda \leq 1$ and write (5.3.1) as

$$x(t) = \lambda a(t) - \int_{-\infty}^{t} D(t,s)\lambda g(s,x(s))ds \qquad (5.3.1_\lambda)$$

so that if $V(t)$ is defined in (5.3.7) with g replaced by λg then we obtain, as in Lemma 5.3.2, that

$$V'(t) \leq -c_1 x(t)\lambda g(t,x(t)) + M\lambda|a(t)|.$$

We now show that there is an *a priori* bound on any T-periodic solution $x(t)$ of $(5.3.1_\lambda)$. If $\lambda = 0$, then $\|x\| = 0$. If $\lambda > 0$, since V is also T-periodic, $0 = V(T) - V(0) \leq -c_1 \int_0^T x(s)\lambda g(s,x(s))ds + G\lambda$ where $G = MT\|a\|$; thus, λ divides out and we have $\int_0^T x(s)g(s,x(s))ds \leq G/c_1$.

Next, let $0 \leq t_1 \leq t_2 \leq T$, $|x(t_1)| = \|x\|$, and note that since $|g(t,x)| \leq M|x|$ we have

$$|x(t_1) - x(t_2)| \leq \lambda |a(t_1) - a(t_2)|$$
$$+ \lambda \left| \int_{-\infty}^{t_1} [D(t_1,s) - D(t_2,s)] g(s, x(s)) ds \right|$$
$$+ \lambda \left| \int_{-\infty}^{t_1} D(t_2,s) g(s, x(s)) ds \right.$$
$$\left. - \int_{-\infty}^{t_2} D(t_2,s) g(s, x(s)) ds \right|$$
$$\leq M|t_1 - t_2| + M^2 |t_1 - t_2| \|x\| + M\|x\| \int_{t_1}^{t_2} |D(t_2,s)| ds$$
$$\leq \left(M + M^2 \|x\| + BM\|x\| \right) |t_1 - t_2| \leq J(1 + \|x\|)|t_1 - t_2|$$

for $|D(t_2,s)| \leq B$ if $0 \leq s \leq T$ and some $J > 0$. If $J|t_1 - t_2| \leq 1/2$, then $|x(t_1)| - |x(t_2)| \leq |x(t_1) - x(t_2)| \leq (1 + \|x\|)/2$ and $|x(t_1)| = \|x\|$ so $\|x\|/2 \leq 1/2 + |x(t_2)|$. If $\|x\| \leq 2$, then this is an *a priori* bound. If $\|x\| \geq 2$, then $\|x\| \leq 1 + 2|x(t_2)| \leq \|x\|/2 + 2|x(t_2)|$ or $\|x\|/2 \leq 2|x(t_2)|$ so that $|x(t_2)| \geq \|x\|/4$ if $|t_1 - t_2| \leq 1/2J$. But

$$\int_0^T x(t) g_1(x(t)) dt \leq \int_0^T x(t) g(t, x(t)) dt \leq G/c_1$$

and $x g_1(x) \to \infty$ as $|x| \to \infty$, while $|x(t_2)| \geq \|x\|/4$ if $|t_1 - t_2| \leq 1/2J$. Thus, the required bound on $\|x\|$ exists for $0 \leq \lambda \leq 1$.

Next, let $(\mathcal{P}, \|\cdot\|)$ be the Banach space of continuous T-periodic functions with the supremum norm. Define a mapping $H: \mathcal{P} \to \mathcal{P}$ by $\phi \in \mathcal{P}$ implies that

$$(H\phi)(t) = a(t) - \int_{-\infty}^t D(t,s) g(s, \phi(s)) ds. \tag{5.3.12}$$

5.3. A REFINED LIAPUNOV FUNCTIONAL

A translation will show that $H : \mathcal{P} \to \mathcal{P}$. To show that $H\phi$ is continuous in t and lies in a compact set we let $\phi \in \mathcal{P}$ with $\|\phi\| \leq K$, where K is an arbitrary positive number. Then

$$|(H\phi)(t_1) - (H\phi)(t_2)| \leq \Big[|a(t_1) - a(t_2)|$$
$$+ \Big|\int_{-\infty}^{t_1} [D(t_1,s) - D(t_2,s)]g(s,\phi(s))ds\Big|$$
$$+ \Big|\int_{t_1}^{t_2} D(t_2,s)g(s,\phi(s))ds\Big|\Big]$$
$$\leq M|t_1 - t_2| + M^2\|\phi\| |t_1 - t_2| + D^*M\|\phi\| |t_2 - t_2|$$

where $D^* = \sup_{0\leq s\leq T, 0\leq t_2\leq T} |D(t_2,s)|$. Hence, $H\phi$ is equicontinuous and bounded by a function of K. To show that $H\phi$ is continuous in ϕ, for fixed t and for $\phi_i \in \mathcal{P}$ we have

$$|(H\phi_1)(t) - (H\phi_2)(t)|$$
$$= \Big|\int_{-\infty}^{t} D(t,s)[g(s,\phi_1(s)) - g(s,\phi_2(s))]ds\Big|$$
$$\leq \int_{-\infty}^{t} |D(t,s)| M |\phi_1(s) - \phi_2(s)|ds$$
$$\leq M\|\phi_1 - \phi_2\| \int_{-\infty}^{t} |D(t,s)|ds$$
$$\leq U\|\phi_1 - \phi_2\| \text{ for some } U > 0.$$

Hence, H is continuous in ϕ. This completes the proof.

Remark. When $B(t,s) = 0$, a more flexible Liapunov functional is

$$H(t) = k\int_{-\infty}^{t}\int_{t-s}^{\infty} |D(u+s,s)|du|g(s,x(s))|ds$$

with

$$H'(t) \leq -\delta\Big[|g(t,x)| + \int_{-\infty}^{t} |C(t,s)g(s,x(s))|ds\Big] + |a(t)|$$

and $\delta > 0$. From this we conclude that if $a \in L^1$, then $|g(t,x)|$ and $|x|$ are L^1. In the previous corollaries we did not yet use the term $-(k-1)\int_{-\infty}^{t}|C(t,s)g(s,x(s))|ds$ in the derivative of V. But here it can be used effectively and we see that under suitable assumptions relating D to one of its integrals we can obtain

$$H'(t) \leq -\gamma H(t) + |a(t)|, \quad \gamma > 0$$

and

$$\mu\big[|x| - |a(t)|\big] \leq H(t), \quad \mu > 0.$$

These are old and useful relations for deriving qualitative properties.

5.3.1 A linear vector equation

Let D be a continuous $n \times n$ matrix with $\int_{-\infty}^{t} |D(t,s)| ds$ continuous, $a : \Re \to \Re^n$ be continuous, and consider the equation

$$x(t) = a(t) - \int_{-\infty}^{t} D(t,s) x(s) ds. \tag{5.3.13}$$

It turns out that all of the work on the scalar equation can be done for (5.3.13) except that we have been unable to obtain a counterpart of (5.3.8). Thus, we readily prove that solutions are L^2, that they converge to $a(t)$, and that there are periodic solutions. But we must rely on techniques independent of (5.3.8) to show boundedness. Formal counterparts of (5.3.2)–(5.3.6) are needed. The symbol $|\cdot|$ will denote absolute value as well as compatible vector and matrix norms.

Suppose there are continuous matrix functions B and Q with

$$B(t,s) = D(t,s) + Q(t,s), \quad B^T(t,s) = B(t,s), \tag{5.3.14}$$

$$x^T B_s(t,s) x \geq 0, \quad x^T B_{st}(t,s) x \leq 0, \tag{5.3.15}$$

$$\int_{-\infty}^{t} \big[|B(t,s)| + |B_s(t,s)|(t-s)^2 + |B_{st}(t,s)| + |Q(t,s)| \big] ds \tag{5.3.16}$$

is continuous,

$$\lim_{s \to -\infty} |t-s| |B(t,s)| = 0 \text{ for fixed } t, \tag{5.3.17}$$

and

$$\int_{0}^{\infty} |Q(u+t,t)| du + \int_{-\infty}^{t} \int_{t-s}^{\infty} |Q(u+s,s)| du\, ds \tag{5.3.18}$$

exists for $t \geq 0$.

5.3. A REFINED LIAPUNOV FUNCTIONAL

Theorem 5.3.2. *If $x(t)$ is continuous on $\Re$ and is a solution of (5.3.13) on $[t_0, \infty)$ with bounded initial function, then the functional*

$$V(t) = \int_{-\infty}^{t} \left\{ \left[\int_{s}^{t} x^T(q)dq \right] B_s(t,s) \int_{s}^{t} x(q)dq \right\} ds$$

$$+ k \int_{-\infty}^{t} \int_{t-s}^{\infty} |Q(u+s,s)|du|x(s)|^2 ds \qquad (5.3.19)$$

satisfies

$$V'(t) \leq 2|a(t)|\,|x(t)|$$
$$- \left[2 - \int_{-\infty}^{t} |Q(t,s)|ds - k \int_{0}^{\infty} |Q(u+t,t)|du \right] |x(t)|^2$$
$$- (k-1) \int_{-\infty}^{t} |Q(t,s)|\,|x(s)|^2 ds. \qquad (5.3.20)$$

Proof. We have

$$V'(t) \leq \int_{-\infty}^{t} 2x^T(t) B_s(t,s) \int_{s}^{t} x(q)dq\, ds$$
$$+ k \int_{0}^{\infty} |Q(u+t,t)|du|x(t)|^2 - k \int_{-\infty}^{t} |Q(t,s)|\,|x(s)|^2 ds$$
$$= 2x^T(t) \left[B(t,s) \int_{s}^{t} x(q)dq \Big|_{s=-\infty}^{s=t} + \int_{-\infty}^{t} B(t,s)x(s)ds \right]$$
$$+ k \int_{0}^{\infty} |Q(u+t,t)|du|x(t)|^2 - k \int_{-\infty}^{t} |Q(t,s)|\,|x(s)|^2 ds$$
$$= 2x^T(t) \left[a(t) - x(t) + \int_{-\infty}^{t} Q(t,s)x(s)ds \right]$$
$$+ k \int_{0}^{\infty} |Q(u+t,t)|du|x(t)|^2 - k \int_{-\infty}^{t} |Q(t,s)|\,|x(s)|^2 ds$$
$$\leq 2|a(t)|\,|x(t)| - 2|x(t)|^2 + |x(t)|^2 \int_{-\infty}^{t} |Q(t,s)|ds$$
$$+ \int_{-\infty}^{t} |Q(t,s)|\,|x(s)|^2 ds + k \int_{0}^{\infty} |Q(u+t,t)|du|x(t)|^2$$
$$- k \int_{-\infty}^{t} |Q(t,s)|\,|x(s)|^2 ds$$
$$\leq - \left[2 - \int_{-\infty}^{t} |Q(t,s)|ds - k \int_{0}^{\infty} |Q(u+t,t)|du \right] |x(t)|^2$$

$$+ 2|a(t)| \, |x(t)| - (k-1) \int_{-\infty}^{t} |Q(t,s)| \, |x(s)|^2 ds,$$

as required.

At this point we do not have a lower bound parallel to (5.3.8); but for linear systems this is not so crucial since solutions can always be defined for all future time. We can prove results for the system parallel to the ones for (5.3.1) as follows. In Corollaries 5.3.1 and 5.3.2 we conclude only that $x \in L^2[0, \infty)$. Corollaries 5.3.3 and 5.3.5 say little about the system. Corollary 5.3.4 and Corollary 5.3.6 hold exactly as they did for (5.3.1).

The material in this section was taken from Burton (1994a), published by Tohoku University.

5.4 A Priori Bounds

This section concerns a very intriguing problem. Consider a linear ordinary differential equation

$$x' = Ax, \quad \det A \neq 0,$$

where A is an $n \times n$ constant matrix. If a solution is bounded for $t \leq 0$, then it might happen that it is periodic, almost periodic, or unbounded for $t > 0$, as is the case when λ is a characteristic root with positive real part. In short, knowing that solutions are bounded for $t < 0$ does not portend well for obtaining boundedness for $t > 0$.

Of course, when $A = A(t)$ the entire matter changes, while the work here is now a commentary on the case with memory.

Consider the equation

$$x(t) = a(t) - \int_{-\infty}^{t} D(t,s) g(x(s)) ds \qquad (5.4.1)$$

with a, D, D_{st}, and g continuous. Suppose also that there exist $A > 0$, $B > 0$, and $G > 0$ so that

$$|a(t)| \leq A, \, xg(x) > 0 \text{ if } x \neq 0, G = \max_{0 \leq |x| \leq 1} |g(x)|, \qquad (5.4.2)$$

$$D(t,s) \geq 0, D_s(t,s) \geq 0, D_{st}(t,s) \leq 0, D(t,t) \leq B, \qquad (5.4.3)$$

$$\lim_{s \to -\infty} (t-s) D(t,s) = 0 \text{ for each fixed } t, \qquad (5.4.4)$$

and

$$\int_{-\infty}^{t} [D(t,s) + \left(D_s(t,s) - D_{st}(t,s)\right)(t-s)^2] ds. \qquad (5.4.5)$$

is continuous.

5.4. A PRIORI BOUNDS

The result which motivates the section is that there is a constant L such that if ϕ is a solution of (5.4.1) on $(-\infty, \infty)$ and if for $t < 0$ it is bounded, then $|\phi(t)| < L$ for $-\infty < t < \infty$.

In the following we will study solutions of (5.4.1) on several different intervals: $(-\infty, 0]$, $(-\infty, \infty)$, $[t_0, \infty)$. We always mean that the solution satisfies Equation (5.4.1) on these intervals. In particular:

a) In the case of $[t_0, \infty)$, we begin with an initial function on $(-\infty, t_0)$; from that initial function we construct a solution satisfying (5.4.1) on $[t_0, \infty)$, while agreeing with the initial function on $(-\infty, t_0)$. There may be a discontinuity at t_0 and the initial function need not satisfy (5.4.1).

b) In the other two cases, the solution is its own initial function on any interval $(-\infty, t_0]$.

We prove our results in several steps. First we show that if there is a solution on $(-\infty, 0]$ then it can be extended to $(-\infty, \infty)$ and it is bounded by a constant L. Then we find a solution of (5.4.1) on $(-\infty, 0]$ using a fixed point theorem. Finally we obtain a solution on $(-\infty, \infty)$ in one step.

Thus, if $(X, \|\cdot\|)$ is the space of bounded continuous functions on $(-\infty, 0]$ with the supremum norm and if

$$M = \{\phi \in X \,|\, \|\phi\| \leq L\}, \tag{5.4.6}$$

then to find a solution in that space is to find a fixed point of the mapping $P : M \to X$ defined by $\phi \in M$ implies

$$(P\phi)(t) = a(t) - \int_{-\infty}^{t} D(t,s)g(\phi(s))ds,$$
$$-\infty < t \leq 0. \tag{5.4.7}$$

When we consider fixed point theorems which are not of contraction type, we are led to the hypothesis that PM be contained in a compact subset of X. And that seems impossible, even though PM may be equicontinuous. There then arises a very interesting interplay between continuity of P and compactness of PM, as well as several other consequences which we now enumerate.

(i) Under general conditions including (5.4.5), it is possible to find a continuous function $h : (-\infty, 0] \to [1, \infty)$ with $h(t)$ tending monotonically to ∞ as $t \to -\infty$ so that P is continuous on M in the norm defined by

$$|\phi|_h := \sup_{t \leq 0} |\phi(t)|/h(t). \tag{5.4.8}$$

If Y is the set of continuous functions ψ on $(-\infty, 0]$ for which $|\psi|_h < \infty$, then PM may be contained in a compact subset of $(Y, |\cdot|_h)$.

(ii) If we consider the classical fixed point theorem of Schaefer we see that it can be extended in a simple way to cover the present situation. Immediately, we find that P has a fixed point in M so that (5.4.1) has a solution ϕ which is bounded on $(-\infty, 0]$ and so $|\phi(t)| < L$ on $(-\infty, \infty)$.

(iii) But there is another useful conclusion. Given any property S of bounded continuous functions, if a convex and complete (in the topology of $(Y, |\cdot|_h)$) subset M^* of M is required to satisfy property S and if $\phi \in M^*$ implies $P\phi$ also has property S, then the fixed point also has property S. Under the conditions of our basic theorem, two interesting examples are:

(a) If there is a $T > 0$ with $a(t+T) = a(t)$ and $D(t+T, s+T) = D(t, s)$ then the fixed point is also T-periodic.

(b) If $a(t)$ is almost periodic in the space $(Y, |\cdot|_h)$ and if $D(t, s) = D(t - s)$, then the fixed point is almost periodic. This result is not given here, but a reference is provided.

5.4.1 A Fixed Point Theorem

Let $h : (-\infty, 0] \to [1, \infty)$ be a continuous decreasing function with $h(0) = 1$ and $h(r) \to \infty$ as $r \to -\infty$. Let

$$(Y, |\cdot|_h)$$

be the Banach space of continuous $\phi : (-\infty, 0] \to \Re^n$ for which

$$|\phi|_h := \sup_{t \leq 0} |\phi(t)|/h(t) < \infty,$$

and let

$$(X, \|\cdot\|)$$

be the Banach space of bounded continuous $\phi : (-\infty, 0] \to \Re^n$ with the supremum norm.

We will need a fixed point theorem patterned after the result of Schaefer, which will fail us here owing to problems with continuity, compactness, norms, and global considerations. The following modification fits our circumstances.

Theorem 5.4.1. Let $L > 0$ and $J > 0$ be constants,

$$M = \{\phi \in X | \|\phi\| \leq L\},$$

and let $P : M \to X$ satisfy:
(i) P is continuous in $|\cdot|_h$.

5.4. A PRIORI BOUNDS

(ii) For each $\phi \in M$, $|(P\phi)(t_1) - (P\phi)(t_2)| \leq J|t_1 - t_2|$ for $-\infty < t_2 < t_1 \leq 0$.

(iii) If $0 \leq \lambda \leq 1$ and if $\phi = \lambda P\phi$ for $\phi \in M$, then $\|\phi\| < L$.

Then P has a fixed point in M.

Proof. Define $H: M \to M$ by

a) $H\phi = P\phi$ if $\|P\phi\| \leq L$

and

b) $H\phi = \lambda P\phi$, $0 < \lambda < 1$, $\|\lambda P\phi\| = L$ if $\|P\phi\| > L$.

Since P is continuous in $|\cdot|_h$, so is H. Since $h(r) \to \infty$ as $r \to -\infty$ and (ii) holds, HM is contained in a compact subset of M in the space $(Y, |\cdot|_h)$ (see Burton (2005b; p.172)). By Schauder's second theorem (see Smart (1980; p. 25)) H has a fixed point: $H\phi = \phi$ or $\lambda P\phi = \phi$. If $\|P\phi\| \leq L$ then $\lambda = 1$ and P has the fixed point as claimed. If $\|P\phi\| > L$ then $\|\lambda P\phi\| = L = \|\phi\|$, and this contradicts (iii).

Corollary 5.4.1. *Let the conditions of Theorem 5.4.1 hold and let M^* be a convex subset of M. Suppose that for H defined in the proof of Theorem 5.4.1 we have $H: M^* \to M^*$. Finally, suppose that if f_n is a sequence in M^* and if there is an $f \in Y$ such that*

$$\lim_{n \to \infty} |f_n - f|_h = 0$$

then $f \in M^*$. Under these conditions P has a fixed point in M^*.

The function h plays a central role in what can be said about solutions and the first thing we want to do is show what its properties must be.

Lemma 5.4.1. *Suppose there are positive constants C and L so that $|x_i| \leq L$ for $i = 1, 2$ imply that $|g(x_1) - g(x_2)| \leq C|x_1 - x_2|$. Suppose also that there is an $E > 0$ and a continuous $h: (-\infty, 0] \to [1, \infty)$ such that $h(0) = 1$, $h(t) \to \infty$ as $t \to -\infty$ so that*

$$\sup_{-\infty < t \leq 0} \int_{-\infty}^{t} [D(t,s)h(s)/h(t)]ds \leq E. \tag{5.4.9}$$

holds. Let P be defined in (5.4.7) and $M = \{\phi \in X \mid \|\phi\| \leq L\}$. Then P is continuous on M in $|\cdot|_h$.

Proof. For a given $\epsilon > 0$ we must find $\delta > 0$ such that $[\phi, \psi \in M, |\phi - \psi|_h < \delta]$ imply that $|P\phi - P\psi|_h < \epsilon$. As $\phi, \psi \in M$ we have

$$|g(\phi(s)) - g(\psi(s))| \leq C|\phi(s) - \psi(s)|$$

for $-\infty < s \leq 0$. Hence, $-\infty < t \leq 0$ implies that

$$|(P\phi)(t) - (P\psi(t))|/h(t) \leq \int_{-\infty}^{t} D(t,s) C |\phi(s) - \psi(s)|/h(t) ds$$

$$\leq [C/h(t)] |\phi - \psi|_h \int_{-\infty}^{t} D(t,s) h(s) ds$$

$$\leq C |\phi - \psi|_h E.$$

For the given ϵ we take $\delta \leq \epsilon/CE$. This completes the proof.

Given (5.4.1), we must find an appropriate h and constant E so that (5.4.9) holds. The reader should experience little difficulty in constructing a suitable h when D is an elementary function satisfying (5.4.5). In the Appendix we show that it can always be done in the convolution case, $0 \leq D(t,s) = D(t-s)$.

The fixed point is a bounded continuous function. A main goal is to refine the solution and those refinements take place in the space $(Y, |\cdot|_h)$ as indicated in the corollary above. We are interested in periodic solutions which must be defined in that norm.

5.4.2 A Uniform Bound

Equation (5.4.1) can have an initial function $\phi : (-\infty, t_0] \to R$; in that case, in order for the solution to be continuous at t_0 it is required that

$$\phi(t_0) = a(t_0) - \int_{-\infty}^{t_0} D(t_0, s) g(\phi(s)) ds.$$

If ϕ is bounded and continuous, then (5.4.5) implies that this integral exists. On the other hand, (5.4.1) may have a solution ψ on all of R; in that case, ψ is its own initial function on any interval $(-\infty, t_0]$. Suitable existence theory is found in Chapter 3. Moreover, as we have mentioned several times before, given ϕ we can find ϕ^* arbitrarily close to ϕ with continuity holding.

5.4. A PRIORI BOUNDS

Theorem 5.4.2. *Let (5.4.2) - (5.4.5) hold. Then there exists a positive constant J such that if $x(t)$ is a continuous solution of (5.4.1) on an interval $[t_0, \infty)$ with bounded continuous initial function ϕ, then for*

$$V(t) = \int_{-\infty}^{t} D_s(t,s) \left(\int_{s}^{t} g(x(v)) dv \right)^2 ds, \tag{5.4.10}$$

$$V'(t) \leq -x(t)g(x(t)) + J|a(t)|, \tag{5.4.11}$$

and

$$(a(t) - x(t))^2 \leq D(t,t) V(t). \tag{5.4.12}$$

We have seen a proof of this result in both Section 5.2 and 5.3.

Theorem 5.4.3. *Let (5.4.2) - (5.4.5) hold and let $x(t)$ be a bounded, continuous solution of (5.4.1) on an interval $(-\infty, t_1]$. If, in addition, there is a $K > 0$ with*

$$\int_{-\infty}^{t} D_s(t,s)[(t-s)^2 + 1] ds \leq K \qquad (t \in (-\infty, \infty)) \tag{5.4.13}$$

then $x(t)$ can be defined on all of $\Re$ and

$$(a(t) - x(t))^2 \leq B[1 + 2((G + JA)^2 + 1)K]. \tag{5.4.14}$$

As $|a(t)| \leq A$, $|x(t)| \leq L$, for some L.

Proof. If we obtain the bound (5.4.14) on the solution so long as it can be defined, it will follow from existence theory of Chapter 3 that it can be defined for all t.

Since x is bounded and (5.4.13) holds, $V(t)$ is bounded on the interval $(-\infty, t_1]$. Then we have two cases.

1. If $V(t)$ is bounded as long as $x(t)$ can be defined, then there exists a t^* such that $V(s) \leq V(t^*) + 1$ for all s where $x(s)$ is defined.

2. If $V(t)$ is not bounded, then we have $\limsup_{s \to \infty} V(s) = \infty$, and hence for all t we can find a $t^* \geq t$ such that $V(s) \leq V(t^*)$ for all $s \leq t^*$.

In both cases we obtained, that for all t we can find a t^* such that $V(t) \leq V(t^*) + 1$ and $V(s) \leq V(t^*) + 1$ for all $s \leq t^*$. Then we have

$$-1 \leq V(t^*) - V(s) \leq -\int_{s}^{t^*} |x(u)g(x(u))| du + JA(t^* - s),$$

and hence (for G defined in (5.4.2))

$$(a(t) - x(t))^2 \leq D(t,t)V(t)$$
$$\leq D(t,t)[V(t^*) + 1]$$
$$\leq B\left[1 + \int_{-\infty}^{t^*} D_s(t^*,s)\left(\int_s^{t^*}[G + |x(u)g(x(u))|]du\right)^2 ds\right]$$
$$\leq B\left[1 + \int_{-\infty}^{t^*} D_s(t^*,s)(G(t^* - s) + JA(t^* - s) + 1)^2 ds\right]$$
$$= B\left[1 + 2(G + JA)^2\int_{-\infty}^{t^*} D_s(t^*,s)(t^* - s)^2 ds + 2\int_{-\infty}^{t^*} D_s(t^*,s)ds\right]$$
$$\leq B[1 + 2[(G + JA)^2 + 1]K].$$

5.4.3 A Summary

It is easy to verify that if in (5.4.1) we replace D by λD and $a(t)$ by $\lambda a(t)$, then (5.4.10) - (5.4.12) and (5.4.14) contain only those changes. Thus, the conclusion of Theorem 5.4.3 is valid for

$$x(t) = \lambda[a(t) - \int_{-\infty}^{t} D(t,s)g(x(s))ds] \tag{1_λ}$$

for $0 \leq \lambda \leq 1$.

I. Let the conditions of Theorem 5.4.3 hold: (5.4.2) - (5.4.5). Then from (5.4.14) we conclude that there is an L such that if (1_λ) has a solution ϕ bounded on $(-\infty, 0]$, then $|\phi(t)| \leq L$ on all of $\Re$. Condition (iii) of Theorem 5.4.1 is satisfied.

II. Let the conditions of Lemma 5.4.1 hold. That is, (5.4.9) is valid and there is a constant $C > 0$ so that $|x_i| \leq L$ imply that $|g(x_1) - g(x_2)| \leq C|x_1 - x_2|$. Then P defined by (5.4.7) is continuous on

$$M = \{\phi \in X \mid \|\phi\| \leq L\} \tag{5.4.16}$$

in the norm $|\cdot|_h$. Condition (i) of Theorem 5.4.1 is satisfied.

III. Let (5.4.2) - (5.4.5) hold and suppose that a satisfies a Lipschitz condition. In addition, let

$$\int_{-\infty}^{t} |D_t(t,s)|ds \tag{5,4.17}$$

be bounded. Then we can obtain a bound on the derivative of the integral in (5.4.7) for $\phi \in M$ and conclude that (ii) of Theorem 5.4.1 holds: $P\phi$ is Lipschitz on M. Condition (ii) of Theorem 5.4.1 is satisfied.

5.4. A PRIORI BOUNDS

Thus, if I, II, III hold, then (5.4.1) has a solution in M; to this point, we then say that (5.4.1) has a solution bounded on R.

We now focus on the Corollary 5.4.1 of Theorem 5.4.1. The idea is to refine the result that we have a bounded solution. If we add a condition S to M and if $\phi \in M$ implies $P\phi$ satisfies S then there may be a solution of (5.4.1) also satisfying S. When we do this we frequently want a condition to hold on all of $\Re$ instead of just on $(-\infty, 0]$. We can revise the above work in a simple way. In Section 5.4.1 leave h as stated, except that it be an even function. Then

$$(Y, |\cdot|_h) \tag{5.4.18}$$

is the Banach space of continuous $\phi : \Re \to \Re^n$ for which

$$|\phi|_h := \sup_{t \in \Re} |\phi(t)|/h(t) < \infty \tag{5.4.19}$$

and $(X, \|\cdot\|)$ is the Banach space of bounded continuous $\phi : \Re \to \Re^n$ with the supremum norm. Then M and P are defined with these changes and we get a fixed point in M.

Example 5.4.1. Let I, II, III hold. If there is a $T > 0$ so that $a(t+T) = a(t)$ and if $D(t+T, s+T) = D(t,s)$, then (5.4.1) has a T-periodic solution.

With the Corollary 5.4.1 of Theorem 5.4.1 in mind, we take M^* as the subset of M consisting of continuous $T-$periodic functions. A sequence in M^* with limit in the norm $|\cdot|_h$ is certainly periodic. It is easy to verify that $\phi \in M$ implies $P\phi$ is periodic.

In Burton Makay (2002) we continue these ideas and obtain an almost periodic solution.

5.4.4 Appendix

We now show the details for constructing h when $0 \leq D(t,s) = D(t-s)$ with $\int_0^\infty D(u)du < \infty$ so that (5.4.9) will be satisfied. Under these conditions on D, in Burton and Grimmer (1973; pp. 207-8) there is constructed a continuous increasing function $p_2(t)$ tending to infinity with t so that $\int_0^\infty D(u)p_2(u)du < \infty$. Referring to that argument, we can construct a sequence $\{t_n\} \to \infty$ (with $t_0 = 0$) and a function h so that

$t_{n+1} - t_n \to \infty$ with n, $h(t_n) = 2^n$, h is linear on the intervals $[t_n, t_{n+1}]$ and $\int_0^\infty D(u)h(u)du < \infty$. First of all we see that

$$t_{k+l} = t_k + (t_{k+1} - t_k) + \ldots + (t_{k+l} - t_{k+l-1})$$
$$\geq t_k + (t_1 - t_0) + \ldots + (t_l - t_{l-1}) = t_k + t_l.$$

Let $a \in (t_{k-1}, t_k]$ and $b \in (t_{l-1}, t_l]$ be arbitrary numbers, then

$$h(a+b) \leq h(t_k + t_l) \leq h(t_{k+l})$$
$$= 2^{k+l} \leq 4 \cdot 2^{k-1} \cdot 2^{l-1}$$
$$= 4h(t_{k-1})h(t_{l-1}) \leq 4h(a)h(b).$$

Now, let h be an even function and have

$$\int_{-\infty}^t [D(t-s)h(s)/h(t)]ds = \int_0^\infty [D(u)h(t-u)/h(t)]du$$
$$\leq \int_0^\infty [D(u)4h(-u)h(t)/h(t)]du < \infty.$$

The material in this section was taken from Burton and Makay (2002), published by the University of Szeged.

Chapter 6

The Krasnoselskii-Schaefer Theorem

6.1 Method and Problems

In Chapter 4 we began by discussing the work of Krasnoselskii and his most perceptive view of the inversion of a perturbed differential operator, obtaining the sum of a contraction and compact map. This led to his theorem which is a combination of Banach's contraction mapping principle and Schauder's fixed point theorem. We noted the difficulties in establishing the exact conditions of Krasnoselskii's theorem and the ease with which we applied Schaefer's theorem. That let us to a combination Krasnoselskii-Schaefer theorem which was well suited to Liapunov's direct method. That result required a proposition whose proof will be needed in the work here. Thus, we will repeat both the proposition and the theorem for ready reference.

Proposition 6.1.1. *If* $(\mathcal{B}, \|\cdot\|)$ *is a normed space, if* $0 < \lambda < 1$, *and if* $B : \mathcal{B} \to \mathcal{B}$ *is a contraction mapping with contraction constant* α, *then* $\lambda B \frac{1}{\lambda} : \mathcal{B} \to \mathcal{B}$ *is also a contraction mapping with contraction constant* α, *independent of* λ; *in particular*

$$\|\lambda B(x/\lambda)\| \leq \alpha \|x\| + \|B0\|.$$

Proof. To see that $\lambda B \frac{1}{\lambda}$ is a contraction, $x \in \mathcal{B} \Rightarrow x/\lambda \in \mathcal{B} \Rightarrow B(x/\lambda) \in \mathcal{B} \Rightarrow \lambda B(x/\lambda) \in \mathcal{B}$; moreover, $x, y \in \mathcal{B} \Rightarrow$

$$\|\lambda B(x/\lambda) - \lambda B(y/\lambda)\| = \lambda \|B(x/\lambda) - B(y/\lambda)\|$$
$$\leq \lambda \alpha \|(x/\lambda) - (y/\lambda)\| = \alpha \|x - y\|.$$

To obtain the bound, for any $x \in \mathcal{B}$ we have

$$\begin{aligned}
\|\lambda B(x/\lambda)\| &= \lambda \|B(x/\lambda)\| \\
&= \lambda(\|B(x/\lambda) - B0 + B0\|) \\
&\leq \lambda(\|B(x/\lambda) - B0\| + \|B0\|) \\
&\leq \lambda(\alpha\|(x/\lambda) - 0\| + \|B0\|) \\
&= (\lambda\alpha/\lambda)\|x\| + \|B0\|,
\end{aligned}$$

as required.

With this proposition we are able to prove the following result which we call the Krasnoselskii-Schaefer theorem. See Burton and Kirk (1998).

Theorem 6.1.1. *Let $(\mathcal{B}, \|\cdot\|)$ be a Banach space, $A, B : \mathcal{B} \to \mathcal{B}$, B a contraction with contraction constant $\alpha < 1$, and A continuous with A mapping bounded sets into compact sets. Either*

(i) $x = \lambda B(x/\lambda) + \lambda Ax$ *has a solution in $\mathcal{B}$ for $\lambda = 1$, or*

(ii) *the set of all such solutions, $0 < \lambda < 1$, is unbounded.*

In this chapter we will look at a number of problems which can be solved quite simply with that theorem. These problems are all in some sense finite dimensional. Examples of application to infinite dimensional problems can be seen in recent work of Belmekki, Benchohra, Ezzinbi, and Ntouyas (2008) and Garcia-Falset (2008).

6.1.1 Duration of Heredity

Volterra emphasized that the integrals in equations of the second kind may take the forms

$$\int_{-\infty}^{t}, \int_{t-r}^{t}, \int_{t_0}^{t}$$

depending on what he called the duration of "heredity:" the solution "inherits" the characteristics of its past. Given an equation

$$x(t) = a(t) - \int_{-\infty}^{t} C(t,s)g(s, x(s))ds \qquad (6.1.1.1)$$

we may seemingly shorten the duration of heredity in several ways, some being more mathematically sound than others.

6.1. METHOD AND PROBLEMS

First, given an initial function $\phi : (-\infty, t_0] \to \Re$ we can write

$$x(t) = \left[a(t) - \int_{-\infty}^{t_0} C(t,s)g(s,\phi(s))ds\right] - \int_{t_0}^{t} C(t,s)g(s,x(s))ds$$

$$=: b(t) - \int_{t_0}^{t} C(t,s)g(s,x(s))ds$$

so that the past heredity is hidden in $b(t)$, an entirely mathematically valid procedure.

Next, if $C(t,s) = C(t-s)$ and if $C(t) \equiv 0$ for $t \geq r$ for some $r > 0$, then the integral is $\int_{t-r}^{t} C(t-s)g(s,x(s))ds$.

Grimmer (1979) noted that if $C(t,s) = C(t-s)$ then we may telescope the integral into a sum so that under periodicity conditions on $a(t)$ and $g(t,x)$ the search for a periodic solution can be conducted on an equation

$$x(t) = a(t) - \int_{t-T}^{t} H(t-s)g(s,x(s))ds.$$

Here is the way that is accomplished. Use the change of variable $s = u - nT$ and assume uniform convergence in the work below to obtain

$$\int_{-\infty}^{t} C(t-s)x(s)ds = \sum_{n=0}^{\infty} \int_{t-(n+1)T}^{t-nT} C(t-s)x(s)ds$$

$$= \int_{t-T}^{t} \sum_{n=0}^{\infty} C(t-u+nT)x(u-nT)du.$$

Then notice that

$$x(t) = a(t) - \int_{-\infty}^{t} C(t-s)x(s)ds$$

has a T-periodic solution if and only if

$$y(t) = a(t) - \int_{t-T}^{t} H(t-u)y(u)du$$

also has such a solution where $H(t-u) = \sum_{n=0}^{\infty} C(t-u+nT)$.

6. THE KRASNOSELSKII-SCHAEFER THEOREM

Truncation can also occur naturally from differential equations. Suppose we want to prove that the scalar equation

$$x' = -a(t)x(t) - g(t, x(t))$$

has a periodic solution under periodic assumptions on a and $g(t,x)$. We could write

$$x(t) = x_0 - \int_0^t [a(s)x(s) + g(s, x(s))]ds$$

and set up a mapping

$$(P\phi)(t) = x_0 - \int_0^t [a(s)\phi(s) + g(s, \phi(s))]ds$$

for ϕ periodic. Two difficulties occur. First, we do not know how to choose x_0 and it is far too much to suppose there will be a periodic solution for every x_0. But what is initially worse is that the mapping will map ϕ into a non-periodic function.

If

$$-\int_0^T a(s)ds < 0 \tag{6.1.1.2}$$

then we can avoid these problems by writing the equation as

$$\left(x \exp \int_0^t a(s)ds\right)' = -g(t,x)\exp \int_0^t a(s)ds.$$

We have two choices and both are good ones. First, we can integrate both sides from $-\infty$ to t, under the assumption that we are dealing with a bounded solution, and obtain an integral equation with infinite delay of the type considered in Chapter 5. There is no x_0 to consider and a translation shows the required periodicity. But the infinite delay introduces other questions and we may note that there is another way to handle the equation. Integrate it from $t - T$ to t and obtain

$$x(t) = x(t-T)e^{-\int_{t-T}^t a(s)ds} - \int_{t-T}^t g(u, x(u))e^{-\int_u^t a(s)ds}du. \tag{6.1.1.3}$$

We now have a finite delay equation and the integral can define a compact mapping, while $x(t-T)e^{-\int_{t-T}^t a(s)ds}$ will define a contraction since that last product term is smaller than one when (6.1.1.2) holds. Moreover, it

6.1. METHOD AND PROBLEMS

is perfect for Theorem 6.1.1 and it allows us to demonstrate several ideas concerning Liapunov functions for integral equations.

Compare (6.1.1.3) with (4.1.6.2) which was derived from the equation of advanced type considered below.

Such equations will be studied in two steps. In Section 6.2 we treat the linear convolution case which has several nice properties not seen in the nonlinear case treated in Section 6.3. It is left as an exercise and the references to examine the case of a contraction plus an infinite delay. It is essentially a combination of the work in Section 6.3 and that of Chapter 5.

6.1.2 An Equation of Advanced Type

We now come to an entirely new kind of singular equation which will be discussed in Section 6.4. An introduction to make it meaningful may be a distraction so we are going to give it here. We will then be able to move right into the mathematical aspects in Section 6.4 without such distractions. Let $|\alpha| < 1$, $a > 0$, $h > 0$, q be continuous and consider the neutral functional differential equation

$$x' = \alpha x'(t-h) + ax - q(t, x, x(t-h)). \tag{6.1.2.1}$$

Such equations have been studied with varying degrees of vigor since the middle of the last century (cf. Driver (1965), Gopalsamy (1992), Gopalsamy and Zhang (1988), El'sgol'ts (1966), Kuang (1993a,b), (1991)). Recently investigators have given heuristic arguments to support their use in describing biological phenomena and much of this is formalized in the final chapter of each of the books by Gopalsamy (1992) and Kuang (1993). In such studies a specific type of solution is sought, such as a periodic solution as the attractor for a logistic equation. And effective search for particular types of solutions is often accomplished by inverting the differential equation to an integral equation and then applying fixed point theory. Thus, with the Krasnoselskii-Schaefer theorem we see that Liapunov functions can play a central role.

Equations like (6.1.2.1) are extremely unstable and that will introduce far different work than we have seen to this point. Notice that the ode part is $x' = ax$ with $a > 0$ so that solutions are exponentially unbounded. Add to this the term $\alpha x'(t-h)$ where α will characteristically be positive and we have massive increase in growth of the solution. Any boundedness must come from the nonlinear term q. It is, of course, a generalization of the stock logistic equation $x' = ax - bx^2$ with a and b positive constants. The nonnegative solutions of that equation consist of two constant solutions, $x = 0$ and $x = a/b$, which are trivially periodic, together with solutions approaching $x = a/b$ asymptotically.

6. THE KRASNOSELSKII-SCHAEFER THEOREM

Equations of this type are going to present great challenges and it is only fair to give some heuristic reasons for being so determined to study them. And let us be clear that the logistic equation itself is founded mainly on heuristic arguments, in spite of the introduction of first principle ideas like the law of mass action. A short survey of those ideas is discussed in Burton (2006b).

Traditionally one begins with the Malthusian assumption that a certain biomass increases at a rate proportional to itself, but then decreases at a rate proportional to the biomass squared: $x' = ax - bx^2$. Such growth has been noted in the literature since 1599. But every parent and every gardener has seen such very different growth. A field of corn is planted in a hot and humid region of the Midwestern United States, for example. The plants emerge in a few days and grow so slowly that every year there is worry that it will not mature before frost and, thus, be mainly ruined. Early farmers developed slogans to reassure themselves: Corn should be knee high by the fourth of July, for example. Suddenly, this seemingly stunted corn starts to grow quickly. And growth begets growth so that after a good rain in the first week of August the corn is growing so fast that, in the quiet hour of midnight, one can actually hear it grow. Quite reasonably, one advances the idea that $x' = \alpha x'(t - h) + ax$; it is the recent growth as much as the size which is causing the growth. But, just as quickly, a trigger point is reached and the brakes are applied in the form of $-q(t, x, x(t - h))$. It is the age, the size, and the recent size which signal a halt to the growth. The growth stops and all the energy goes into filling out that great ear of corn containing more than 750 kernels which are packed in so tightly that, when mature and dry, the ear is as solid as a stick of fire wood.

The growth of a child can be described in a similar way and the clear evidence is that trousers which were too long in September when school started are well up the calf by Christmas and the embarrassed child prays that Santa will bring something more appropriate.

Simple analysis of the characteristic quasi-polynomial for $x' = ax + \alpha x(t - h)$ shows how the growth can move from e^{at} to e^{bt} with b much larger than a, particularly when h is small and α is near 1.

If our discussion of (6.1.2.1) were restricted to such problems as just described, only the region $x \geq 0$ would be of interest. We deal with that at the end of Section 6.4. Moreover, in such a case, a change of variable is often made to map the problem into a space where all points are significant. Such an example is found in Burton (1983b; pp. 110-111), for example.

We now review the work in Chapter 4 of inverting a differential equation to get an integral equation. Write (6.1.2.1) as

$$(x - \alpha x(t - h))' = a(x - \alpha x(t - h)) + a\alpha x(t - h) - q(t, x, x(t - h)),$$

multiply by e^{-at}, and group terms as

$$[(x - \alpha x(t - h))e^{-at}]' = [a\alpha x(t - h) - q(t, x, x(t - h))]e^{-at}.$$

We search for a solution having the property that

$$(x(t) - \alpha x(t - h))e^{-at} \to 0 \text{ as } t \to \infty$$

so that an integration from t to infinity yields

$$-(x(t) - \alpha x(t - h))e^{-at} = \int_t^\infty [a\alpha x(s - h) - q(s, x(s), x(s - h))]e^{-as}ds$$

and, finally,

$$x(t) = \alpha x(t-h) + \int_t^\infty [q(s, x(s), x(s-h)) - a\alpha x(s-h)]e^{a(t-s)}ds. \quad (6.1.2.2)$$

A general form for such equations is

$$x(t) = f(x(t - h)) + \int_t^\infty Q(s, x(s), x(s - h))C(t - s)ds + p(t). \quad (6.1.2.3)$$

We call this a neutral delay integral equation of advanced type and it is a very interesting equation. It may have a solution on all of $\Re$ or it may have a solution on $[0, \infty)$ generated by an initial function φ on $[-h, 0]$. In the latter case, notice that we can not obtain a local solution: we must get the full solution on $[0, \infty)$. Thus, we will need to employ a fixed point theorem to get existence and that means that we will get a fixed point in the solution space; hence, we must know in advance the form of the solution space. We study this problem in Burton (1998b) and find that the solution will have discontinuities at $t = nh$, but the jumps will tend to zero as $n \to \infty$. (This is parallel to solutions of functional differential equations smoothing (cf. El'sgol'ts (1966)).) Equally important is the need to know in advance the growth of the solution so that functions in the solution space will have a weighted norm allowing such growth.

6.2 A Linear Truncated Equation

In an effort to introduce the subject slowly we will begin with a truncated convolution equation and a delay term which will generate a contraction.

6. THE KRASNOSELSKII-SCHAEFER THEOREM

As discussed in Section 6.1, the delay term is typical of the example in which we integrated an ordinary differential equation from $t - r$ to t. The truncated integral is typical of that discussed concerning Grimmer's telescoping of the infinite integral into a truncated integral.

Thus, we consider the scalar equation

$$x(t) = \alpha x(t - h) + p(t) - \int_{t-T}^{t} C(t - s)x(s)\, ds, \tag{6.2.1}$$

where p and C are continuous, and there is a $T > 0$ with

$$p(t + T) = p(t), \quad \int_0^T p(s)\, ds = 0,$$

$h > 0$, h constant,

$$C''(t) \geq 0, |\alpha| < 1. \tag{6.2.2}$$

Theorem 6.2.1. *If (6.2.2) holds then Equation (6.2.1) has a unique T-periodic solution.*

Proof. With the parameter λ of Theorem 6.1.1 we write (6.2.1) as

$$x(t) = \alpha x(t - h) + \lambda \left[p(t) - \int_{t-T}^{t} x(s)C(t - s)\, ds \right] \tag{6.2.1$_\lambda$}$$

$0 < \lambda < 1$. A fixed point for $\lambda = 1$ will solve (6.2.1). Let $(\mathcal{P}_T^0, \|\cdot\|)$ be the Banach space of continuous T-periodic functions having mean value zero with the supremum norm. Define $P : \mathcal{P}_T^0 \to \mathcal{P}_T^0$ by $\varphi \in \mathcal{P}_T^0$ implies that

$$(P\varphi)(t) = \alpha \varphi(t - h) + \lambda \left[p(t) - \int_{t-T}^{t} \varphi(s) C(t - s)\, ds \right] \tag{6.2.3}$$

and define $A : \mathcal{P}_T^0 \to \mathcal{P}_T^0$ by $\varphi \in \mathcal{P}_T^0$ implies that

$$(A\varphi)(t) = p(t) - \int_{t-T}^{t} \varphi(s) C(t - s)\, ds. \tag{6.2.4}$$

Thus, in Theorem 6.1.1 we have $(B\varphi)(t) = \alpha \varphi(t - h)$. To see that $P : \mathcal{P}_T^0 \to \mathcal{P}_T^0$, note that $\varphi \in \mathcal{P}_T^0$ implies that $P\varphi$ is T-periodic and

$$\int_0^T \int_{u-T}^u \varphi(s) C(u - s)\, ds\, du = \int_0^T \int_{-T}^0 \varphi(s + u) C(-s)\, ds\, du$$

$$= \int_{-T}^0 \int_0^T \varphi(s + u)\, du\, C(-s)\, ds = 0$$

for each fixed s; hence $P\varphi \in \mathcal{P}_T^0$.

6.2. A LINEAR TRUNCATED EQUATION

We may note that A is continuous and maps bounded sets into equicontinuous sets. Clearly, $(B\varphi)(t) = \alpha\varphi(t-h)$ is a contraction. Thus, all that remains to prove that (6.2.1$_\lambda$) has a solution in $\mathcal{P}_T^0$ is to find an *a priori* bound on fixed points of P in $\mathcal{P}_T^0$. Thus, let $x \in \mathcal{P}_T^0$ solve (6.2.1$_\lambda$) and define

$$V(t) = \lambda \int_{t-T}^{t} C_s(t-s) \left(\int_s^t x(u)\,du \right)^2 ds \qquad (6.2.5)$$

so that

$$V'(t) = -\lambda C_s(T) \left(\int_{t-T}^{t} x(u)\,du \right)^2$$
$$+ \lambda \int_{t-T}^{t} C_{st}(t-s) \left(\int_s^t x(u)\,du \right)^2 ds$$
$$+ 2\lambda x \int_{t-T}^{t} C_s(t-s) \int_s^t x(u)\,du\,ds.$$

The first term in V' is zero because $x \in \mathcal{P}_T^0$, while the second term is not positive. Thus, we have

$$V'(t) \leq 2\lambda x \int_{t-T}^{t} C_s(t-s) \int_s^t x(u)\,du\,ds$$
$$= 2\lambda x \left[C(t-s) \int_s^t x(u)\,du \Big|_{t-T}^{t} + \int_{t-T}^{t} C(t-s)x(s)\,ds \right].$$

We now have

$$V'(t) \leq 2x\lambda \int_{t-T}^{t} C(t-s)x(s)\,ds$$
$$= 2x\left[-x + \alpha x(t-h) + \lambda p(t)\right]$$

from (6.2.1$_\lambda$). If $\varepsilon > 0$ and $-2 + 2|\alpha| + \varepsilon < 0$, then

$$V'(t) \leq -2x^2 + |\alpha|(x^2 + x^2(t-h)) + \varepsilon x^2 + p^2/\varepsilon$$
$$= [-2 + |\alpha| + \varepsilon]x^2 + |\alpha|x^2(t-h) + p^2(t)/\varepsilon.$$

It is clear that $x \in \mathcal{P}_T^0$ implies that $V(T) = V(0)$ (write $V(t+T)$ and then $s = v + T$) and so

$$0 = V(T) - V(0) \leq \left[-2 + |\alpha| + \varepsilon\right] \int_0^T x^2(s)\, ds$$

$$+ |\alpha| \int_0^T x^2(s-h)\, ds + T\|p\|^2/\varepsilon$$

$$= [-2 + 2|\alpha| + \varepsilon] \int_0^T x^2(s)\, ds + T\|p\|^2/\varepsilon$$

and there is an $M > 0$ with

$$\int_0^T x^2(s)\, ds \leq M^2. \tag{6.2.6}$$

From (6.2.1$_\lambda$) and (6.2.6) we have

$$|x(t)| \leq |\alpha||x(t-h)| + \|p\| + \left[\int_{t-T}^t C^2(t-s)\, ds \int_{t-T}^t x^2(s)\, ds\right]^{1/2}$$

so that for t satisfying $|x(t)| = \|x\|$ we have

$$\|x\|(1 - |\alpha|) \leq \|p\| + M \left[\int_{-T}^0 C^2(-s)\, ds\right]^{1/2},$$

a suitable *a priori* bound. This proves that (6.2.1) has a T-periodic solution. To see that it is unique, if there are two, then the difference is periodic and satisfies

$$x(t) = \alpha x(t-h) - \int_{t-h}^t \lambda C(t-s)x(s)\, ds, \quad \lambda = 1.$$

We may proceed to get an *a priori* bound on all possible periodic solutions of this equation just as before and see that in (6.2.6) we can let $M \to 0$.

Conjecture 6.2.1. In (6.2.1) let $\alpha \uparrow 1$ and let $h = T$. Notice that under the conditions of Theorem 6.2.1 there is a periodic solution for each such value of α and that those solutions satisfy a uniform bound. We conjecture that there is also a T-periodic solution for $\alpha = 1$. If so, then in (5.2.1) the $x(t)$ on each side of the equation will cancel and we will have obtained a periodic solution of the singular Volterra equation of the first kind.

We have shown that (6.2.1) has a unique periodic solution. Does it have other solutions and do they converge to the periodic solution? We will answer questions of that type in later sections.

The material in this section is from Burton (1997), published by Fuzhou University.

6.3 A Nonlinear Truncated Equation

Motivation for this type of problem was given in Section 6.1.1. Let $(\mathcal{P}_T, \|\cdot\|)$ be the Banach space of continuous T-periodic functions $\varphi : \Re \to \Re$ with the supremum norm. Consider

$$x(t) = f(t, x(t)) - \int_{t-h}^{t} D(t, s)g(s, x(s))ds \qquad (6.3.1)$$

and suppose there is a $T > 0$ and an $\alpha \in (0, 1)$ with:

$$\begin{aligned} f(t+T, x) &= f(t, x), \\ D(t+T, s+T) &= D(t, s), \\ g(t+T, x) &= g(t, x), \end{aligned} \qquad (6.3.2)$$

$$t - h \leq s \leq t \text{ implies that } D_s(t, t-h) \geq 0,$$
$$D_{st}(t, s) \leq 0, \quad D(t, t-h) = 0, \qquad (6.3.3)$$

$$|f(t, x) - f(t, y)| \leq \alpha|x - y|, \quad xg(t, x) \geq 0, \qquad (6.3.4)$$

$$\forall k > 0 \, \exists P > 0 \, \exists \beta > 0 \text{ with} \qquad (6.3.5)$$
$$2\lambda[-(1-\alpha)xg(t, x) + k|g(t, x)|\,] \leq \lambda[P - \beta|g(t, x)|\,],$$

$f, g,$ and D_{st} are continuous. $\qquad (6.3.6)$

Liu and Li (2007) offer the same problem with the same techniques and use a Krasnoselskii-Schaefer theorem in which (6.3.4) is replaced by a separate contraction. Liu and Li (2006) then obtain a periodic solution for a partial differential equation with the same technique. We view the latter as a major result which should lead to many more interesting problems. See also Belmekki, Benchohra, Ntouas (2006) and Garcia-Falset (2008) for other infinite dimensional work using Theorem 6.1.1.

We are going to use these conditions and Theorem 6.1.1 to prove that (6.3.1) has a T-periodic solution. Condition (6.3.3) bears some study and comparison to the conditions of the previous section. In that linear case we did not need $D(t, t-h) = 0$. Condition (6.3.3) shows that the duration of heredity is being dictated by the vanishing of the weighting function. It was not needed in the last section and in that case the kernel could be obtained from telescoping an infinite integral. But the conditions here definitely do not suppose convexity.

6. THE KRASNOSELSKII-SCHAEFER THEOREM

Theorem 6.3.1. *If (6.3.2)–(6.3.6) hold, then (6.3.1) has a T-periodic solution.*

Proof. Define a mapping $B : \mathcal{P}_T \to \mathcal{P}_T$ by $\varphi \in \mathcal{P}_T$ implies

$$(B\varphi)(t) = f(t, \varphi(t)). \tag{6.3.7}$$

Lemma 6.3.1. *If B is defined by (6.3.7), then B is a contraction mapping from $\mathcal{P}_T$ into $\mathcal{P}_T$ with contraction constant α of (6.3.4).*

Proof. If $\varphi, \psi \in \mathcal{P}_T$, then

$$\begin{aligned}
\|B\varphi - B\psi\| &= \sup_{t \in [0,T]} |(B\varphi)(t) - (B\psi)(t)| \\
&= \sup_{t \in [0,T]} |f(t, \varphi(t)) - f(t, \psi(t))| \\
&\leq \sup_{t \in [0,T]} \alpha|\varphi(t) - \psi(t)| = \alpha\|\varphi - \psi\|,
\end{aligned}$$

as required.

Define a mapping $A : \mathcal{P}_T \to \mathcal{P}_T$ by $\varphi \in \mathcal{P}_T$ implies

$$(A\varphi)(t) = -\int_{t-h}^{t} D(t,s)g(s, \varphi(s))\, ds. \tag{6.3.8}$$

Lemma 6.3.2. *If A is defined by (6.3.8) then $A : \mathcal{P}_T \to \mathcal{P}_T$, A is continuous, A maps bounded sets into compact sets.*

Proof. Let $k > 0$ be given, $\varphi \in \mathcal{P}_T$ be an arbitrary element with $\|\varphi\| \leq k$. By the continuity of D, if $0 \leq t \leq T$ and $-h \leq s \leq T$, there is an $M > 0$ with

$$|D(t,s)| \leq M. \tag{I}$$

By the uniform continuity of D on $[0,T] \times [-h,T]$, for each $\varepsilon > 0$ there is a $\delta_1 > 0$ such that $u, v \in [0,T]$, $s, t \in [-h,T]$, $|u-v|+|s-t| \leq \delta_1$ implies

$$|D(u,s) - D(v,t)| \leq \varepsilon. \tag{II}$$

Next, since g is continuous on $[0,T] \times [-k,k]$ and periodic in t there is an $L > 0$ with

$$|g(s,x)| \leq L \text{ for } s \in \Re \text{ and } x \in [-k,k]. \tag{III}$$

In fact, by the uniform continuity of g on that set, for any $\varepsilon > 0$ there is a positive $\delta_2(\varepsilon) < T$ such that if $s, t \in [0,T]$ and $x, y \in [-k,k]$ with $|s-t|+|x-y| \leq \delta_2$, then $|g(s,x) - g(t,y)| \leq \varepsilon$ and, by the periodicity,

$$|g(s,x) - g(t,y)| \leq \varepsilon \text{ for } s, t \in \Re \text{ and } x, y \in [-k,k] \tag{IV}$$

with $|s-t|+|x-y| < \delta_2$.

6.3. A NONLINEAR TRUNCATED EQUATION

The assertions about A will now follow. If $\varphi \in \mathcal{P}_T$, a change of variable shows that $A\varphi$ is T-periodic. Clearly, if $\varphi \in \mathcal{P}_T$, then $A\varphi$ is continuous. Thus, $A\varphi \in \mathcal{P}_T$.

We now show that A maps bounded sets into compact sets. First,

$$\{A\varphi : \varphi \in \mathcal{P}_T \text{ and } \|\varphi\| \leq k\} \text{ is equicontinuous.} \tag{V}$$

To see this, note that if $u < v$, then

$$(A\varphi)(u) - (A\varphi)(v) = -\int_{u-h}^{v-h} D(u,s)g(s,\varphi(s))\,ds$$
$$-\int_{v-h}^{u}[D(u,s) - D(v,s)]g(s,\varphi(s))\,ds$$
$$+\int_{u}^{v} D(v,s)g(s,\varphi(s))\,ds.$$

By (I)-(III), for all $u, v \in [0, T]$ with $|u - v| < \delta_1$, if $\varphi \in \mathcal{B}$ and $\|\varphi\| \leq k$, then

$$|(A\varphi)(u) - (A\varphi)(v)| \leq ML|u-v| + L|u-v+h|\varepsilon$$
$$+ML|u-v|$$
$$\leq 2ML\delta_1 + L(\delta_1 + h)\varepsilon. \tag{VI}$$

Next, for $\varphi \in \mathcal{P}_T$ and $\|\varphi\| \leq k$, it follows from (I) and (III) that $|(A\varphi)(u)| \leq LMh$ so that

$$\|A\varphi\| \leq LMh \text{ for } \varphi \in \mathcal{P}_T \text{ and } \|\varphi\| \leq k. \tag{VII}$$

By Ascoli's theorem A maps bounded sets into compact sets.

To see that A is continuous, fix φ and $\psi \in \mathcal{P}_T$ with $\|\varphi - \psi\| < \delta_2$, $\|\varphi\| \leq k$, $\|\psi\| \leq k$. Then for $0 \leq u \leq T$ we have by (I) and (IV) that

$$|(A\varphi)(u) - (A\psi)(u)| \leq \int_{u-h}^{u} |D(u,s)|\,|g(s,\varphi(s)) - g(s,\psi(s))|\,ds$$
$$\leq Mh\varepsilon.$$

This completes the proof of Lemma 6.3.2.

Next, notice that if $B : \mathcal{P}_T \to \mathcal{P}_T$ is defined by

$$(Bx)(t) = f(t, x(t)), \text{ then } \left(\lambda B\frac{x}{\lambda}\right)(t) = \lambda f\left(t, \frac{x(t)}{\lambda}\right).$$

302   6. THE KRASNOSELSKII-SCHAEFER THEOREM

Lemma 6.3.3. *There is a $K \geq 0$ such that if $0 < \lambda < 1$ and if $x \in \mathcal{P}_T$ solves*

$$x(t) = \lambda f\left(t, \frac{x}{\lambda}\right) - \lambda \int_{t-h}^{t} D(t,s)g(s, x(s))\, ds \qquad (6.3.1_\lambda)$$

then $\|x\| \leq K$.

Proof. Let $x \in \mathcal{P}_T$ solve $(3.6.1_\lambda)$ and define

$$V(t) = \lambda^2 \int_{t-h}^{t} D_s(t,s) \left(\int_s^t g(v, x(v))\, dv\right)^2 ds.$$

Now $D_{st}(t,s) \leq 0$ so

$$V'(t) \leq -\lambda^2 D_s(t, t-h) \left(\int_{t-h}^{t} g(v, x(v))\, dv\right)^2$$

$$+ 2\lambda^2 g(t,x) \int_{t-h}^{t} D_s(t,s) \int_s^t g(v, x(v))\, dv\, ds.$$

The first term on the right-hand-side is not positive by (6.3.3); if we integrate the last term by parts and use (6.3.3) again we have

$$V'(t) \leq 2\lambda g(t,x) \int_{t-h}^{t} \lambda D(t,s) g(s, x(s))\, ds$$

$$= 2\lambda g(t,x) \left[\lambda f\left(t, \frac{x}{\lambda}\right) - x(t)\right]$$

from $(6.3.1_\lambda)$. But by the reasoning in the proof of Proposition 6.1.1,

$$\left|\lambda f\left(t, \frac{x}{\lambda}\right)\right| \leq \alpha |x(t)| + |f(t,0)| \leq \alpha |x(t)| + k$$

for some $k > 0$. Thus,

$$V'(t) \leq 2\lambda\{|g(t,x)|[\alpha|x| + k] - xg(t,x)\}$$
$$= 2\lambda[|\alpha x g(t,x)| + k|g(t,x)| - xg(t,x)]$$
$$\leq 2\lambda[-(1-\alpha)xg(t,x) + k|g(t,x)|].$$

As $\alpha < 1$, from (6.3.5) we have

$$V'(t) \leq \lambda[-\beta|g(t,x)| + P].$$

Thus, $x \in \mathcal{P}_T$ implies $V \in \mathcal{P}_T$ so that

$$0 = V(T) - V(0) \leq \lambda\left[-\beta \int_0^T |g(t, x(t))|\, dt + PT\right]$$

or

$$\int_0^T |g(t, x(t))| \, dt \leq PT/\beta \tag{6.3.9}$$

since $\lambda > 0$. As $g(t, x(t)) \in \mathcal{P}_T$, there is an $n > 0$ with

$$\int_{t-h}^t |g(t, x(t))| \, dt \leq n. \tag{6.3.10}$$

Taking $M = \max_{-h \leq s \leq t \leq T} |D(t, s)|$, then from the proposition, $(6.3.1_\lambda)$, and (6.3.10) we have

$$|x(t)| \leq \left|\lambda f\left(t, \frac{x}{\lambda}\right)\right| + \lambda \left|\int_{t-h}^t D(t, s) g(s, x(s)) \, ds\right|$$
$$\leq \alpha |x(t)| + k + Mn$$

or

$$\|x\| \leq (Mn + k)/(1 - \alpha),$$

as required. Application of Theorem 6.1.1 completes the proof.

This material is taken from Burton and Kirk (1998).

6.4 Neutral Integral Equations

In Section 6.1.2 we discussed the motivation for the following equation and we suggest that the reader reviews that before proceeding.

We motivate our work by starting with an equation of interest from biology in the form of a generalized logistic equation

$$x' = ax + \alpha x'(t - h) - q(t, x, x(t - h)) \tag{6.4.1}$$

where $a > 0$, $0 \leq |\alpha| < 1$, $h > 0$, all are constant. We are interested in a solution for $t \geq 0$, possibly arising from a given initial function, or for

6. THE KRASNOSELSKII-SCHAEFER THEOREM

a solution on $(-\infty, \infty)$, possibly a periodic solution. In Section 6.1.2.1 we worked with (6.4.1) and obtained

$$x(t) = \alpha x(t-h) + \int_t^\infty [q(s, x(s), x(s-h)) \\ - a\alpha x(s-h)] e^{a(t-s)} ds. \quad (6.4.1\text{I})$$

This form motivates our study and we focus on an equation

$$x(t) = f(x(t-h)) \\ + \int_t^\infty Q(s, x(s), x(s-h)) C(t-s) ds + p(t). \quad (6.4.2)$$

Sometimes we let $C = C(t, s)$. Here, the forcing function $p(t)$ can be critical. If the equation has an equilibrium point without $p(t)$, then the subsequent struggle to prove the existence of a bounded or periodic solution might simply yield that obvious equilibrium point.

We will ask that

$$|f(x) - f(y)| \le \alpha |x - y|, \quad 0 \le \alpha < 1, \quad (6.4.3)$$

and for some fixed k with $0 \le k \le 1$

$$|Q(t, x, y) - Q(t, w, z)| \\ \le (k|x - w| + (1-k)|y - z|), \quad (6.4.4)$$

$$|Q(t, 0, 0)| \le 1, \quad (6.4.5)$$

Q, p, and C are continuous,

$$\int_0^\infty |C(-u)| du =: C_0 < \infty. \quad (6.4.6)$$

Equation (6.4.2) can be called a neutral delay integral equation of advanced type. In keeping with Krasnoselskii's observation, note that it may generate the sum of a contraction and compact map.

While the integral from t to ∞ is not common, it can be found in many places. Coddington and Levinson (1955; p. 331) write an ordinary

6.4. NEUTRAL INTEGRAL EQUATIONS

differential equation in this way (without h) when studying an unstable manifold. Also, investigators have long written differential equations

$$x' = F(t, x)$$

as

$$x(t) = x(t_0) + \int_{t_0}^{t} F(s, x(s))ds.$$

Then, if it can be determined that every solution converges to zero as $t \to \infty$, it is permissible to bypass $x(t_0)$ and write

$$x(t) = -\int_{t}^{\infty} F(s, x(s))ds.$$

While we see many equations of the form of (6.4.2) (without h), such equations usually are studied only after existence of solutions has been established. That is a central question here.

For this equation, we must immediately establish existence on $[t_0, \infty)$. And because of this, the central problem for (6.4.2) is to determine the space in which solutions reside. Once that is done, the actual proof of existence follows from application of a fixed point theorem. Moreover, unlike the case just mentioned of a differential equation, the existence of a solution of (6.4.2) in a given space endows that solution with the properties of that space. Thus, proof of existence yields important qualitative properties as well. We illustrate that in the examples.

Equation (6.4.2) holds many surprises. For functional differential equations of neutral type, it may be deduced from Driver (1965) that repeated discontinuities of a solution on $[t_0, \infty)$ must be expected as a result of a given initial function. We note that this is true for (6.4.2); but we also show that the magnitude of the jumps in the discontinuities tends to zero as $t \to \infty$. This is parallel to an interesting phenomenon for delay-differential equations. El'sgol'ts (1966; p. 7) discusses solutions of

$$x' = H(t, x(t), x(t-h))$$

in which H is in C^∞ and there is an initial function $\bar{\phi} : [-h, 0] \to R$ yielding a solution $x(t, 0, \bar{\phi})$ on $[0, \infty)$. He notes that $x^{(k)}(t, 0, \bar{\phi})$ may have a discontinuity at $(k-1)h$, but will be continuous for $t > kh$; the solution smooths with time.

When we obtain a solution on all of $\Re$ then we avoid these discontinuities. Consider again

$$x(t) = f(x(t-h)) + \int_{t}^{\infty} Q(s, x(s), x(s-h))C(t-s)ds + p(t) \quad (6.4.2)$$

with (6.4.3)–(6.4.6) holding.

6. THE KRASNOSELSKII-SCHAEFER THEOREM

Theorem 6.4.1. *Let p be bounded on $\Re$, let (6.4.3)–(6.4.6) hold, and suppose that*

$$\mu := \alpha + C_0 < 1, \tag{6.4.7}$$

where α and C_0 are in (6.4.3) and (6.4.6). Then (6.4.2) has a unique bounded continuous solution on $-\infty < t < \infty$.

Proof. Let $(\mathcal{B}, \|\cdot\|)$ be the Banach space of bounded continuous functions $\phi : \Re \to \Re$ with the supremum norm. Define $P : \mathcal{B} \to \mathcal{B}$ by $\phi \in \mathcal{B}$ implies that

$$(P\phi)(t) = f(\phi(t-h)) + \int_t^\infty Q(s, \phi(s), \phi(s-h))C(t-s)ds + p(t). \tag{6.4.8}$$

As ϕ is bounded and continuous, so is $Q(t, \phi(t), \phi(t-h))$ using (6.4.4)–(6.4.6); since C is $L^1[0, \infty)$ by (6.4.6) it follows from a theorem of Hewitt and Stromberg (1971; p. 398) that the integral is uniformly continuous. That integral is bounded because of (6.4.6). Hence, $P\phi$ is bounded and continuous.

Next, if $\phi, \psi \in \mathcal{B}$ then

$$|(P\phi)(t) - (P\psi)(t)| \leq \alpha\|\phi - \psi\|$$
$$+ \int_t^\infty [k|\phi(s) - \psi(s)| + (1-k)|\phi(s-h) - \psi(s-h)|]|C(t-s)|ds$$
$$\leq \alpha\|\phi - \psi\| + \|\phi - \psi\| \int_t^\infty |C(t-s)|ds$$
$$= \mu\|\phi - \psi\|.$$

As $\mu < 1$, P is a contraction with unique fixed point. This completes the proof.

Corollary 6.4.1. *If the conditions of Theorem 6.4.1 hold and if there is a $T > 0$ such that $Q(t, x, y) = Q(t+T, x, y)$ and $p(t+T) = p(t)$, then (6.4.2) has a unique T-periodic solution.*

Proof. A change of variable shows that if $(\mathcal{B}, \|\cdot\|)$ is the Banach space of continuous T-periodic functions with the supremum norm, always denoted by

$$(\mathcal{P}_T, \|\cdot\|),$$

then $\phi \in \mathcal{P}_T$ implies that $P\phi \in \mathcal{P}_T$. Thus, the fixed point is in $\mathcal{P}_T$.

6.4. NEUTRAL INTEGRAL EQUATIONS

It is natural to ask if, under the conditions of Theorem 6.4.1, (6.4.2) might also have an unbounded solution, $D(t)$, on $(-\infty, \infty)$. A study of the work up to Theorem 6.4.2 will show this is possible only if $\int_t^\infty |C(t-s)||D(s)|ds$ is large. We continue this question after Example 6.4.1.

The solutions just obtained are continuous and satisfy (6.4.2) on $\Re$. A different point of view studies (6.4.2) and an initial condition. Suppose that there is a given initial function $\bar\phi : [-h, 0] \to \Re$ which is continuous. The initial value problem for (6.4.2) asks that we find a function $x : [-h, \infty) \to \Re$, denoted by $x(t, 0, \bar\phi)$, with $x(t, 0, \bar\phi) = \bar\phi(t)$ on $[-h, 0)$ and x satisfies (6.4.2) on $[0, \infty)$.

Remark 6.4.1. If $x(t) = x(t, 0, \bar\phi)$ is to be continuous at $t = 0$, then we must have

$$x(0) = \bar\phi(0) = f(\bar\phi(-h)) + \int_0^\infty Q(s, x(s), x(s-h))C(-s)ds + p(0). \quad (6.4.9)$$

Thus, for an arbitrary continuous $\bar\phi$, a finite jump discontinuity at $t = 0$ must be expected. Under conditions to be given, the integral in (6.4.2) will be continuous whenever x is piecewise continuous and so, from (6.4.2), we must expect x to have jump discontinuities at nh, for $n = 1, 2, \ldots$ This shows us how to search for a solution on $[0, \infty)$.

First, examine $p(t)$ and select a constant $K > 0$ and a continuous function
$D : [-h, \infty) \to [1, \infty)$ with

$$\sup_{t \geq -h} |p(t)/D(t)| < \infty \text{ and } D(t-h)/D(t) \leq K. \quad (6.4.10)$$

The function D will be the weight for a norm on a Banach space and notation is simplified if D is increasing, but we do not ask that. We will see that we want $D(t)$ to be as small as possible.

Next, for D to be compatible with $C(t-s)$ we will need

$$\sup_{t \geq 0} \int_t^\infty |C(t-s)|[D(s)/D(t)]ds < \infty. \quad (6.4.11)$$

If we can not satisfy (6.4.10) and (6.4.11), then we are unable to say anything here about (6.4.2).

308   6. THE KRASNOSELSKII-SCHAEFER THEOREM

Finally, we will want a mapping induced by (6.4.2) to map piecewise continuous functions into piecewise continuous functions. This will lead us to ask that

$$\int_t^\infty Q(s, \phi(s), \phi(s-h))C(t-s)ds$$

be continuous whenever ϕ is piecewise continuous and $\phi(t)/D(t)$ is bounded. We have already noted in the proof of Theorem 6.4.1 that if ϕ is bounded and continuous then a classical theorem says that this integral is uniformly continuous. We ask that

$$\begin{cases} \forall J > 0, \text{ if } 0 \leq t_1 < t_2 \leq J, \text{ then} \\ \int_{t_2}^\infty |C(t_1 - s) - C(t_2 - s)|D(s)ds \to 0 \text{ as } |t_1 - t_2| \to 0. \end{cases} \quad (6.4.12)$$

In order to have a contraction mapping we also need

$$\alpha K + (k + (1-k)K) \sup_{t \geq 0} \int_t^\infty |C(t-s)|[D(s)/D(t)]ds$$

$$=: \mu < 1. \quad (6.4.13)$$

Define

$$(\mathcal{B}, |\cdot|_D) \quad (6.4.14)$$

as the Banach space of functions $\phi : [-h, \infty) \to \Re$ which are continuous on $[(n-1)h, nh)$ with left-hand limits existing at nh and with the property that

$$|\phi|_D := \sup_{t \geq -h} |\phi(t)/D(t)| \text{ exists.} \quad (6.4.15)$$

Theorem 6.4.2. *Let (6.4.3)–(6.4.6) and (6.4.10)–(6.4.13) hold. Let $\bar{\phi}$: $[-h, 0] \to \Re$ be a given continuous function. Then there is a unique $x \in \mathcal{B}$ satisfying (6.4.2) for $t \geq 0$ and $x(t) = \bar{\phi}(t)$ on $[-h, 0)$.*

Proof. Let

$$\mathcal{B}^* = \{\phi \in \mathcal{B} \mid \phi(t) = \bar{\phi}(t) \text{ on } [-h, 0)\}.$$

Then $(\mathcal{B}^*, |\cdot|_D)$ is a complete metric space. Define $P : \mathcal{B}^* \to \mathcal{B}^*$ by $\phi \in \mathcal{B}^*$ implies that

$$\begin{cases} (P\phi)(t) = \bar{\phi}(t) \text{ if } -h \leq t < 0, \\ (P\phi)(t) = f(\phi(t-h)) \\ \quad + \int_t^\infty Q(s, \phi(s), \phi(s-h))C(t-s)ds + p(t), \quad t \geq 0. \end{cases} \quad (6.4.16)$$

6.4. NEUTRAL INTEGRAL EQUATIONS

Notice that

$$|f(x)| \leq |f(0)| + \alpha|x|$$

and

$$|Q(t,x,y)| \leq |Q(t,0,0)| + k|x| + (1-k)|y|$$
$$\leq 1 + k|x| + (1-k)|y|$$

by (6.4.3)–(6.4.5) and so using (6.4.10) and (6.4.11) we have

$$|(P\phi)(t)/D(t)| \leq [|f(0)|/D(t)] + \alpha K[|\phi(t-h)|/D(t-h)]$$
$$+ \int_t^\infty [(1/D(s)) + k(|\phi(s)|/D(s))$$
$$+ (1-k)K(|\phi(s-h)|/D(s-h))]|C(t-s)|(D(s)/D(t))ds$$
$$+ |p(t)|/D(t).$$

Taking the supremum for $t \geq 0$ we see that $|P\phi|_D$ exists. To see that the integral in (6.4.16) is continuous, if $J > 0$ and if $0 \leq t_1 < t_2 \leq J$, then

$$\left| \int_{t_1}^\infty Q(s,\phi(s),\phi(s-h))C(t_1-s)ds \right.$$
$$\left. - \int_{t_2}^\infty Q(s,\phi(s),\phi(s-h))C(t_2-s)ds \right|$$
$$\leq \left| \int_{t_1}^{t_2} Q(s,\phi(s),\phi(s-h))C(t_1-s)ds \right|$$
$$+ \int_{t_2}^\infty |Q(s,\phi(s),\phi(s-h))||C(t_1-s) - C(t_2-s)|ds.$$

The first term on the R-H-S can be made small if t_1 or t_2 is fixed and then $|t_1 - t_2|$ is made small. The last term is bounded by

$$\int_{t_2}^\infty [|Q(s,\phi(s),\phi(s-h))|/D(s)]|C(t_1-s) - C(t_2-s)|D(s)ds.$$

The first factor in the integrand is bounded because of (6.4.4), (6.4.5), and (6.4.10). By (6.4.12) the integral tends to zero as $|t_1 - t_2| \to 0$. Hence, the integral is continuous and $P\phi \in \mathcal{B}^*$.

6. THE KRASNOSELSKII-SCHAEFER THEOREM

Finally, we show that P is a contraction. If $\phi, \psi \in \mathcal{B}^*$, then

$$|(P\phi)(t) - (P\psi)(t)|/D(t)$$
$$\leq \alpha K|\phi - \psi|_D + \int_t^\infty \{k[|\phi(s) - \psi(s)|/D(s)]$$
$$+ (1-k)K[|\phi(s-h) - \psi(s-h)|/D(s-h)]\}|C(t-s)|[D(s)/D(t)]ds$$
$$\leq \alpha K|\phi - \psi|_D + \left(k|\phi - \psi|_D\right.$$
$$\left. + (1-k)K|\phi - \psi|_D\right) \int_t^\infty |C(t-s)|[D(s)/D(t)]ds$$
$$\leq \mu|\phi - \psi|_D$$

by (6.4.13), as required. Hence, P is a contraction with unique fixed point $\phi \in \mathcal{B}^*$ and the proof is complete.

The first candidates which come to mind for $C(t)$ are βe^{at} and $\beta/(1 + t^{2n})$, $n \geq 1$. We illustrate the results with the first one. Consider the equation

$$x(t) = \alpha x(t-h) + \beta \int_t^\infty x(s)e^{a(t-s)}ds + p(t) \qquad (6.4.17)$$

so that $Q(t, x, y) = x$ and $k = 1$. Clearly, if p is differentiable, it could be reduced to a neutral functional differential equation. We now give three forms for $p(t)$.

Example 6.4.1. If $p(t)$ is bounded and continuous on $\mathfrak{R}$, then (6.4.17) has a unique bounded continuous solution on $\mathfrak{R}$ provided that

$$\mu = \alpha + |\beta| \int_t^\infty e^{a(t-s)}ds = \alpha + (|\beta|/a) < 1. \qquad (6.4.7^*)$$

Question: Is it possible that with $p(t)$ bounded, then (6.4.17) can have an unbounded solution? By following the next examples we can see that the solution of Example 6.4.1 may be the only solution of (6.4.17) which is bounded by $Me^{\theta at}$, $0 < \theta < 1$, for $t > 0$ and some $M > 0$. If there are larger solutions, questions of convergence will arise.

Example 6.4.2. If $p(t) = t$, then $D(t) = t + 1 + h$, $D(t-h)/D(t) \leq 1 =: K$, and

$$\mu = \alpha K + |\beta| \sup_{t \geq 0}[at + 1 + a + ah]/a^2(t + 1 + h). \qquad (6.4.7^{**})$$

If, for example, $a = 1$, then $\mu < 1$ if and only if $\alpha + |\beta| < 1$. When $\mu < 1$, then for any continuous initial function $\bar{\phi} : [-h, 0] \to \mathfrak{R}$, there is a unique

6.4. NEUTRAL INTEGRAL EQUATIONS

solution $x(t) = x(t, 0, \bar{\phi})$ of (6.4.17) on $[0, \infty)$ satisfying $\sup_{t \geq 0} |x(t)/(t + 1 + h)| < \infty$.

Proof. Clearly, (6.4.11) will be satisfied. To verify (6.4.12) we have

$$\int_{t_2}^{\infty} |e^{a(t_1-s)} - e^{a(t_2-s)}|(s + 1 + h)ds$$

$$= \int_{t_2}^{\infty} e^{-as}|e^{at_1} - e^{at_2}|(s + 1 + h)ds$$

so that if $J > 0$ is given and if $0 \leq t_1 < t_2 \leq J$ then this quantity tends to zero as $|t_1 - t_2| \to 0$. For (6.4.13) we examine

$$\int_t^{\infty} e^{-as}[s + 1 + h]ds = e^{-at}[at + 1 + a + ah]/a^2$$

so that

$$\int_t^{\infty} |C(t - s)|[D(s)/D(t)]ds$$
$$= |\beta|[e^{at}/(t + 1 + h)]e^{-at}[at + 1 + a + ah]/a^2$$
$$= |\beta|[at + 1 + a + ah]/a^2(t + 1 + h),$$

as required.

Example 6.4.3. Let $0 < c < a$, $p(t) = e^{ct}$ and $D(t) = e^{ct}e^{ch}$. Then $D(t - h)/D(t) = e^{ct}/e^{ct}e^{ch} = e^{-ch} = K$ and

$$\mu = \alpha e^{-ch} + (|\beta|/(a - c)). \tag{6.4.7***}$$

If $\mu < 1$, then for any given continuous $\bar{\phi} : [-h, 0] \to \Re$, there is a unique solution $x(t) = x(t, 0, \bar{\phi})$ of (6.4.17) satisfying $\sup_{t \geq 0} |x(t)/e^{ct}| < \infty$. The proof is a routine calculation.

Remark 6.4.2. Theorem 6.4.1 and Example 6.4.1 can be stated far more strongly. When (6.4.3)–(6.4.6) hold, if (6.4.2) has a bounded solution then the integral in (6.4.2) is bounded for $t \geq 0$ and $x(t) - f(x(t - h))$ is bounded for $t \geq 0$. Hence, if we suppose that (6.4.3)–(6.4.7) hold, then (6.4.2) has a solution bounded for $t \geq 0$ if and only if $p(t)$ is bounded. Parallel remarks hold for the other results.

Theorem 6.4.3. *Let $\mathcal{B}^*$ be defined in the proof of Theorem 6.4.2 and let (6.4.12) hold. If $0 \leq \alpha < 1$ and $x \in \mathcal{B}^*$ satisfies (6.4.2) for $t \geq 0$, then $x(nh^-) - x(nh^+) \to 0$ as $n \to \infty$.*

6. THE KRASNOSELSKII-SCHAEFER THEOREM

Proof. In the proof of Theorem 6.4.2 we noted that the integral in (6.4.2) is continuous. Since p is continuous, so is $x(t) - f(x(t-h))$. Thus,

$$x(nh^-) - f(x((n-1)h^-)) = x(nh^+) - f(x((n-1)h^+))$$

or

$$\begin{aligned}
|x(nh^-) - x(nh^+)| &= |f(x((n-1)h^-)) - f(x((n-1)h^+))| \\
&\leq \alpha |x((n-1)h^-) - x((n-1)h^+)| \\
&\leq \alpha |f(x((n-2)h^-)) - f(x((n-2)h^+))| \\
&\leq \alpha^2 |x((n-2)h^-) - x((n-2)h^+)| \\
&\leq \cdots \\
&\leq \alpha^{n-1} |x(h^-) - x(h^+)| \\
&\to 0 \text{ as } n \to \infty,
\end{aligned}$$

completing the proof.

We have been looking at small kernels and now we consider large kernels and Liapunov functionals yielding bounds. Let

$$x(t) = f(x(t-h)) + \int_t^\infty [g(x(s)) + r(x(s-h))] C(t,s) ds + p(t) \quad (6.4.18)$$

in which

$$|f(x) - f(y)| \leq \alpha |x-y|, \ 0 \leq \alpha < 1, \ xg(x) > 0 \text{ if } x \neq 0, \quad (6.4.19)$$
$$g, r, C, C_s, C_{st}, p \text{ are continuous, } p(t+T) = p(t), \quad (6.4.20)$$
$$C(t+T, s+T) = C(t,s),$$

and

$$\sup_{t \geq 0} \left[\int_t^\infty |C(t,s)| ds \right.$$
$$\left. + \int_t^\infty (|C_s(t,s)| + |C_{st}(t,s)|)[1 + (t-s)^2] ds \right] < \infty, \quad (6.4.21)$$
$$|C(t,s)||t-s| \to 0 \text{ as } s \to \infty.$$

6.4. NEUTRAL INTEGRAL EQUATIONS

In addition to (6.4.18) we consider the equation

$$x(t) = \lambda \bigg[f(x(t-h)/\lambda) \\ + \int_t^\infty [g(x(s)) + r(x(s-h))]C(t,s)ds + p(t) \bigg], \\ 0 < \lambda \leq 1. \qquad (6.4.22)$$

We will obtain an *a priori* bound on all T-periodic solutions of (6.4.22) for $0 < \lambda \leq 1$ and then use Theorem 6.1.1 to prove that (6.4.22) has a T-periodic solution.

Lemma 6.4.1. *If $x(t)$ is a continuous T-periodic solution of (6.4.22) and if V is defined by*

$$V(t) = \int_t^\infty \lambda^2 C_s(t,s) \left(\int_s^t [g(x(u)) + r(x(u-h))] du \right)^2 ds, \qquad (6.4.23)$$

then

$$V'(t) = \lambda^2 \int_t^\infty C_{st}(t,s) \left(\int_s^t [g(x(u)) + r(x(u-h))] du \right)^2 ds \\ + 2\lambda \big(g(x) + r(x(t-h))\big)(x(t) - \lambda f(x(t-h)/\lambda) - \lambda p(t)). \qquad (6.4.24)$$

Proof. The first term is clear. In addition we have

$$2\lambda^2 [g(x) + r(x(t-h))] \int_t^\infty C_s(t,s) \int_s^t [g(x(u)) + r(x(u-h))] du\, ds \\ = 2\lambda^2 [g(x) + r(x(t-h))] \bigg\{ C(t,s) \int_s^t [g(x(u)) + r(x(u-h))] du \Big|_t^\infty \\ + \int_t^\infty C(t,s) [g(x(s)) + r(x(s-h))] ds \bigg\} \\ = 2\lambda [g(x) + r(x(t-h))] [x - \lambda f(x(t-h)/\lambda) - \lambda p(t)],$$

as required.

Remark 6.4.4. As we have done before, we are going to use V' alone (not V) to obtain *a priori* bounds on solutions of (6.4.22). The goal is to show that a modification of V, say W, satisfies

$$W'(t) \leq \lambda(-K|xg(x)| + M), \text{ some } K > 0,\ M > 0,$$

or

$$W'(t) \geq \lambda(K|xg(x)| - M).$$

As $x \in \mathcal{P}_T$ implies that $W \in \mathcal{P}_T$ we obtain $W(T) = W(0)$ and so an integration of *either* inequality yields

$$\int_0^T |x(t)g(x(t))|dt \leq M/K.$$

This inequality is then parlayed into an *a priori* bound on the supremum of x using (6.4.22), (6.4.21), and a Schwarz inequality. A fixed point theorem then yields a T-periodic solution of (6.4.18).

We now specialize further and write (6.4.22) as

$$x(t) = \alpha x(t-h) + \lambda \left[\int_t^\infty (\beta x^{2n+1}(s) + \gamma x^m(s-h)) C(t,s) ds + p(t) \right] \quad (6.4.25)$$

so that (6.4.24) becomes

$$V'(t) = \lambda^2 \int_t^\infty C_{st}(t,s) \left(\int_s^t [\beta x^{2n+1}(u) + \gamma x^m(u-h)] du \right)^2 ds$$
$$+ 2\lambda(\beta x^{2n+1} + \gamma x^m(t-h))(x - \alpha x(t-h) - \lambda p(t)). \quad (6.4.26)$$

We will need to work with the last term which we write as

$$Q := 2\lambda[\beta x^{2n+2} - \alpha\beta x^{2n+1} x(t-h) + \gamma x x^m(t-h)$$
$$- \gamma\alpha x^{m+1}(t-h) - \lambda\beta x^{2n+1} p - \gamma\lambda x^m(t-h)p]. \quad (6.4.27)$$

We now give two lemmas with different relations between n and m which will yield (6.4.28).

Lemma 6.4.2. *If $C_{st} \leq 0$, if $\beta < 0$, if m is odd and if $2n > m$, then there is a constant X so that if x is a T-periodic solution of (6.4.25), $0 < \lambda \leq 1$, then*

$$\int_0^T x^{2n+2}(s) ds \leq X. \quad (6.4.28)$$

6.4. NEUTRAL INTEGRAL EQUATIONS

Proof. We need two basic inequalities for Q:

$$|x^{2n+1}x(t-h)| \leq [(2n+1)x^{2n+2} + x^{2n+2}(t-h)]/(2n+2)$$

and

$$|x^m(t-h)x| \leq [m|x(t-h)|^{m+1} + |x|^{m+1}]/(m+1).$$

With these, from (6.4.27) we get

$$Q \leq 2\lambda \bigg[\beta x^{2n+2} + \frac{|\alpha\beta|}{2n+2}(2n+1)x^{2n+2}$$
$$+ \frac{|\alpha\beta|}{2n+2}x^{2n+2}(t-h) + \frac{|\gamma|m}{m+1}|x(t-h)|^{m+1}$$
$$+ \frac{|\gamma|}{m+1}|x|^{m+1} - \gamma\alpha|x(t-h)|^{m+1} + |\lambda\beta|\,\|p\|\,|x|^{2n+1}$$
$$+ |\lambda\gamma|\,\|p\|\,|x(t-h)|^m \bigg].$$

Next, define the function

$$W(t) = 2\lambda \bigg[\bigg(\frac{|\gamma|m}{m+1} + |\gamma\alpha| \bigg) \int_{t-h}^{t} |x(s)|^{m+1} ds$$
$$+ |\gamma\lambda|\,\|p\| \int_{t-h}^{t} |x(s)|^m ds + \frac{|\alpha\beta|}{2n+2} \int_{t-h}^{t} x^{2n+2}(s) ds \bigg]$$

with derivative

$$W'(t) = 2\lambda \bigg[\bigg(\frac{|\gamma|m}{m+1} + |\gamma\alpha| \bigg) \bigg(|x|^{m+1} - |x(t-h)|^{m+1} \bigg)$$
$$+ |\gamma\lambda|\|p\| \bigg(|x|^m - |x(t-h)|^m \bigg)$$
$$+ \frac{|\alpha\beta|}{2n+2} \bigg(x^{2n+2} - x^{2n+2}(t-h) \bigg) \bigg]. \tag{6.4.29}$$

If we now form $V + W$, then the derivative has the first term in (6.4.26), while (6.4.29) added to Q yields a quantity

$$\overline{Q} \leq 2\lambda \bigg[\bigg\{ \beta + |\alpha\beta|\bigg(\frac{2n+1}{2n+2}\bigg) + \frac{|\alpha\beta|}{2n+2} \bigg\} x^{2n+2}$$
$$+ \bigg(\frac{|\gamma|}{m+1} + \frac{m|\gamma|}{m+1} + |\gamma\alpha| \bigg) |x|^{m+1} + |\lambda\beta|\,\|p\|\,|x|^{2n+1}$$
$$+ |\gamma\lambda|\,\|p\|\,|x|^m \bigg]$$

6. THE KRASNOSELSKII-SCHAEFER THEOREM

or

$$\overline{Q} \leq 2\lambda[\{\beta + |\alpha\beta|\}x^{2n+2} + (|\gamma| + |\gamma\alpha|)|x|^{m+1} \\ + |\lambda\beta| \, \|p\| \, |x^{2n+1}| + |\gamma\lambda| \, \|p\| \, |x|^m]. \tag{6.4.30}$$

Since $m < 2n$, and since $\beta < 0$ and $|\alpha| < 1$, there is an $M > 0$ with

$$(V + W)' \leq 2\lambda \left[-\frac{|\beta|(1 - |\alpha|)}{2} x^{2n+2} + M \right]. \tag{6.4.31}$$

But $x \in \mathcal{P}_T$ yields $V + W \in \mathcal{P}_T$ so

$$0 = V(T) + W(T) - V(0) - W(0)$$

$$\leq 2\lambda \left[-\frac{|\beta|(1 - |\alpha|)}{2} \int_0^T x^{2n+2}(s)ds + MT \right].$$

As $\lambda > 0$ we get (6.4.28). This proves Lemma 6.4.2.

Lemma 6.4.3. *If $C_{st} \leq 0$, if $\beta < 0$, if $\gamma\alpha > 0$, if $2n + 2 = m + 1$, and if $|\beta| > |\gamma|$, then (6.4.28) holds.*

Proof. Proceeding with V and Q, as before in the proof of Lemma 6.4.2, we get

$$Q \leq 2\lambda \left[\beta x^{2n+2} + |\alpha\beta| \left(\frac{2n+1}{2n+2} \right) x^{2n+2} \right. \\ + \frac{|\alpha\beta|}{2n+2} x^{2n+2}(t-h) + |\gamma| \left(\frac{2n+1}{2n+2} \right) x^{2n+2}(t-h) \\ + \frac{|\gamma|}{2n+2} x^{2n+2} - \gamma\alpha x^{2n+2}(t-h) + |\lambda\beta| \, \|p\| \, |x|^{2n+1} \\ \left. + |\gamma\lambda| \, \|p\| \, |x(t-h)|^{2n+1} \right].$$

Define a function

$$Z(t) = 2\lambda \left[\left\{ \frac{|\alpha\beta|}{2n+2} - \gamma\alpha + |\gamma| \left(\frac{2n+1}{2n+2} \right) \right\} \int_{t-h}^t x^{2n+2}(s)ds \\ + |\gamma\lambda| \, \|p\| \int_{t-h}^t |x(s)|^{2n+1}ds \right]$$

with derivative

$$Z'(t) = 2\lambda \left[\left\{ \frac{|\alpha\beta|}{2n+2} - \gamma\alpha + |\gamma| \left(\frac{2n+1}{2n+2} \right) \right\} \{x^{2n+2}(t) - x^{2n+2}(t-h)\} \\ + |\gamma\lambda| \, \|p\| \left(|x|^{2n+1} - |x(t-h)|^{2n+1} \right) \right].$$

6.4. NEUTRAL INTEGRAL EQUATIONS

Thus,
$$Q + Z' \leq 2\lambda[(\beta + |\alpha\beta| - \gamma\alpha + |\gamma|)x^{2n+2} \\ + (|\lambda\beta|\,\|p\| + |\gamma\lambda|\,\|p\|)|x|^{2n+1}]. \tag{6.4.32}$$

Now, $\beta < 0$ and $\gamma\alpha > 0$ so
$$\beta + |\alpha\beta| + |\gamma| - \gamma\alpha = -|\beta|(1 - |\alpha|) + |\gamma|(1 - |\alpha|)$$
$$= (-|\beta| + |\gamma|)(1 - |\alpha|) < 0.$$

This means that there are positive constants U and R with
$$(V + Z)' \leq 2\lambda[-Ux^{2n+2} + R]$$
and (6.4.28) will now follow, proving Lemma 6.4.3.

Theorem 6.4.4. *Let (6.4.20), (6.4.21) hold, $C_s \leq 0$, $C_{st} \leq 0$, $|\alpha| < 1$, and suppose that (6.4.28) holds for any T-periodic solution x of (6.4.25). Let $m = 2n + 1$. Then (6.4.25) has a T-periodic solution for $\lambda = 1$.*

Proof. In the notation of Theorem 6.1.1,
$$(Bx)(t) = \alpha x(t - h)$$
and
$$(Ax)(t) = \int_t^\infty (\beta x^{2n+1}(s) + \gamma x^m(s-h))C(t,s)ds + p(t),$$
while $(\mathcal{B}, \|\cdot\|)$ is $(\mathcal{P}_T, \|\cdot\|)$, the Banach space of continuous T-periodic functions with the supremum norm. By the conditions in (6.4.20)–(6.4.21) we can show that A is continuous. Computations will show that bounded sets are mapped into equicontinuous sets. We will use (6.4.28) to show that (ii) of Theorem 6.1.1 can not hold, and that will complete the proof.

From (6.4.25) we get (integrating by parts)
$$(x - \alpha x(t-h) - \lambda p)^2$$
$$\leq \left[\int_t^\infty (\beta x^{2n+1}(s) + \gamma x^{2n+1}(s-h))C(t,s)ds\right]^2$$
$$= \left[-C(t,s)\int_s^t (\beta x^{2n+1}(u) + \gamma x^{2n+1}(u-h))du\Big|_t^\infty\right.$$
$$\left. - \int_t^\infty -C_s(t,s)\int_s^t (\beta x^{2n+1}(u) + \gamma x^{2n+1}(u-h)du\,ds\right]^2$$

(The first term on the right is zero by (6.4.21) if $x \in \mathcal{P}_T$.)

$$\leq \left[\int_t^\infty -C_s(t,s) \int_s^t (\beta x^{2n+1}(u) + \gamma x^{2n+1}(u-h)) du ds \right]^2$$

$$\leq \left[\int_t^\infty -C_s(t,s) \int_t^s K(x^{2n+2}(u) + x^{2n+2}(u-h) + M) du ds \right]^2$$

for some positive constants K and M. Now, divide $[t,s]$ into intervals of length T. There will be no more than $1 + (s-t)/T$ of them. On each such interval the inner integration will yield no more than $K(2X + MT)$ since $x \in \mathcal{P}_T$ and (6.4.28) holds. So our integral here is bounded by

$$\left[\int_t^\infty -C_s(t,s)[1 + (t-s)/T][2X + MT] ds \right]^2$$
$$\leq Y^2$$

for some $Y > 0$ by (6.4.21).

This means that

$$|x - \alpha x(t-h) - \lambda p| \leq Y$$

and

$$|x(t)| \leq Y + \|p\| + |\alpha||x(t-h)|$$
$$=: \delta + |\alpha||x(t-h)|.$$

If $0 \leq t \leq T$ and n is a positive integer, then

$$|x(t+nh)| \leq \delta + |\alpha||x(t+(n-1)h)|$$
$$\leq \delta + |\alpha|(\delta + |\alpha||x(t+(n-2)h)|)$$
$$\leq \delta(1 + |\alpha|) + \alpha^2(\delta + |\alpha||x(t+(n-3)h)|)$$
$$\leq \cdots$$
$$\leq \delta \sum_{n=1}^\infty |\alpha|^{n-1} + |\alpha|^{n-1}|x(t-h)|$$
$$\to \delta/(1 - |\alpha|) \text{ as } n \to \infty.$$

This gives the *a priori* bound and proves the theorem.

We conclude this section with some remarks about (6.4.1)

$$x' = ax + \alpha x'(t-h) - q(t, x, x(t-h)).$$

Our results have all been of a global nature, while (6.4.1) is usually of interest for only a certain range of x values, namely $x > 0$, since (6.4.1)

6.4. NEUTRAL INTEGRAL EQUATIONS

is often a population problem. Moreover, we have dealt with a periodic forcing function. There is a vast gulf between that search for a periodic solution of (6.4.25) with nontrivial $p(t)$ and the search for a periodic solution of (6.4.1). If we seek periodic solutions of (6.4.1), then under common assumptions one can be found by inspection as an equilibrium solution; this means that the struggle to prove that there is a periodic solution may simply yield an obvious one.

Equation (6.4.18) is quite general and, in that form, the infinite integral seems to be unavoidable. But (6.4.1) is special and, if one is interested only in periodic solutions, then (6.4.1) can be converted to a neutral integral equation of finite delay type as follows. In Section 6.1.2 we wrote (6.4.1) as

$$[(x - \alpha x(t-h))e^{-at}]' = [a\alpha x(t-h) - q(t,x,x(t-h))]e^{-at}.$$

If $x \in \mathcal{P}_T$ solves that equation, integrate from $t-T$ to t and use $x(t+T) = x(t)$ to obtain

$$(x(t) - \alpha x(t-h))e^{-at} - (x(t) - \alpha x(t-h))e^{-a(t-T)}$$
$$= \int_{t-T}^{t} [a\alpha x(s-h) - q(s,x(s),x(s-h)]e^{-as} ds$$

so that

$$x(t) = \alpha x(t-h)$$
$$+ \int_{t-T}^{t} \frac{[a\alpha x(s-h) - q(s,x(s),x(s-h))]}{1 - e^{aT}} e^{a(t-s)} ds. \quad (6.4.33)$$

It is now possible to stipulate conditions so that we can find numbers $0 < c < d$ for which the complete metric space of continuous T-periodic functions ϕ with $c \leq \phi(t) \leq d$ will be mapped into itself by

$$(P\phi)(t) = \alpha \phi(t-h)$$
$$+ \int_{t-T}^{t} \frac{[a\alpha \phi(s-h) - q(s,\phi(s),\phi(s-h))]}{1 - e^{aT}} e^{a(t-s)} ds$$

and to obtain a fixed point by Krasnoselskii's theorem. This periodic function will satisfy (6.4.1).

Good discussions of periodic solutions for (6.4.1) are found in Gopalsamy (1992) and Kuang (1993).

This material is taken from Burton (1998b) which was published by Elsevier. The study was continued in Burton and Hering (2000) for equations of retarded type.

Chapter 7

Stability: Convex Memory

7.1 Introduction

In this chapter we explore the idea of stability theory for integral equations. As we have mentioned before, an integral equation can be a primary representation of fading memory. Moreover, a convex kernel can be one of the first approximations to a fading memory. Thus, in this introduction we focus primarily on convex kernels, realizing that this can be extended in ways already discussed in Chapter 5.

In the theory of stability for differential equations we begin with a solution and study the behavior of other solutions which start near the given solution. Frequently, we make a change of variable so that instead of studying the behavior of solutions starting near the given solution, we study the behavior of solutions starting near a constant solution, usually the zero solution. We call it the equilibrium solution because the system is in equilibrium at the zero function. And that is such a great simplification that we quickly forget where we began and we always start with the assumption that for $x' = f(t, x)$ then $f(t, 0) = 0$ so that we have the zero solution. The details will help us greatly in understanding the coming definitions and theorems.

Consider the ordinary differential equation

$$x' = f(t, x) \qquad (7.1.1)$$

where $f : [0, \infty) \times \Re^n \to \Re^n$ is continuous. Thus, for each (t_0, x_0) in the domain there is at least one solution $x : [0, L) \to \Re^n$ and satisfying (7.1.1); if the solution remains bounded, then $L = \infty$. For notation here we will denote any solution with that initial condition as $\phi(t, t_0, x_0)$ or as $x(t, t_0, x_0)$. Stability theory begins as follows.

7.1. INTRODUCTION

Definition 7.1.1. Let ϕ be a solution of (7.1.1) defined on an interval $[t_0, \infty)$ with $t_0 \geq 0$, denoted by $\phi(t, t_0, x_0)$.

(i) We say that ϕ is *Liapunov stable* if for each $t_1 \geq t_0$ and each $\epsilon > 0$ there is a $\delta > 0$ such that $|\phi(t_1) - x_1| < \delta$ and $t \geq t_1$ imply that $|\phi(t) - x(t, t_1, x_1)| < \epsilon$.

(ii) We say that ϕ is *uniformly stable* if ϕ is Liapunov stable and if δ is independent of $t_1 \geq t_0$.

(iii) If ϕ is Liapunov stable and if for each $t_1 \geq t_0$ there exists $\eta > 0$ such that $|x_0 - \phi(t_1)| < \eta$ implies $|x(t, t_1, x_0) - \phi(t)| \to 0$ as $t \to \infty$, then ϕ is *asymptotically stable*.

(iv) If ϕ is Liapunov stable, and if for each $t_1 \geq t_0$ and each $x_0 \in R^n$ we have $\lim_{t \to \infty} |x(t, t_1, x_0) - \phi(t)| = 0$, then $x = \phi$ is *globally asymptotically stable*.

We are confronted with two realities. One is good news, the other is so bad that the stability theory investigator frequently defines away the difficulty and moves ahead without another thought. First, in so many real-world problems there are constant solutions (called equilibrium solutions) which are central. The system is at equilibrium and we must study the behavior of solutions when the equilibrium is disturbed. Thus, rather than study the complicated definition just given, we would translate that constant solution to the origin and study stability of the zero solution; in the definition, ϕ would be replaced by zero for great simplicity. Next, there is also the bitter reality that if the solution is not constant, then it is more than likely so complicated that it can not even be expressed by elementary functions. Thus, the definitions are now far more complicated than they appeared. Our solution is to combine the two cases as follows.

Let $\phi'(t) = f(t, \phi(t))$ and make the substitution $y = x - \phi$ to obtain

$$y' = x' - \phi' = f(t, x) - \phi'(t)$$
$$= f(t, y + \phi(t)) - \phi'(t)$$
$$=: g(t, y)$$

and we see that when $y = 0$ then

$$y' = g(t, 0) = f(t, \phi(t)) - \phi'(t) \equiv 0.$$

Our new equation has the zero solution and we will discuss stability of the zero solution instead of stability of ϕ. With some very significant

7. STABILITY: CONVEX MEMORY

exceptions, mainline stability investigators approach $y' = g(t,y)$ asking that $g(t,0) = 0$ without any focus on that forgotten function ϕ.

Integral equations offer even more challenges. Consider the integral equation

$$x(t) = a(t) - \int_{\alpha(t)}^{t} Q(t,s,x(s))\, ds \qquad (7.1.2)$$

where $\alpha(t) \geq \alpha_0 \geq -\infty$. We focus on functions which are analogous to equilibrium points of ordinary differential equations and obtain results, by way of Liapunov's direct method, concerning the long-time behavior of solutions. So seldom is it the case that $a(t)$ is constant and even more seldom is it the case that there is a constant solution. If we are going to give a stability analysis which will tell us the long-time behavior of all solutions, then we are going to have to learn to focus on functions which are in some sense almost solutions. What we offer here is one possibility and it begins with behavior which we discussed early in this book. The classical view is that if the kernel is nice, then the solution follows $a(t)$. As we know, sometimes it does, and sometimes it does not. The most tractable case is the one in which $a(t) \in L^1[0, \infty)$ which we have discussed at length. Thus, if the integral equation is linear, if $a \in L^1$, and if x follows a, then x is near zero much of the time for large time. Hence, zero is approximately an asymptotic equilibrium for the integral equation and the solution approaches that asymptotic equilibrium.

There is a very fruitful way to formalize and generalize this idea. That is the topic of the rest of the chapter. It provides one type of stability theory for integral equations. As a simple case, we outline how the process works in the linear finite delay equation

$$x(t) = a(t) - \int_{t-h}^{t} C(t,s)x(s)\,ds \qquad (7.1.3)$$

under convexity conditions on $C(t,s)$ of the type discussed several times before.

First, suppose we can find a pair of functions (ψ, Ψ) for (7.1.3) with the property that ψ is an approximate solution of (7.1.3) in the sense that if we substitute ψ into (7.1.3) then it fails to satisfy it by the amount of an L^1-function Ψ. That is

$$\Psi(t) := a(t) - \psi(t) - \int_{t-h}^{t} C(t,s)\psi(s)ds \in L^1[0,\infty). \qquad (7.1.4)$$

This is the analog of trying to find an exact solution of (7.1.1). We will call (ψ, Ψ) an L^1-near equilibrium for (7.1.3). We also consider L^p-near equilibria for $1 \leq p \leq \infty$. The case $p = \infty$ can be treated with some ease.

7.1. INTRODUCTION

If we can find such an equilibrium it will turn out to be absolutely fundamental. We will be able to show that if $x(t, 0, \phi)$ is a continuous solution of (7.1.3) with a continuous initial function ϕ then $|x(t) - \psi(t) - \Psi(t)| \to 0$ as $t \to \infty$. This means that even though ψ is only an approximate solution of (7.1.3), missing by an L^1-function, Ψ, every exact solution of (7.1.3) will converge to $\psi(t)$ to within the same L^1-function, Ψ.

Here is the reason that this is significant. The function $a(t)$ may be large and badly behaved, but if we can find an L^1-near equilibrium (ψ, Ψ) then we never have to deal with ψ or $a(t)$. Let us review the details.

Subtract (7.1.4) from (7.1.3) and obtain

$$x(t) - \psi(t) = \Psi(t) - \int_{t-h}^{t} C(t,s)[x(s) - \psi(s)]ds$$

which we denote by

$$y(t) = \Psi(t) - \int_{t-h}^{t} C(t,s)y(s)ds. \tag{7.1.5}$$

Now, use a Liapunov function and the convexity condition to show that $\Psi \in L^1[0, \infty)$ implies that for every continuous initial function yielding a continuous solution $y(t)$ of (7.1.5) we have

$$|y(t) - \Psi(t)| \quad \text{small in some sense.}$$

See Corollary 7.2.3.

Recall also that we studied many problems

$$x(t) = a(t) - \int_0^t C(t,s)x(s)ds$$

in which $a(t)$ was very large and badly behaved, yet we found that there was a solution in L^p. That solution can be taken as ψ and the stability problem will involve the unforced equation alone; $a(t)$ can be completely ignored.

We give several ways of finding an L^1-near equilibrium, but readily concede that it can be arduous, exactly as is the problem of finding ϕ in the original ode problem described above. On the other hand, the other extreme case of L^∞-near equilibrium is almost trivial to achieve and that yields

$$|x(t) - \psi(t) - \Psi(t)| \to 0$$

where $\Psi(t)$ is a bounded function; every exact solution for every continuous initial function eventually resides in a ball around $\psi(t)$.

7.1.1 Open Problems

This is just one short chapter in one book and we merely introduce one possibility for stability theory of integral equations. We mention only L^p-near equilibria and develop the necessary inequalities in Theorems 7.2.1A,B; those inequalities need to be developed for the other L^p cases; we would expect this to be a significant and rewarding study. The case in which $\Psi(t) \to 0$ as $t \to \infty$ is not even touched and it is as important as anything mentioned here. One would need to examine Theorems 7.2.1A,B and determine the corresponding functions needed in this case. We barely touch on nonlinearities. The L^∞ case yields what can be called uniform boundedness and that is known to promote periodic solutions under certain conditions as may be seen in Burton and Zhang (1990); there is a beautiful theory waiting to be developed here. In our Liapunov theorems we use only convexity; yet, earlier chapters show a number of other kinds of Liapunov functions. Moreover, we have not even mentioned fixed point techniques for stability as developed throughout Burton (2006b).

7.2 Near Equilibria

In this section we offer one choice for equilibrium points and we show that it is a good choice by developing a Liapunov theory around it and use it to obtain new results on limit sets for three problems of classical interest.

In particular, we study three forms of the integral equation

$$x(t) = a(t) - \int_{\alpha(t)}^{t} Q(t,s,x(s)) \, ds \qquad (7.2.1)$$

with a constant α satisfying $\alpha(t) \geq \alpha \geq -\infty$. We focus on functions which are analogous to equilibrium points of ordinary differential equations and obtain results, by way of Liapunov's direct method, concerning the long-time behavior of solutions.

Definition 7.2.1. *A pair of functions* (ψ, Ψ), *each mapping* $[\alpha, \infty) \to R^n$ *with* $\alpha \leq 0$, *is said to be an L^p-near equilibrium for (7.2.1) if there is a* $p \in [1, \infty]$ *with*

$$\Psi(t) := a(t) - \psi(t) - \int_{\alpha(t)}^{t} Q(t,s,\psi(s)) \, ds \in L^p[0,\infty). \qquad (7.2.2)$$

Thus, for $p = 1$, if (7.2.1) is perturbed by the L^1 function $-\Psi$, then ψ is a solution of (7.2.1); in other words, ψ fails to be a solution of (7.2.1) by an amount of an L^1 function.

7.2. NEAR EQUILIBRIA

Example 7.2.1. Here are some L^1-near equilibria.

(i) If $a \in L^1[0, \infty)$ and $Q(t, s, 0) = 0$, then $\psi(t) = 0$ and $\Psi(t) = a(t)$ so $(0, \Psi)$ is a near L^1-equilibrium for (7.2.1).

(ii) If $x(t) = a(t) + \int_{-\infty}^{t} C(t - s)x(s)\, ds$ where $a \in L^1[0, \infty)$ and $\int_0^\infty C(t)\, dt = 1$, then for every constant x_0, $\psi(t) = x_0$ and $\Psi(t) = a(t)$, so $(x_0, a(t))$ is an L^1-near equilibrium for this equation.

(iii) If $x(t) = a + a_1(t) + \int_{-\infty}^{t} C(t - s)x(s)\, ds$ where a is constant, $a_1 \in L^1[0, \infty)$, $\int_0^\infty C(t)\, dt = c \neq 1$, then for β defined by $\beta(1 - c) = a$, it follows that $\psi(t) = \beta$ and $\Psi(t) = a_1(t)$ so (ψ, Ψ) is an L^1-near equilibrium.

(iv) If ψ is an L^1 solution of $x(t) = a(t) + \int_{-\infty}^{t} D(t, s)g(x(s))\, ds$ and if $E \in L^1$, then (ψ, Ψ) is a near equilibrium for

$$x(t) = a(t) + \int_{-\infty}^{t} D(t, s)g(x(s))\, ds + \int_0^t E(t - s)x(s)\, ds$$

where $\Psi(t) := \int_0^t E(t - s)\psi(s)\, ds$.

(v) Finally, it must be remembered that throughout Chapter 2 we worked with

$$x(t) = a(t) - \int_0^t C(t, s)x(s)\, ds$$

in which $a(t)$ was a very large function, yet $x \in L^p[0, \infty)$. Thus, $(x(t), 0)$ is an L^p-near equilibrium and the transformation yielding (7.1.5) will rid us completely of that large $a(t)$; similar results are found in Section 3.3 for nonlinear problems.

Remark. In the next subsection we deal with $p = 1$ in the first two theorems. We frequently ask that not only Ψ, but powers of Ψ be $L^1[0, \infty)$. A number of transformations may be used to achieve this. In the equation

$$x(t) = a(t) + \int_0^t C(t - s)x(s)\, ds$$

with a and C in $L^1[0, \infty)$ and $C(t) \to 0$ as $t \to \infty$, let $y = x - a(t)$ so that

$$y(t) = \int_0^t C(t - s)a(s)\, ds + \int_0^t C(t - s)y(s)\, ds.$$

The first term on the right is $L^1[0, \infty)$ and it tends to zero. Hence, all powers are $L^1[0, \infty)$.

For a given t_0 we require a continuous initial function $\varphi : [\alpha, t_0] \to \Re^n$ and seek a solution $x(t, t_0, \varphi)$ of (7.2.1) with x continuous on $[\alpha, \infty)$, $x(t) = \varphi(t)$ on $[\alpha, t_0]$, and $x(t, t_0, \varphi)$ satisfying (7.2.1) for $t \geq t_0$. While existence theory may be given for (7.2.1) which allows a discontinuity in x at t_0, in most of our work we perform certain integration by parts which requires continuity; thus, φ must be selected with care.

Definition 7.2.2. *A metric space $(\Omega(t_0), \rho)$ of continuous functions $\varphi : [\alpha, t_0] \to R^n$ is said to be admissible for (7.2.1) if for each $\varphi \in \Omega(t_0)$ there is a solution $x(t, t_0, \varphi)$ of (7.2.1) with $x(t, t_0, \varphi) = \varphi(t)$ for $\alpha \leq t \leq t_0$, $x(t, t_0, \varphi)$ satisfies (7.2.1) for $t \geq t_0$ and $x(t, t_0, \varphi)$ is continuous on $[\alpha, \infty)$.*

Thus, given $\varphi \in \Omega(t_0)$, Equation (7.2.1) is usually written as

$$x(t) = a(t) - \int_{\alpha(t)}^{t_0} Q(t, s, \varphi(s)) \, ds - \int_{t_0}^{t} Q(t, s, x(s)) \, ds$$

and the first two terms on the right are taken as the inhomogeneous term. In this form there is much existence theory, as may be seen in Chapter 3 or in Corduneanu (1991) and Gripenberg, Londen, Staffans (1990), for example.

Notation. The symbol $\Omega(t_0)$ will always denote an admissible set. If $\varphi \in \Omega(t_0)$ and $\Psi : [\alpha, \infty) \to R^n$, then $\rho(\varphi, \Psi)$ means Ψ is restricted to $[\alpha, t_0]$.

Clearly, φ must be chosen so that

$$\varphi(t_0) = a(t_0) - \int_{\alpha(t_0)}^{t_0} Q(t_0, s, \varphi(s)) \, ds. \tag{7.2.3}$$

However, if for large t we have $\alpha(t) > \alpha(t_0)$ then (7.2.3) can be avoided, as we will see in the next section.

But what is important here is that any bounded continuous φ on $(-\infty, 0]$ can be approximated arbitrarily well by a function satisfying (7.2.3) with $t_0 = 0$.

Proposition 7.2.1. *Let $Q : \Re \times \Re \times \Re^n \to \Re^n$ be continuous and suppose that $\int_{-\infty}^{0} Q(0, s, \varphi(s)) \, ds$ converges for each bounded and continuous $\varphi : (-\infty, 0] \to \Re^n$. Let $\varphi : (-\infty, 0] \to \Re^n$ be an arbitrary bounded and*

7.2. NEAR EQUILIBRIA

continuous function. For each $\varepsilon > 0$ there is a $t_1 < 0$, t_1 near 0, and $\varphi_1 : (-\infty, 0] \to \Re^n$ which is continuous, which satisfies

$$\varphi_1(0) = a(0) - \int_{-\infty}^0 Q(0, s, \varphi_1(s)) \, ds, \tag{7.2.3*}$$

$$\varphi(t) = \varphi_1(t) \text{ for } -\infty < t \leq t_1, \text{ and } |\tilde{\varphi}(0) - \varphi_1(0)| \leq \varepsilon$$

where

$$\tilde{\varphi}(0) = a(0) - \int_{-\infty}^0 Q(0, s, \varphi(s)) \, ds.$$

Proof. For any $x \in \Re^n$ and any $t_1 < 0$ define

$$\varphi^x = \begin{cases} \varphi(s), & \text{if } s \leq t_1, \\ [(t_1 - s)x + s\varphi(t_1)]/t_1, & \text{if } t_1 < s \leq 0. \end{cases}$$

Now, let t_1 be any number such that for any $x \in R^n$ with $|x - \tilde{\varphi}(0)| \leq \varepsilon$ we have

$$\left| \int_{-\infty}^0 Q(0, s, \varphi(s)) \, ds - \int_{-\infty}^0 Q(0, s, \varphi^x(s)) \, ds \right| \leq \varepsilon. \tag{*}$$

By the continuity of Q and the assumed convergence, $(*)$ can be satisfied. Also, t_1 is as near 0 as we please.

Next, let $S = \{x \in \Re^n : |x - \tilde{\varphi}(0)| \leq \varepsilon\}$ and define $P : S \to S$ by $x \in S$ implies that

$$P(x) = a(0) - \int_{-\infty}^0 Q(0, s, \varphi^x(s)) \, ds.$$

Now P is continuous and, by construction, maps S into S. By Brouwer's theorem, there is a fixed point x_1 and φ^{x_1} is the required function.

Definition 7.2.3. An L^1-near equilibrium (ψ, Ψ) for (7.2.1) is said to be stable relative to Ω if there is a wedge W and continuous functions $\gamma(t)$ and $p(t)$, where $\gamma \in L^1[0, \infty)$, while $p(t) \to 0$ and $W(t) \to \infty$ as $t \to \infty$, and for each $\varepsilon > 0$ and $t_0 \in R$ there is a $\delta > 0$ such that $[\varphi \in \Omega(t_0), \rho(\varphi, \psi + \Psi) < \delta]$ imply that

$$W(|x(t, t_0, \varphi) - \psi(t) - \Psi(t)|) < \varepsilon + p(t_0) + \int_{t_0}^t \gamma(s) \, ds$$

where $x(t, t_0, \phi)$ solves (7.2.1). If, in addition, $|x(t, t_0, \varphi) - \psi(t) - \Psi(t)| \to 0$ as $t \to \infty$, then (ψ, Ψ) is asymptotically stable relative to Ω.

To relate this to differential equations, first note in (7.2.1) that if $a(t) \equiv 0$ and $Q(t, s, 0) \equiv 0$, then ψ and Ψ may be both zero so $(0, 0)$ is an L^1-near equilibrium. If we take W as the identity function and $p(t) = \gamma(t) = 0$ then our definition is the usual one for stability of an integrodifferential equation

$$x'(t) = \int_{\alpha(t)}^{t} Q(t, s, x(s)) \, ds, \qquad Q(t, s, 0) \equiv 0,$$

so that the zero function is a solution (equilibrium point). See, for example, Yoshizawa (1996; pp. 27–31, 183–190), Burton (1983b; pp. 12–3, 33–34, 227–237).

Next, if $\varphi(t)$ is a solution of (7.2.1) and we wish to study the behavior of solutions starting near it, we can write $x = y + \varphi$ so that

$$y(t) = -\int_{\alpha(t)}^{t} [Q(t, s, y(s) + \varphi(s)) - Q(t, s, \varphi(s))] \, ds \qquad (7.2.1^*)$$

has the L^1-near equilibrium $(0, 0)$.

Early in the book we discussed the classical idea that for a nice kernel then the solution follows $a(t)$. We considered cases in which it is true and cases in which it is not true at all. It failed mainly when $a(t)$ is large. Here, we will look again at cases in which it is true. For example, consider

$$x(t) = a(t) + \int_{-\infty}^{t} D(t - s) x(s) \, ds \qquad (7.2.1^{**})$$

where $D \in L^1[0, \infty)$. Three facts are derived by elementary considerations which motivated Definitions 7.2.1 and 7.2.3:

(i) Does (7.2.1**) have any constant solutions?
It does if $a(t)$ is constant and $\int_0^\infty D(u) \, du \neq 1$. It does only if $a(t)$ is constant.

(ii) Does (7.2.1**) have a solution in $L^1[0, \infty)$?
It does only if $a(t) \in L^1[0, \infty)$.

(iii) Does (7.2.1**) have any solutions tending to zero?
It does only if $a(t) \to 0$ as $t \to \infty$.

Part (ii) is the most interesting. We frequently show that there is not only a solution in L^1, but it converges pointwise to $a(t)$ as $t \to \infty$.

7.2.1 A Finite Delay Problem

In our discussions we always consider a pair of equations: one is linear, one nonlinear. The linear equation will be the prototype and will lead us to the

7.2. NEAR EQUILIBRIA

results; in effect, it will be an example. But the basic theory is nonlinear and we provide nonlinear examples.

Let h be a positive constant, Q be continuous, and consider the scalar equations

$$x(t) = a(t) - \int_{t-h}^{t} D(t,s)x(s)\,ds \tag{7.2.4}$$

and

$$x(t) = a(t) - \int_{t-h}^{t} Q(t,s,x(s))\,ds \tag{7.2.4$_N$}$$

with

$$a: \Re \to \Re \text{ being continuous, } a \text{ and } a^2 \in L^1[0,\infty), \tag{7.2.5}$$

and suppose there is a $P > 0$ with

$$\begin{aligned} D(t,t) &\leq P, D(t,s) \geq 0, D_s(t,s) \geq 0, \\ D_{st}(t,s) &\leq 0, D(t,t-h) = 0. \end{aligned} \tag{7.2.6}$$

Let $a(t)$ be bounded and consider the convolution equation

$$x(t) = a(t) - \int_{t-h}^{t} C(t-s)x(s)\,ds.$$

If $C(t) < 0$ and large, for a positive initial function we readily expect $x(t)$ to grow; thus, we ask $C(t) > 0$. But this is an equation with memory and, although the memory is lost on each interval of length h, we still expect the memory to immediately begin to fade with time; thus, we ask that $C'(t) \leq 0$. To ask that $C(t) = 0$ for $t \geq h$ makes it a finite memory problem. Hence, there is an uncontrived reason for $D(t,s) \geq 0$, $D_s(t,s) \geq 0$, and $D(t,t-h) = 0$, and investigators traditionally ask $D_{st} \leq 0$ out of technical necessity, but which is fully consistent with the perceived problems. One of our goals is to reduce $D_{st} \leq 0$.

The discussion here is the same for any t_0 so we take $t_0 = 0$ and $\Omega = \Omega(0)$ to be the set of continuous $\varphi: [-h, 0] \to R$ with

$$\varphi(0) = a(0) - \int_{-h}^{0} D(0,s)\varphi(s)\,ds \tag{7.2.7}$$

for the stability statements. But (7.2.7) will not be needed for the study of limit sets.

The metric ρ on Ω will be the L^2-norm, $|||\cdot|||$. Also, if $q : [-h, A) \to R$, $A > 0$, then $q_t(s) = q(t+s)$ for $-h \leq s \leq 0$ and

$$|||q_t|||^2 = \int_{-h}^{0} q^2(t+s)\, ds. \tag{7.2.8}$$

Clearly, the pair $(0, a(t))$ is an L^1-near equilibrium for (7.2.4) and we will show that it is asymptotically stable relative to Ω. In addition, it will motivate a general theorem. It is convenient to give them in reverse order and to prove Theorem 7.2.1B first.

In preparation for that work we remind the reader that a *wedge* is a scalar function $W : [0, \infty) \to [0, \infty)$, $W(0) = 0$, W is continuous and strictly increasing. Usually, we ask that $W(r) \to \infty$ as $r \to \infty$. Wedges are denoted by W_i.

Theorem 7.2.1A. *Suppose that $x, \Gamma : [-h, \infty) \to \Re$ are continuous and that $p, q : [0, \infty) \to [0, \infty)$ are also continuous with $p(t) \to 0$ as $t \to \infty$ and $q \in L^1[0, \infty)$. If there is a differentiable scalar functional $V(t, x(\cdot))$ and wedges W_i for which*

(i) $W_1(|x(t) - \Gamma(t)|) \leq V(t, x(\cdot)) \leq W_2(|||(x - \Gamma)_t|||) + p(t)$, *and*

(ii) $V'(t, x(\cdot)) \leq -W_3(|x(t) - \Gamma(t)|) + q(t)$

then for $t \geq 0$, $\phi \in \Omega$, and x solving $(7.2.4_N)$ we have

$$W_1(|x(t) - \Gamma(t)| \leq W_2(|||\phi - \Gamma|||_0) + p(0) + \int_0^t q(s)\, ds$$

and $|x(t) - \Gamma(t)| \to 0$ as $t \to \infty$.

In the next theorem there is the tacit assumption that there was a (ψ, Ψ) which was an L^1-near equilibrium and the transformation yielding (7.1.5) has already been made. This is the same as saying that $a \in L^1[0, \infty)$.

Theorem 7.2.1B. *Let (7.2.5), (7.2.6), (7.2.7) hold and let $a \in L^1[0, \infty)$ so that $(0, a(t))$ is an L^1-near equilibrium. Then there exist a constant P, continuous functions $p, q : [0, \infty) \to [0, \infty)$, $p(t) \to 0$ as $t \to \infty$, $q \in L^1[0, \infty)$, a continuous functional $V(t, x(\cdot))$ defined for a solution $x(t) = x(t, 0, \varphi)$ of (7.2.4) with $\varphi \in \Omega$, and wedges W_i such that $W_1(r) \to \infty$ as $r \to \infty$,*

(i) $W_1(|x(t) - a(t)|) \leq PV(t, x(\cdot)) \leq W_2(|||(x - a)_t|||) + p(t)$,
    and

7.2. NEAR EQUILIBRIA

(ii) $V'(t, x(\cdot)) \leq -W_3(|x(t) - a(t)|) + q(t)$.

Thus, the L^1-near equilibrium $(0, a(t))$ of (7.2.4) is asymptotically stable relative to Ω.

Notice in these expressions we have $|x(t) - \psi(t) - a(t)|$ where $\psi = 0$.

Proof. To prove Theorem 7.2.1B, let $\varphi \in \Omega$, $x(t) = x(t, 0, \varphi)$, and define

$$V(t) := V(t, x(\cdot)) = \int_{t-h}^{t} D_s(t, s) \left(\int_s^t x(v)\, dv \right)^2 ds. \qquad (7.2.9)$$

Then (ii) holds because

$$V'(t) =$$
$$- D_s(t, t-h) \left(\int_{t-h}^{t} x(v)\, dv \right)^2 ds + \int_{t-h}^{t} D_{st}(t, s) \left(\int_s^t x(v)\, dv \right)^2 ds$$
$$+ 2x(t) \int_{t-h}^{t} D_s(t, s) \int_s^t x(v)\, dv$$
$$\leq 2x(t) \left[D(t, s) \int_s^t x(v)\, dv \bigg|_{t-h}^{t} + \int_{t-h}^{t} D(t, s) x(s)\, ds \right]$$
$$= 2x(t)[a(t) - x(t)]$$
$$= -x^2(t) - (x(t) - a(t))^2 + a^2(t) \leq -(x(t) - a(t))^2 + a^2(t)$$
$$=: -W_3(|x(t) - a(t)|) + q(t).$$

Next, from (7.2.4) we have, upon finding P with $D(t, t) \leq P$,

$$W_1(|x(t) - a(t)|) := (x(t) - a(t))^2 = \left(-\int_{t-h}^{t} D(t, s) x(s)\, ds \right)^2$$
$$= \left\{ D(t, s) \int_s^t x(v)\, dv \bigg|_{t-h}^{t} - \int_{t-h}^{t} D_s(t, s) \int_s^t x(v)\, dv\, ds \right\}^2$$
$$\leq \int_{t-h}^{t} D_s(t, s)\, ds \int_{t-h}^{t} D_s(t, s) \left(\int_s^t x(v)\, dv \right)^2 ds \leq PV(t)$$
$$\leq P \int_{t-h}^{t} D_s(t, s) 2 \left[\left(\int_s^t |x(v) - a(v)|\, dv \right)^2 + \left(\int_s^t |a(v)|\, dv \right)^2 \right] ds$$
$$\leq 2P \int_{t-h}^{t} D_s(t, s)(t - s) \int_s^t |x(v) - a(v)|^2\, dv$$
$$+ 2P \left(\int_{t-h}^{t} D_s(t, s)\, ds \right) \left(\int_{t-h}^{t} |a(v)|\, dv \right)^2$$

$$\leq 2P^2 h \int_{t-h}^{t} |x(v) - a(v)|^2 \, dv + 2P^2 \left(\int_{t-h}^{t} |a(v)| \, dv \right)^2$$

$$=: W_2(|||(x-a)_t|||) + p(t)$$

where $p(t) \to 0$ as $t \to \infty$; hence, (i) holds and Theorem 7.2.1B will be proved when we have proved Theorem 1A.

To that end, in Theorem 7.2.1A we note that an integration of (ii) yields $V(t)$ bounded and, since $W_1(r) \to \infty$ as $r \to \infty$, in (i) we see that $|x(t) - \Gamma(t)|$ is bounded. This means that (ii) can be sharpened to

$$V'(t) \leq -W_4(|x(t) - \Gamma(t)|^2) + q(t) \tag{ii^*}$$

where W_4 is convex downward. (See Natanson (1960; pp. 36–46) for a good discussion of convexity and Jensen's inequality. In particular, if W is a wedge, then for $0 \leq r \leq 1$ we have

$$W^*(r) = \int_0^r W(s) \, ds = W(\xi) r \leq W(r)$$

for some ξ in $[0, r]$ and W^* is convex downward.)

From (i) and (ii) we have

$$W_1(|x(t) - \Psi(t)|) \leq V(t, x(\cdot)) \leq V(0) + \int_0^t q(s) \, ds$$

$$\leq W_2(|||(\varphi - \Gamma)_0|||) + p(0) + \int_0^t q(s) \, ds,$$

(as we have taken t_0 to be zero for convenience) and this is the required stability.

We now show that $|x(t) - \Gamma(t)| \to 0$ as $t \to \infty$. If it does not, then there is an $\varepsilon > 0$ and $\{t_n\} \uparrow \infty$ with $h < t_n$, $t_{n+1} > t_n + h$, and $|x(t_n) - \Gamma(t_n)| \geq \varepsilon$. Using (i) and the fact that $p(t) \to 0$, we can say that there is a $\delta > 0$ with $|||(x-\Gamma)_{t_n}|||^2 \geq \delta$ for large n, say $n \geq 1$. Using (ii^*) and Jensen's inequality, we take N large, integrate (ii^*) from t_1 to t_N and obtain

$$V(t_N) - V(t_1) \leq -\int_{t_1}^{t_N} W_4(|x(s) - \Psi(s)|^2) \, ds + \int_{t_1}^{t_N} q(s) \, ds$$

$$\leq -\sum_{i=2}^{N} \int_{t_i - h}^{t_i} W_4(|x(s) - \Psi(s)|^2) \, ds + \int_{t_1}^{t_N} q(s) \, ds$$

$$\leq -\sum_{i=2}^{N} h W_4 \left(\frac{1}{h} \int_{t_i - h}^{t_i} |x(s) - \Psi(s)|^2 \, ds \right) + \int_{t_1}^{t_N} q(s) \, ds$$

$$\leq -\sum_{i=2}^{N} h W_4(\delta/h) + \int_{t_1}^{t_N} q(s) \, ds,$$

7.2. NEAR EQUILIBRIA

a contradiction for large N since $V(t) \geq 0$ and $q \in L^1[0, \infty)$. This proves Theorem 7.2.1A, so 7.2.1B is also true.

The only place (7.2.7) was used was in the integration by parts when differentiating V. For any continuous φ there is a solution $x(t, 0, \varphi)$ for $t > 0$ of (7.2.4) which may have a discontinuity at $t = 0$ but V is differentiable for $t > h$. There is the question of stability, but it can be resolved using continuous dependence of solutions on initial conditions in conjunction with the following result.

Corollary 7.2.1. *If (7.2.5) and (7.2.6) hold then there exist continuous functions $p, q : [0, \infty) \to [0, \infty)$, $p(t) \to 0$ as $t \to \infty$, $q \in L^1[0, \infty)$, and wedges W_i such that if $\varphi : [-h, 0] \to R$ is continuous and $x(t) = x(t, 0, \varphi)$ solves (7.2.4), then there is a continuous functional $V(t, x(\cdot))$ satisfying (i) of Theorem 7.2.1B for $t > 0$ and (ii) for $t > h$. In particular, $|x(t) - a(t)| \to 0$ as $t \to \infty$.*

In Burton and Furumochi (1994), Corollary 2, p. 455, it is shown how to reduce $D_{st} \leq 0$.

Remark 7.2.1. **A Reversal of Roles** The terminology we have used about the solution staying near Ψ can be disturbing. We have used it because all of that work centered on $a \in L^1[0, \infty)$ so that $(0, \Psi)$ is a near identity and the solution stays near $\psi = a$. All of that changes when we step out and actually find a nontrivial function ψ which may even be unbounded and then show that $x(t)$ stays close to $\psi + \Psi$. In effect, $x(t)$ will stay Ψ close to ψ and that is exactly what we intuitively draw from the definition.

Corollary 7.2.2. *Suppose that (ψ, Ψ) is an L^1-near equilibrium for (7.2.4) and that (7.2.5–7.2.7) hold. Then*

$$W_1(|x(t) - \psi(t) - \Psi(t))|) \leq PV(t, x(\cdot)) \leq W_2(|||x - \psi - \Psi|||_t) + p(t),$$

$$V'(t, x(\cdot) \leq -W_3(|x(t) - \psi(t) - \Psi(t))|) + q(t),$$

and $|x(t) - \psi(t) - \Psi(t)| \to 0$ as $t \to \infty$.

Proof. We have

$$\Psi(t) = a(t) - \psi(t) - \int_{t-h}^{t} C(t, s)\psi(s)ds \in L^1[0, \infty)$$

and

$$x(t) = a(t) - \int_{t-h}^{t} C(t, s)x(s)ds.$$

Thus,
$$x(t) - \psi(t) = \Psi(t) - \int_{t-h}^{t} C(t,s)[x(s) - \psi(s)]ds.$$

Let
$$y(t) = \Psi(t) - \int_{t-h}^{t} C(t,s)y(s)ds$$

so the inequalities in Theorem 7.2.1B hold with $\Psi(t)$ replacing $a(t)$. Hence, $|y(t) - \Psi(t)| \to 0$ and that is what was to be proved.

Theorem 7.2.1A emphasizes that linearity is not essential; it merely serves as a convenient example with fewer hypotheses. We now give examples of superlinear and sublinear cases. The wedges in the theorems still arise in a natural way.

Consider the equation
$$x(t) = a(t) - \int_{t-h}^{t} D(t,s)g(s,x(s))\,ds \tag{7.2.4*}$$

with (7.2.5), (7.2.6) holding and with g bounded for x bounded. Let the initial function $\phi \in \Omega$ satisfy
$$\varphi(0) = a(0) - \int_{-h}^{0} D(0,s)g(s,\varphi(s))\,ds. \tag{7.2.7*}$$

Then for $x(t,0,\phi)$ a solution of (7.2.4*) and
$$V(t,x(\cdot)) = \int_{t-h}^{t} D_s(t,s)\left(\int_s^t g(v,x(v))\,dv\right)^2 ds$$

we have
$$(x(t) - a(t))^2 \leq D(t,t)V(t,x(\cdot))$$

and
$$V'(t,x(\cdot)) \leq -2g(t,x)[x - a(t)].$$

Example 7.2.2. It is easily verified that Theorem 7.2.1A is still true if in (i) we have $|x - \Gamma|^k$ and in (ii) we have $|x(t) - \Gamma(t)|^k$ for $k > 0$. If (7.2.5), (7.2.6), and (7.2.7*) hold and if both a^3 and $a^4 \in L^1[0,\infty)$, then conditions (i) and (ii) of Theorem 7.2.1A hold when $g(t,s) = x^3$ in (4*) and $|x(t) - a(t)| \to 0$ as $t \to \infty$.

Proof. We have just defined V and we have

$$V'(t) \leq -2x^4 + 2a(t)x^3$$
$$= -x^4 - (x - a(t))^4 - 2a(t)x^3 + 6x^2a^2(t) - 4xa^3(t) + a^4(t).$$

Use Hölder's inequality to parlay that into

$$V'(t) \leq -(x - a(t))^4 + Ma^4(t) - Nx^4$$

for some positive constants M and N. This will take care of (ii) in Theorem 7.2.1A. Moreover, it yields $\int_0^\infty x^4(t)dt < \infty$. From that we can argue that $\int_{t-h}^t |x(s)|^3 ds \to 0$ as $t \to \infty$.

We now work toward (ii). If we take

$$r(s) = 3(x(s) - a(s))^2|a(s)| + 3|x(s) - a(s)|a^2(s) + |a(s)|^3,$$

we may note that $|x|^3 = |x - a + a|^3 \leq |x - a|^3 + r$. Working with the terms of r, we have

$$\int_{t-h}^t x^2(s)|a(s)|\, ds \leq \left(\int_{t-h}^t |x(s)|^3\, ds\right)^{2/3} \left(\int_{t-h}^t |a(s)|^3\, ds\right)^{1/3}$$
$$\leq h^{1/6}\left(\int_{t-h}^t |x(s)|^4\, ds\right)^{1/2} \left(\int_{t-h}^t |a(s)|^3\, ds\right)^{1/3}$$

$$\to 0 \text{ as } t \to \infty$$

and

$$\int_{t-h}^t |x(s)|a^2(s)\, ds \leq \left(\int_{t-h}^t |x(s)|^3\, ds\right)^{1/3} \left(\int_{t-h}^t |a(s)|^3\, ds\right)^{2/3}$$

$$\to 0 \text{ as } t \to \infty.$$

Clearly,

$$\int_{t-h}^t |a(s)|^3\, ds \to 0 \text{ as } t \to \infty.$$

7. STABILITY: CONVEX MEMORY

Thus, we have, for r defined above,

$$V(t, x(\cdot)) = \int_{t-h}^{t} D_s(t,s) \left(\int_s^t x^3(v)\, dv \right)^2 ds$$

$$\leq \int_{t-h}^{t} D_s(t,s) 2 \left\{ \left(\int_s^t |x(v) - a(v)|^3\, dv \right)^2 \right.$$

$$\left. + \left(\int_s^t r(v)\, dv \right)^2 \right\} ds$$

$$\leq \int_{t-h}^{t} D_s(t,s) 2 \left\{ h^{1/4} \left(\int_{t-h}^{t} |x(v) - a(v)|^4\, dv \right)^{3/4} \right.$$

$$\left. + \left(\int_{t-h}^{t} r(v)\, dv \right)^2 \right\} ds$$

$$\leq 2 D(t,t) h^{1/4} \left(\int_{t-h}^{t} (x(v) - a(v))^4\, dv \right)^{3/4} + p(t)$$

where we have verified that $p(t) \to 0$ as $t \to \infty$, so that (i) of Theorem 7.2.1A is satisfied and the conclusion follows.

Example 7.2.3. If (7.2.5), (7.2.6), and (7.2.7*) hold for (7.2.4*) and if $g(t, x) = x^{1/3}$, while $a(t)$ is bounded, then the conditions of Theorem 7.2.1A hold and $|x(t) - a(t)| \to 0$ as $t \to \infty$.

Proof. We have $V(t) := V(t, x(\cdot))$ and

$$V'(t) \leq -2x^{4/3} + 2x^{1/3} a(t)$$

so that

$$0 \leq V(t) \leq V(0) - 2 \int_0^t x^{4/3}(s)\, ds$$

$$+ 2 \left(\int_0^t x^{4/3}(s)\, ds \right)^{1/4} \left(\int_0^t a^{4/3}(s)\, ds \right)^{3/4}$$

and so the terms in V' are $L^1[0, \infty)$. Moreover, familiar arguments yield (i). Hence, V is bounded so $(x(t) - a(t))^2$ is bounded; but $a(t)$ bounded yields $x(t)$ bounded. Thus, there exists $M > 0$ with

$$\int_0^t x^2(s)\, ds = \int_0^t x^{2/3}(s) x^{4/3}(s)\, ds \leq M \int_0^t x^{4/3}(s)\, ds.$$

Hence, $(x - a)^2 = x^2 - 2ax + a^2$ is in $L^1[0, \infty)$ and we can write

$$V'(t) \leq -(x(t) - a(t))^2 + q(t)$$

so that (ii) of Theorem 7.2.1A holds and the proof is complete.

7.2. NEAR EQUILIBRIA

Exercise 7.2.1. The ideas in Corollary 7.2.3 also work for (7.2.4*) if
$$g(t,x) - g(t,y) = \frac{\partial g}{\partial x}(t,\xi)(x-y),$$
where we are using the mean value theorem for derivatives, and where that partial derivative is bounded and continuous. In particular, assume that $0 < \gamma \leq \frac{\partial g}{\partial x}(t,s) \leq \Gamma$. Suppose that (ψ, Ψ) is a near equilibrium for (7.2.4*). We then have
$$\psi(t) = a(t) - \Psi(t) - \int_{t-h}^{t} D(t,s)g(s,\psi(s))ds.$$
Thus, there is a fixed continuous function $G(s)$ with
$$g(s,x(s)) - g(s,\psi(s)) = \frac{\partial g(s,\xi(s))}{\partial x}(x(s) - \psi(s))$$
and $G(s) = \frac{\partial g(s,\xi(s))}{\partial x}$ lies in the interval $[\gamma, \Gamma]$. Then for
$$y(t) = x(t) - \psi(t)$$
we have
$$y(t) = \Psi(t) - \int_{t-h}^{t} D(t,s)G(s)y(s)ds.$$
Use
$$V(t,y(\cdot)) = \int_{t-h}^{t} D_s(t,s)\left(\int_s^t G(v)y(v)dv\right)^2 ds$$
and obtain the conditions of Theorem 7.2.1B. State the appropriate conclusion.

Theorem 7.2.1A is predicated on finding a near equilibrium; once that is found, the limit set of all solutions is specified by Corollary 7.2.1. To find a near equilibrium is to find a function which fails to solve (7.2.4) only by an amount of an L^1-function. If we can find a function which fails to solve (7.2.4) only by an amount of a bounded function, then we can locate a bounded set which contains the limit set of all solutions of (7.2.4). When the conditions of this theorem hold, then we are assured that all stable near equilibria are a bounded distance from that function.

Theorem 7.2.2A. *Let $x(t)$ solve ($7.2.4_N$) with $x(t) = x(t,0,\varphi)$ and $\varphi : [-h,0] \to \Re$ be continuous. Suppose there is a continuous function $\Psi : [-h,\infty) \to \Re$, positive constants Q and L, wedges W_i with $W_1(r) \to \infty$ as $r \to \infty$, and a continuous functional $V(t,x(\cdot))$ so that for $t > h$*

(i) $W_1(|x(t) - \Psi(t)|) \leq V(t, x(\cdot)) \leq W_2(|||(x - \Psi)_t|||) + Q$

and

(ii) $V'(t, x(\cdot)) \leq -W_3(|x(t) - \Psi(t)|^2) + L$

with W_3 convex downward. Then there is a number B independent of φ with $|x(t)| \leq B$ for large t.

Proof. Consider the intervals $I_n = [(n-1)h, nh]$ for $n = 2, 3, \ldots$. Let $V(t) := V(t, x(\cdot))$. Either
  (a) $V(nh) \geq V((n-1)h) - 1$ so that from (ii)

$$-1 \leq V(nh) - V((n-1)h) \leq -hW_3\left(\frac{1}{h}|||(x - \Psi)_{nh}|||^2\right) + Lh$$

or

$$|||(x - \Psi)_{nh}|||^2 \leq hW_3^{-1}\left(L + \frac{1}{h}\right)$$

so from (i)

$$V(nh) \leq W_2((hW_3^{-1}(L + \frac{1}{h}))^{1/2}) + Q =: C \qquad (*)$$

or
  (b) $V(nh) \leq V((n-1)h) - 1$.

Since (b) can not hold for all n, there is a k with (*) holding for $n = k$:

$$V(kh) \leq C.$$

From (ii) we have

$$V(t) \leq C + Lh \text{ if } kh \leq t \leq (k+1)h.$$

But by the arguments in (a) and (b), either

$$V((k+1)h) \leq V(kh) - 1 < C \text{ by (b)}$$

or

$$V((k+1)h) \leq C \text{ by } (*).$$

Hence,

$$W_1(|x(t) - \Psi(t)|) \leq V(t) \leq C + Lh$$

for all large t and we take

$$B = W_1^{-1}(C + Lh).$$

This completes the proof.

7.2. NEAR EQUILIBRIA

Suppose there is an $A > 0$ with

$$a : \Re \to \Re \text{ is continuous and } |a(t)| \leq A \text{ for } t \geq 0. \qquad (7.2.5^*)$$

When $(7.2.5^*)$ holds then $(0, a(t))$ is an L^∞-near equilibrium for (7.2.4). In effect, there was an L^∞-near equilibrium, say $(\psi, a(t))$ and the transformation yielding (7.1.5) has already been made.

Theorem 7.2.2B. *Let $(7.2.5^*)$ and (7.2.6) hold. Then there are constants Q and L, wedges W_i with $W_1(r) \to \infty$ as $r \to \infty$ and a continuous functional V with the following properties. If $\varphi : [-h, 0] \to \Re$ is continuous and if $x(t) = x(t, 0, \varphi)$ satisfies (7.2.4) then*

(i) $W_1(|x(t) - a(t)|) \leq PV(t, x(\cdot)) \leq W_2(|||(x - a)_t|||) + Q$

and for $t > h$

(ii) $V'(t, x(\cdot)) \leq -W_3(|x(t) - a(t)|^2) + L$

where W_3 is convex downward. Thus, there is a $B > 0$, independent of φ, with $|x(t)| \leq B$ for large t.

Proof. The proof of (i) proceeds by familiar arguments. We have

$$Q = 2P^2h^2A^2 \text{ and in (ii) } L = A^2.$$

Remark 7.2.2. When we study the proof of Theorem 7.2.2A, part (b), we see that for each $B_1 > 0$ there is a $B_2 > 0$ such that $|||(\varphi - \Psi)_0||| < B_1$ and $t \geq 0$ imply $|x(t, 0, \varphi)| < B_2$. Also, for each $B_3 > 0$ there is a $T > 0$ such that $|||(\varphi - \Psi)_0||| < B_3$ and $t \geq T$ imply $|x(t)| \leq B$. This may be called uniform boundedness and uniform ultimate boundedness.

The material for this section was taken from Burton and Furumochi (1994) which was published by the Rocky Mountain Consortium. The study is continued there with consideration of

$$x(t) = a(t) - \int_{-\infty}^{t} D(t, s)x(s)\, ds \qquad (7.2.10)$$

and

$$x(t) = a(t) - \int_{-\infty}^{t} Q(t, s, x(s))\, ds \qquad (7.2.10_N)$$

as well as for

$$x(t) = a(t) - \int_0^t D(t,s)x(s)ds \qquad (7.2.11)$$

and

$$x(t) = a(t) - \int_0^t Q(t,s,x(s))ds. \qquad (7.2.11_N)$$

Finally, the study is continued for some partial integral equations in Burton, Furumochi, Huang (1995) illustrating stability in several measures.

Chapter 8

Appendix: Preparing the Kernel

8.1 Introduction

In order to present a unified treatment of a large class of problems we have consistently started with

$$x(t) = a(t) - \int_0^t C(t,s)x(s)ds, \qquad (8.1.1)$$

asking that C have a set of derivatives of certain signs, as in Theorem 2.1.10 or Theorem 2.2.7, or that $\int_0^\infty |C(u+t,t)|du \leq \alpha < 1$ as in Theorem 2.1.7, or that $\int_0^\infty |C_t(u+t,t)|du < C(t,t)$ as in Theorem 2.2.5. Parallel conditions were asked concerning the resolvent equation

$$R(t,s) = C(t,s) - \int_s^t C(t,u)R(u,s)du \qquad (8.1.2)$$

in Section 2.6.

It can happen that all of these conditions fail and yet investigators are able to make certain changes which bring the problem in line with the standard theory. In the next section we will present an example illustrating one such modification. The methods here concern fully nonconvolution problems, but the work becomes so much more transparent in the convolution case that good communication would seem preferable to generality. We will offer a technique applied to the kernel $C(t,s) = (1+t-s)^{-2}$ which, together with specified $a(t)$, offer difficulties for all the Liapunov functionals previously discussed. We will show how to bring it into line with earlier work.

8.2 Adding x and x'

Consider the scalar equation

$$x(t) = a(t) - \int_0^t (1+t-s)^{-2} x(s)\,ds \qquad (8.2.1)$$

where we first suppose that

$$a \in L^1[0,\infty) \text{ and } a' \in L^1[0,\infty). \qquad (8.2.2)$$

Observe that

$$\int_0^\infty (1+u)^{-2}\,du = 1$$

and that

$$\int_0^\infty 2(1+u)^{-3}\,du = 1 = C(t,t).$$

A study of these relations suggests that our Liapunov theorems will not yield $x \in L^1$ and neither will our Razumikhin techniques. We show how to rectify the situation.

From (8.2.1) we have

$$x'(t) = a'(t) - x(t) - \int_0^t -2(1+t-s)^{-3} x(s)\,ds$$

and for $k > 0$ we have

$$kx(t) = ka(t) - \int_0^t k(1+t-s)^{-2} x(s)\,ds$$

so that

$$x'(t) + kx(t) = a'(t) + ka(t) - x(t)$$
$$- \int_0^t \{k(1+t-s)^{-2} - 2(1+t-s)^{-3}\} x(s)\,ds$$

8.2. ADDING X AND X'

or

$$x'(t) = a'(t) + ka(t) - (1+k)x(t)$$
$$- \int_0^t \{k(1+t-s)^{-2} - 2(1+t-s)^{-3}\}x(s)ds. \qquad (8.2.3)$$

While there are other ways to proceed, it is easiest here to observe that if we select $k = 2$ then the kernel does not change sign; the second term in the new kernel continually subtracts from the first term, decreasing the integral of the kernel and, at the same time, the ode part of $-x$ is increased to $-3x$ so that the fact that we multiplied by 2 is balanced in the ode term. In view of our Liapunov theory, we win in both crucial places.

Thus, for $k = 2$ we define a Liapunov functional for (8.2.3) as

$$V(t) = |x(t)| + \int_0^t \int_{t-s}^\infty |2C(u+s,s) + C_t(u+s,s)|du|x(s)|ds$$
$$= |x(t)| + \int_0^t \int_{t-s}^\infty \{2(1+u)^{-2} - 2(1+u)^{-3}\}du|x(s)|ds$$

and that integrand is non-negative. We have

$$V'(t) \leq |2a(t) + a'(t)| - 3|x(t)|$$
$$+ \int_0^t \{2(1+t-s)^{-2} - 2(1+t-s)^{-3}\}|x(s)|ds$$
$$+ \int_0^\infty \{2(1+u)^{-2} - 2(1+u)^{-3}\}du|x(t)|$$
$$- \int_0^t \{2(1+t-s)^{-2} - 2(1+t-s)^{-3}\}|x(s)|ds$$
$$= |2a(t) + a'(t)| - 3|x(t)| + |x(t)|.$$

Theorem 8.2.1. *If (8.2.2) holds then the solution of (8.2.1) and any solution of (8.2.3) for $k = 2$ satisfies x is bounded, $x \in L^1[0,\infty)$, and $x' \in L^1[0,\infty)$.*

Proof. A solution of (8.2.1) also solves (8.2.3). From V and V' we have

$$|x(t)| \leq V(t) \leq V(0) + \int_0^t |2a(s) + a'(s)|ds - 2\int_0^t |x(s)|ds.$$

This yields the first two conclusions. As $x \in L^1$, it is readily verified from (8.2.3) that $x' \in L^1$.

Theorem 8.2.2. *If*

$$2a(t) + a'(t) \text{ is bounded} \tag{8.2.4}$$

then the solution of (8.2.1) and any solution of (8.2.3) on $[0,\infty)$ for $k = 2$ is bounded.

Proof. We use the Razumikhin function $V(t) = |x|$ and notice that if $x(t)$ is any fixed unbounded solution of (8.2.3) then there is a sequence $\{t_n\} \uparrow \infty$ with $|x(t)| \leq |x(t_n)|$ for $0 \leq t \leq t_n$ and $|x(t_n)| \uparrow \infty$. Let $|2a(t) + a'(t)| \leq M$ and note that for $0 \leq t \leq t_n$ we have

$$V'(t) \leq |2a(t) + a'(t)| - 3|x(t)| + |x(t_n)|$$

and at $t = t_n$ we must have $V'(t_n) \geq 0$. But $V'(t_n) \leq M - 3|x(t_n)| + |x(t_n)| < 0$ if $M < 2|x(t_n)|$. This is a contradiction and completes the proof.

Recall that we were often interested in moving from an $a(t) = \ln(t+1)$ to $a'(t) \in L^2[0,\infty)$. We can avoid the properties of $a(t)$ itself by performing these same kinds of operations on the resolvent equation which we treated in Section 2.6 by means of Liapunov functionals, thereby obtaining qualitative properties of R which, in turn, can then be applied directly to the original equation by means of the variation of parameters formula. We can also avoid $a(t)$ itself by adding the pair of equations from $x''(t)$ and $kx'(t)$.

For further results both with $x' + kx$ and $x'' + kx'$ see Burton and Haddock (2009). The technique is also effective when the kernel is convex, as may be seen in Burton (2009), especially in the nonlinear case. A fourth application involves the existence of periodic solutions.

Notice! This process is producing a uniformly asymptotically ode with an integral perturbation.

Here is a very enlightening exercise. Take $d > 0$ and

$$C(t,s) = de^{-(t-s)} \cos(t-s).$$

Form $x' + kx$ with $k = 2$ and divide $[0,\infty)$ into $[0,\pi]$ and $[\pi,\infty)$. Approximate the new kernel on $[\pi,\infty)$ by the exponential part. Determine how large d can be chosen so that our Liapunov functional given above will satisfy

$$V'(t) \leq |a'(t) + 2a(t)| - \beta|x(t)|$$

for some $\beta > 0$. This will show how understated our theorems are unless the kernel is prepared by careful considerations.

8.3 Uniform Boundedness

In the first ten pages of Chapter 1 we outlined the following material. For the equation

$$x(t) = a(t) - \int_0^t C(t,s)x(s)ds \qquad (8.3.1)$$

there is the resolvent equation with solution $R(t,s)$ and the variation of parameters formula

$$x(t) = a(t) - \int_0^t R(t,s)a(s)ds. \qquad (8.3.2)$$

Two of the major goals were to give conditions under which

$$\sup_{t \geq 0} \int_0^t |R(t,s)|ds < \infty \qquad (8.3.3)$$

and/or

$$\sup_{0 \leq s \leq t < \infty} \int_s^t |R(u,s)|du < \infty \qquad (8.3.4)$$

which are useful in so many contexts discussed in this book.

Moreover, when we differentiate (8.3.1) we obtain

$$x'(t) = a'(t) - C(t,t)x(t) - \int_0^t C_t(t,s)x(s)ds \qquad (8.3.5)$$

and we construct another resolvent equation with solution $Z(t,s)$ and obtain the variation of parameters formula for (8.3.5) given by

$$x(t) = Z(t,0)x(0) + \int_0^t Z(t,s)a'(s)ds. \qquad (8.3.6)$$

We were then able to write (1.2.14), stated here as

$$x(t) = a(t) - \int_0^t Z_s(t,s)a(s)ds, \qquad (8.3.7)$$

and sought the counterpart of (8.3.3) in the form

$$\sup_{t \geq 0} \int_0^t |Z_s(t,s)|ds < \infty. \qquad (8.3.8)$$

which was exploited in Theorems 2.1.3, 2.2.10, and 2.6.2.1.

Relation (8.3.8) was achieved mainly by using a Razumikhin technique by integrating the second coordinate of C, of C_t, or of C_s, followed by an argument with Perron's theorem. But our Liapunov functionals tend to integrate the first coordinate of C or its derivatives and yield (8.3.4).

If we could replace $a'(t)$ in (8.3.5) by an arbitrary bounded continuous function $b(t)$ and prove that every solution of the resulting equation is bounded for every bounded continuous function $b(t)$, then it would follow from (8.3.6) and Perron's theorem that

$$\sup_{t \geq 0} \int_0^t |Z(t,s)| ds < \infty. \tag{8.3.9}$$

This could then be parlayed into (8.3.8) as we did in Theorem 2.2.10(ii). We now present a theorem which allows us to use most of our Liapunov functionals to do exactly that.

Definition 8.3.1. *Solutions of an equation*

$$x'(t) = b(t) + A(t)x(t) + \int_0^t C(t,s)x(s)ds \tag{8.3.10}$$

are said to be uniformly bounded at $t = 0$ if for each $B_1 > 0$ there is a $B_2 > 0$ such that $[\|x_0\| < B_1, t \geq 0]$ imply that $|x(t, 0, x_0)| < B_2$.

In the next result the W_i are wedges and it is assumed that every $W_i(r) \to \infty$ as $r \to \infty$. Here is part of Theorem 4.4.4 from p. 321 of Burton (2005b), modified for (8.3.10). The reader may consult the reference for a proof and many related results.

Theorem 8.3.1. *Let $V(t, x(\cdot))$ be a scalar Liapunov functional. Suppose there is a continuous function $\Phi : [0, \infty) \to [0, \infty)$ which is in $L^1[0, \infty)$ and $\Phi'(t) \leq 0$ with*

(a) $W_1(|x(t)|) \leq V(t, x(\cdot)) \leq W_2(|x(t)|) + W_3(\int_0^t \Phi(t-s) W_4(|x(s)|) ds$

and

(b) $V'_{(8.3.10)}(t, x(\cdot)) \leq -W_4(|x(t)|) + M$

for some $M > 0$.

Then solutions of (8.3.10) are uniformly bounded at $t = 0$.

It is critical to understand that W_4 is the same in both places.

8.4. ANOTHER UNUSUAL EQUATION

For a typical application refer back to Theorem 2.2.5, as well as the suggestions for preparation of the kernel in Section 8.2, with

$$|x(t)| \leq V(t, x(\cdot)) = |x(t)| + \int_0^t \int_{t-s}^{\infty} |C_1(u+s,s)| du |x(s)| ds$$

and

$$V'(t, x(\cdot)) \leq -\alpha |x(t)| + |a'(t)|.$$

To obtain properties of $Z(t,s)$ let $M > 0$ be arbitrary and take $a'(t)$ to be any bounded continuous function with $|a'(t)| \leq M$. The only additional assumption we need is that there is a continuous function Φ with $\Phi \in L^1$, $\Phi' \leq 0$, and $\Phi(t-s) \geq \int_{t-s}^{\infty} |C_1(u+s,s)| du$.

Thus, if $B_1 > 0$ is given then there is a $B_2 > 0$ such that if $|x(0)| < B_1$ then $|x(t, 0, x(0))| < B_2$ for $t \geq 0$. Each solution of (8.3.5) is bounded for each bounded and continuous $a'(t)$; by Perron's theorem $\int_0^t |Z(t,s)| ds$ is bounded and this can be parlayed into (8.3.8) under the kind of conditions used in the proof of Theorem 2.2.10.

A major goal is to advance Theorem 8.3.1 to cover Liapunov functionals of the Levin type, as seen in the proof of Theorem 2.2.7.

8.4 Another Unusual Equation

In Chapter 2 we cleansed the kernel through differentiation with respect to t and also with respect to s. That led us to a number of results based on a variety of Liapunov functionals, Razumikhin functions, and contraction mappings. The work here does not seem to offer as many avenues of investigation, but it yields some very different kinds of results.

If the kernel has an additive constant, then we can remove that constant without differentiation. At the same time, we can work with $a'(t)$ instead of $a(t)$ and show that $a' \in L^2$ implies $x \in L^2$. Interestingly, we can then show that $a \in L^2$ implies that $x \in L^2$ under almost the same conditions. This work follows some seen in Burton (2005b), Section 1.6, and can actually be done for vector systems. Consider the scalar equation

$$x(t) = a(t) - \int_0^t [k + C(t,s)] x(s) ds \tag{8.4.1}$$

with a' and C continuous and k a positive constant. Then write (8.4.1) as

$$x'(t) = a'(t) - kx(t) - \frac{d}{dt} \int_0^t C(t,s) x(s) ds. \tag{8.4.2}$$

8. APPENDIX: PREPARING THE KERNEL

Theorem 8.4.1. *Suppose there are positive constants α and ϵ with*

$$\epsilon - 2k + (\epsilon + k) \int_t^\infty |C(u,t)| du + k \int_0^t |C(t,s)| ds \leq -\alpha. \qquad (8.4.3)$$

If $a' \in L^2[0,\infty)$, so is x.

Proof. Let $J = \epsilon + k$ and define

$$V(t) = \left(x(t) + \int_0^t C(t,s)x(s) ds \right)^2 + J \int_0^t \int_t^\infty |C(u,s)| du\, x^2(s) ds.$$

We can find $M > 0$, depending on ϵ, so that the derivative of V along a solution of (8.4.2) satisfies

$$V'(t) = 2\left(x(t) + \int_0^t C(t,s)x(s) ds \right)(a'(t) - kx(t))$$

$$+ J \int_t^\infty |C(u,t)| du\, x^2(t) - J \int_0^t |C(t,s)| x^2(s) ds$$

$$\leq M(a'(t))^2 + \epsilon x^2(t) + \int_0^t |C(t,s)|(\epsilon x^2(s) + M(a'(t))^2) ds$$

$$- 2kx^2(t) + k \int_0^t |C(t,s)|(x^2(s) + x^2(t)) ds$$

$$+ J \int_t^\infty |C(u,t)| du\, x^2(t) - J \int_0^t |C(t,s)| x^2(s) ds$$

$$\leq M\left(1 + \int_0^t |C(t,s)| ds \right)(a'(t))^2 - \alpha x^2(t).$$

An integration completes the proof.

Theorem 8.4.2. *Suppose there are $\epsilon > 0$ and $\alpha > 0$ with*

$$\epsilon - 2k + k \int_0^t |C(t,s)| ds + k \int_t^\infty |C(u,t)| du \leq -\alpha. \qquad (8.4.4)$$

If $a \in L^2[0,\infty)$, so is x.

Proof. Write (8.4.1) as
$$x'(t) = -kx(t) + \frac{d}{dt}\left(a(t) - \int_0^t C(t,s)x(s)ds\right) \qquad (8.4.5)$$

and define
$$V(t) = \left(x(t) - a(t) + \int_0^t C(t,s)x(s)ds\right)^2 + k\int_0^t \int_t^\infty |C(u,s)|du\, x^2(s)ds.$$

Then for the $\epsilon > 0$ we can find $M > 0$ so that
$$V'(t) = 2\left(x(t) - a(t) + \int_0^t C(t,s)x(s)ds\right)(-kx(t))$$
$$+ k\int_t^\infty |C(u,t)|du\, x^2(t) - k\int_0^t |C(t,s)|x^2(s)ds$$
$$\leq -2kx^2(t) + Ma^2(t) + \epsilon x^2(t) + k\int_0^t |C(t,s)|(x^2(s) + x^2(t))ds$$
$$+ k\int_t^\infty |C(u,t)|du\, x^2(t) - k\int_0^t |C(t,s)|x^2(s)ds$$
$$= Ma^2(t) + x^2(t)\left(\epsilon - 2k + k\int_0^t |C(t,s)|ds + k\int_t^\infty |C(u,t)|du\right)$$
$$\leq Ma^2(t) - \alpha x^2(t),$$

from which the result follows.

We leave it to the reader to treat
$$x'(t) = a'(t) + \frac{d}{dt}\left(b(t) - \int_0^t C(t,s)x(s)ds\right)$$
where $a' \subset L^2$ and $b \in L^2$.

Concluding Principle If $a^{(n)}$ and $C^{(n)}$ are continuous and if $C(t)$ satisfies a linear n-th order homogeneous ordinary differential equation with constant coefficients, then $x(t) = a(t) - \int_0^t C(t-s)x(s)ds$ can be reduced to an n-th order linear ordinary differential equation with constant coefficients.

References

Avramescu, C. (2003). Some remarks on a fixed point theorem of Krasnoselskii. *E. J. Qualitative Theory of Diff. Equ.* **No. 5**, 1-15. (http://www.math.u-szeged.hu/ejqtde/2003/200305.html)

Avramescu, C. and Vladimirescu, C. (2003). Some remarks on Krasnoselskii's fixed point theorem. *Fixed Point Theory* **4**, 3-13.

Avramescu, C. and Vladimirescu, C. (2004). Fixed point theorems of Krasnoselskii type in a space of continuous functions. *Fixed Point Theory* **5**, 181-195.

Avramescu, C. and Vladimirescu, C. (2005). Asymptotic stability results for certain integral equations. *E. J. Differential Equations* **126**, 1-10.(http://ejde.math.txstate.edu)

Banach, S. (1932). "Théorie des Opérations Linéairs" (reprint of the 1932 ed.). Chelsea, New York.

Barbashin, E. A. (1968). The construction of Lyapunov functions. *Differential Equations* **4**, 1097–1112.

Barroso, C. S. (2003). Krasnoselskii's fixed point theorem for weakly continuous maps. *Nonlinear Anal.* **55**, 25-31.

Barroso, C. S. and Teixeira, E. V. (2005). A topological and geometric approach to fixed point results for sum of operators and applications. *Nonlinear Anal.* **60**, 625-650.

Becker, L. C. (1979). Stability considerations for Volterra integro-differential equations. Ph.D. dissertation. Southern Illinois University, Carbondale, Illinois.

Becker, Leigh C. (2006). Principal matrix solutions and variation of parameters for a Volterra integro-differential equation and its adjoint. *E. J. Qualitative Theory of Diff. Equ.* **14**, 22 pp.

Becker, Leigh C. (2007). Function bounds for solutions of Volterra equations and exponential asymptotic stability. *Nonlinear Anal.* **67**, 382-397.

Becker, L. C., Burton, T. A., and Krisztin, T. (1988). Floquet theory for a Volterra equation. *J. London Math. Soc.* **37**, 141–147.

Becker, L. C., Burton, T. A., and Zhang, S. (1989). Functional differential equations and Jensen's inequality. *J. Math. Anal. Appl.* **138**, 137–156.

Belmekki, M., Benchohra, M., and Ntouyas, S. K. (2006). Existence results for semilinear perturbed functional differential equations with nondensely defined operators. *Fixed Point Theory Appl.* **Art. ID 43696**, 13pp.

REFERENCES

Belmekki, M., Benchohra, M., Ezzinbi, K., and Ntouyas, S. K. (2008) Existence results for some partial functional differential equations with infinite delay. *Nonlinear Studies* **15**, 373–385.

Burton, T. A. (1978). Uniform asymptotic stability in functional differential equations. *Proc. Amer. Math. Soc.* **68**, 195–199.

Burton, T. A. (1980c). An integrodifferential equation. *Proc. Amer. Math. Soc.* **79**, 393–399.

Burton, T. A. (1983a). Volterra equations with small kernels. *J. Integral Equations* **5**, 271–285.

Burton, T. A. (1983b). "Volterra Integral and Differential Equations." Academic Press, Orlando.

Burton, T. A. (1985). "Stability and Periodic Solutions of Ordinary and Functional Differential Equations." Academic Press, Orlando.

Burton, T. A. (1991). The nonlinear wave equation as a Liénard equation. *Funkcialaj Ekvacioj* **34**, 529-545.

Burton, T. A., (1993). Boundedness and periodicity in integral and integrodifferential equations. *Differential Equations and Dynamical Systems* **1**, 161-172.

Burton, T. A. (1994a). Liapunov functionals and periodicity in integral equations. *Tohoku Math. J.* **46**, 207-220.

Burton, T. A. (1994b). Differential inequalities and existence theory for differential, integral, and delay equations, pp. 35-56. In "Comparison Methods and Stability Theory." Xinzhi Liu and David Siegel, eds, Dekker, New York.

Burton, T. A. (1996a). Examples of Lyapunov functionals for non-differentiated equations. Proc. First World Congress of Nonlinear Analysts, 1992. V. Lakshmikantham, ed. Walter de Gruyter, New York. pp. 1203–1214.

Burton, T. A. (1996b). Integral equations, implicit functions, and fixed points. *Proc. Amer. Math. Soc.* **124**, 2383-2390.

Burton, T. A., (1997). Linear integral equations and periodicity. *Annals of Differential Equations* **13**, 313-326.

Burton, T. A. (1998a). A fixed point theorem of Krasnoselskii. *Appl. Math. Lett.* **11**, 85-88.

Burton, T. A. (1998b). Basic neutral integral equations of advanced type. *Nonlinear Anal.* **31**, 295-310.

Burton, T. A. (2005b). "Stability and Periodic Solutions of Ordinary and Functional Differential Equations." Dover, New York. (This is a slightly corrected reprint of the 1985 edition published by Academic Press.)

Burton, T. A. (2005c). "Volterra Integral and Differential Equations, Second Edition." Elsevier, Amsterdam.

Burton, T. A. (2006a). Integral equations, Volterra equations, and the remarkable resolvent. *E. J. Qualitative Theory of Diff. Equ.* **No. 2**, 1-17. (http://www.math.u-szeged.hu/ejqtde/2006/200602.html)

Burton, T. A. (2006b). "Stability by Fixed Point Theory for Functional Differential Equations." Dover, New York.

Burton, T. A., (2007a). Integral equations, L^p-forcing, remarkable resolvent: Liapunov functionals. *Nonlinear Anal.* **68**, 35-46.

Burton, T. A., (2007b). Integral equations, large and small forcing functions: periodicity. *Math. Computer Modelling* **45**, 1363-1375.

Burton, T. A., (2007c). Integral equations, large forcing, strong resolvents. *Carpathian J. of Math.* **23**, 1-10.

Burton, T. A., (2007d). Scalar nonlinear integral equations. *Tatra Mountains Mathematical Publications* **38**, 41–56.

Burton, T. A., (2008a). Integral equations with contrasting kernels. *E. J. Qualitative Theory of Diff. Equ.*
http:/www.math.u-szeged.hu/ejqtde/2008/200802.html

Burton, T. A., (2008b). Integral equations, periodicity, and fixed points. *Fixed Point Theory* **9**, 47-65.

Burton, T. A., (2009). Liapunov functionals, convex kernels, and strategy. Preprint.

Burton, T. A. and Dwiggins, D. P., (2009). Resolvents of integral equations with continuous kernels. *Nonlinear Studies*, to appear.

Burton, T. A., Eloe, P. W., and Islam, M. N. (1990). Periodic solutions of linear integrodifferential equations. *Math. Nach.* **147**, 175-184.

Burton, T. A. and Furumochi, Tetsuo (1994). A stability theory for integral equations. *J. Integral Equations Appl.* **6**, 445–477.

Burton, T. A. and Furumochi, Tetsuo (1995). Periodic solutions of a Volterra equation and robustness. *Nonlinear Anal.* **25**, 1199-1219.

Burton, T. A. and Furumochi, Tetsuo (1996). Periodic and asymptotically periodic solutions of Volterra integral equations. *Funkcialaj Ekvacioj* **39**, 87-107.

Burton, T. A. and Furumochi, Tetsuo (2001). A note on stability by Schauder's theorem. *Funkcialaj Ekvacioj* **44**, 73-82.

Burton, T. A., Furumochi, Tetsuo, and Huang, Qichang (1995). Stability in several measures and a differential inequality for a partial integral equation. *Nonlinear Anal.* **25**, 885–898.

Burton, T. A. and Grimmer, R. (1973). Oscillation, continuation, and uniqueness of solutions of retarded differential equations. *Trans. Amer. Math. Soc.* **179**, 193–209.

Burton, T. A. and Haddock, J. (2009). Qualitative properties of solutions of integral equations. *Nonlinear Anal.*, **71**, 5712–5723.

Burton, T. A. and Hatvani, L. (1991). On the existence of periodic solutions of some nonlinear functional differential equations with unbounded delay. *Nonlinear Anal.* **16**, 389-398.

Burton, T. A. and Hering, R. H. (2000). Neutral integral equations of retarded type. *Nonlinear Anal.* **41**, 545-572.

Burton, T. A. and Kirk, Colleen (1998). A fixed point theorem of Krasnoselskii-Schaefer Type. *Math. Nachr.* **189**, 23-31.

Burton, T. A. and Makay, G. (2002). Continuity, compactness, fixed points, and integral equations. *E. J. Qualitative Theory of Diff. Equ.* **No. 14**, 1-13. (http://www.math.u-szeged.hu/ejqtde/2002/)

REFERENCES

Burton, T. A. and Somolinos, A. (2007). The Lurie control saisfies a Liénard equation. *Dyn. Contin., Discrete Impuls. Syst. Ser. B Appl. Algorithms* **14**, 625-640.

Burton, T. A. and Zhang, B. (1990). Uniform ultimate boundedness and periodicity in functional differential equations. *Tohoku Math. J.* **42**, 93–100.

Burton, T. A. and Zhang, B. (2004). Fixed points and stability of an integral equation: nonuniqueness. *Applied Math. Letters* **17**, 839-846.

Cain, G. L. and Nashed, M. Z. (1971). Fixed points and stability for a sum of two operators in locally convex spaces. *Pacific J. Math.* **39**, 581-592.

Calvert, B. (1977). Fixed points for $U+C$ where U is Lipschitz and C is compact. *Yokohama Math. J.* **25**, 1-4.

Cao, Jinde, Li, H. X., and Ho, Daniel W. C. (2005). Synchronization criteria of Luré systems with time-delay feedback control. *Chaos, Solitons, and Fractals* **23:4**, 1285-1298.

Coddington, E.A. and Levinson, N. (1955). "Theory of Ordinary Differential Equations." McGraw-Hill, New York.

Corduneanu, C. (1971). "Principles of Differential and Integral Equations." Allyn and Bacon, Rockledge, New Jersey.

Corduneanu, C. (1973). "Integral Equations and Stability of Feedback Systems." Academic Press, New York.

Corduneanu, C. (1977). "Principles of Differential and Integral Equations." Chelsea Publishing Co., New York.

Corduneanu, C. (1991). "Integral Equations and Applications." Cambridge Univ. Press, Cambridge, U.K.

Corduneau, C. (1997). Neutral functional equations of Volterra type. *Functional Differential Equations* **4**, 265-270.

Corduneanu, C. (2000). Existence of solutions for neutral functional differential equations with causal operators. *J. Differential Equations* **168**, 93-101.

Corduneanu, C. (2001). Some existence results for functional equations with causal operators. *Nonlinear Anal.* **47**, 709-716.

Dhage, B. C. (2002). On a fixed point theorem of Krasnoselskii-Schaefer type. *E. J. Qualitative Theory of Diff Equ.* **No. 6**, 1-9. (http://www.math.u-szeged.hu/ejqtde/2002/200206.html)

Dhage, B. C. (2003a). Remarks on two fixed-point theorems involving the sum and the product of two operators. *Comput. Math. Appl.* **46**, 1779 -1785.

Dhage, B. C. (2003b). Local fixed point theory for the sum of two operators in Banach spaces. *Fixed Point Theory* **4**, 49-60.

Dhage, B. C. and Ntouyas, S. K. (2002). Existence results for nonlinear functional integral equations via a fixed point theorem of Krasnoselskii-Schaefer type. *Nonlinear Studies* **9**, 307-317.

Driver, R. D. (1962). Existence and stability of solutions of a delay-differential system. *Arch. Rational Mech. Anal.* **10**, 401–426.

Driver, R. D. (1963). Existence theory for a delay-differential system. *Contrib. Differential Equations* **1**, 317–335.

Driver, R. D. (1965). Existence and continuous dependence of solutions of a neutral functional-differential equation. *Archive Rat. Mech. Anal.* **19**, 149-186.

El'sgol'ts, L. E. (1966). "Introduction to the Theory of Differential Equations with Deviating Arguments." Holden-Day, San Francisco.

Epstein, I. J. (1962). Periodic solutions of systems of differential equations. *Proc. Amer. Math. Soc.* **13**, 690-694.

Feller, W. (1941). On the integral equation of renewal theory. *Ann. Math. Statist.* **12**, 243–267.

Garcia-Falset, J. (2008). Existence fixed points for the sum of two operators. *Math. Nachr.* to appear.

Gopalsamy, K. (1992). "Stability and Oscillations in Delay Differential Equations of Population Dynamics." Kluwer, Dordrecht.

Gopalsamy, K. and Zhang, B.G. (1988). On a neutral delay logistic equation. *Dynamics Stability Systems* **2**, 183-195.

Graef, J. R., Qian, Chuanxi, and Zhang, Bo. (2004). Formulas of Liapunov Functions for systems of linear difference equations. *Proc. London Math. Soc.* **88**, 185-203.

Graves, L. M. (1946). "The Theory of Functions of Real Variables." McGraw-Hill, New York.

Grimmer, R. (1979). Existence of periodic solutions of functional differential equations. *J. Math. Anal. Appl.* **72**, 666–673.

Gripenberg, G., Londen, S. O., and Staffans, O. (1990). "Volterra Integral and Functional Equations." Cambridge Univ. Press, Cambridge, U.K.

Grossman, S. I., and Miller, R. K. (1970). Perturbation theory for Volterra integrodifferential systems. *J. Differential Equations* **8**, 457–474.

Hartman, P. (1964). "Ordinary Differential Equations." Wiley, New York.

Henry, D. (1981). "Geometric Theory of Semilinear Parabolic Equations." Springer, Berlin.

Hewitt, E. and Stromberg, K. (1971). "Real and Abstract Analysis." Springer, Berlin.

Hoa, L. H. and Schmitt, K. (1994). Fixed point theorems of Krasnosel'skii type in locally convex spaces and applications to integral equations. *Results in Math.* **25**, 290-314.

Hurewicz, Witold (1958). "Lectures on Ordinary Differential Equations." The M.I.T. Press, Cambridge.

Islam, M. N. and Neugebauer, J. T. (2008). Qualitative properties of nonlinear Volterra integral equations. *E. J. Qualitative Theory of Diff Equ.* **12**, pp. 1-16 (http://www.math.u-szeged.hu/ejqtde/2008/200812.html)

James, Glenn and James, Robert C. (1959) Mathematics Dictionary, 2nd ed. Van Nostrand, New York.

Kato, J. (1994). Stability criteria for difference equations related with Liapunov functions for delay-differential equations. *Dynamic Systems and Applications* **3**, 75–84.

REFERENCES

Kato, J., and Strauss, A. (1967). On the global existence of solutions and Liapunov functions. *Ann. Math. Pura. Appl.* **77**, 303–316.

Kirk, C. M. and Olmstead,W. E. (2000). The influence of two moving heat sources on blow-up in a reactive-diffusive medium. *Z. Angew. Math. Phys.* **51**, 1-16.

Kirk, C. M. and Olmstead, W. E. (2002). Blow-up in a reactive-diffusive medium with a moving heat source. *Z. Angew. Math. Phys.* **53**, 147-159.

Kirk, C. M. and Olmstead, W. E. (2005). Blow-up solutions of the two-dimensional heat equation due to a localized moving source. *Analysis and Appl.* **3**, 1-16.

Kirk, C. M. and Roberts, Catherine A. (2002). A quenching poblem for the heat equation. *J. Integral Equations Appl.* **14**, 53-72.

Krasnoselskii, M. A. (1958). Some problems of nonlinear analysis. *Amer. Math. Soc. Transl.* **(2) 10**, 345-409.

Krasovskii, N. N. (1963). "Stability of Motion." Stanford Univ. Press.

Kuang, Y. (1993). "Delay Differential Equations with Applications to Population Dynamics." Academic Press, Boston.

Kuang, Y. (1993). Global stability in one or two species neutral delay population models." *Canadian Appl. Math. Quart.* **1**, 23–45.

Kuang, Y. (1991). On neutral delay logistic Gause-type predator-prey systems. *Dynamics Stability Systems* **6**, 173–189.

Lakshmikantham, V. and Leela, S. (1969). "Differential and Integral Inequalities," Vol. I. Academic Press, New York.

LaSalle, J. P. (1968). Stability theory for ordinary differential equations. *J. Differential Equations* **4**, 57-65.

LaSalle, J. P. and Lefschetz, Solomon (1961). "Stability by Liapunov's Direct Method with Applications." Academic Press, New York.

Lefschetz, Solomon (1965). "Stability of Nonlinear Control Systems." Academic Press, New York.

Levin, J. J. (1963). The asymptotic behavior of a Volterra equation. *Proc. Amer. Math. Soc.* **14**, 434–451.

Levin, J. J. (1964). On a nonlinear delay equation. *J. Math. Anal. Appl.* **8**, 31-44.

Levin, J. J. (1965). The qualitative behavior of a nonlinear Volterra equation. *Proc. Amer. Math. Soc.* **16**, 711-718.

Levin, J. J. (1972). On a nonlinear Volterra equation. *J. Math. Anal. Appl.* **39**, 458-476.

Levin, J. J. (1968). A nonlinear Volterra equation not of convolution type. *J. Differential Equations* **4**, 176–186.

Levin, J. J., and Nohel, J. A. (1960). On a system of integrodifferential equations occuring in reactor dynamics. *J. Math. Mechanics* **9**, 347–368.

Levinson, N. (1960). A nonlinear Volterra equation arising in the theory of superfluidity. *J. Math. Anal. Appl.*, **1**, 1-11.

Liu, Yicheng and Li, Zhiaxiang (2007). Krasnoselskii type fixed point theorems and applications. *Proc. Amer. Math. Soc.* to appear.

Liu, Yicheng and Li, Zhixiang (2006). Schaefer type theorem and periodic solutions of evolution equations. *J. Math. Anal. Appl* **316**, 237-255.

Lurie, A. I. (1951). "On some nonlinear problems in the theory of automatic control." H. M. Stationery Office, London.
Lyapunov, A. M. (1992). The general problem of the stability of motion. *Int. J. Control.* **55**, 531-773. (This is a modern translation by A. T. Fuller of the 1892 monograph.)
Massera, J. L. (1949). On Liapunov's condition of stability. *Ann. of Math.* **50**, 705–721.
Massera, J. L. (1950). The existence of periodic solutions of a system of differential equations. *Duke Math. J.* **17**, 457–475.
Melville, Herman (1950). "Moby Dick." The Modern Library, New York.
Miller, R. K. (1968). On the linearization of Volterra integral equations. *J. Math. Anal. Appl.* **23**, 198–208.
Miller, R. K. (1971a). "Nonlinear Volterra Integral Equations." Benjamin, New York.
Miller, R. K. (1971b). Asymptotic stability properties of linear Volterra integrodifferential equations. *J. Differential Equations* **10**, 485–506.
Nashed, M. Z. and Wong, J. S. W. (1969). Some variants of a fixed point theorem of Krasnoselskii and applications to nonlinear integral equations. *J. Math. Mech.* **18**, 767-777.
Natanson, I. P. (1960). "Theory of Functions of a Real Variable," Vol. II. Ungar, New York.
Olmsted, John M. H. (1959). "Real Variables." Appleton-Century-Crofts, Inc., New York.
Padmavally, Komarath (1958). On a non-linear integral equation. *J. Math. Mech.* **7**, 533-555.
Park, Sehie (2007). Generalizations of the Krasnoselskii fixed point theorem. *Nonlinear Anal.* **67**, 3401-3410.
Perron, O. (1930). Die Stabilitätsfrage bei Differentialgleichungen. *Math. Z.* **32**, 703–728.
Purnaras, I. K. (2006). A note on the existence of solutions to some nonlinear functional integral equations. *E. J. Qualitative Theory of Diff. Equ.* **No. 17**, 1-24. (http://www.math.u-szeged.hu/ejqtde/2006/200617.html)
Reynolds, David W. (1984). On linear singular Volterra integral equations of the second kind. *J. Math. Anal. Appl* **103**, 230-262.
Ritt, Joseph Fels (1966). "Differential Algebra." Dover, New York.
Sansone, G. and Conti, R. (1964). "Non-linear Differential Equations." Macmillan, New York.
Schaefer, H. (1955). Über die Methode der a priori Schranken. *Math. Ann.* **129**, 415–416.
Sehgal, V. M. and Singh, S. P. (1978). A fixed point theorem for the sum of two mappings. *Math. Japonica* **23**, 71-75.
Smart, D. R. (1980). "Fixed Point Theorems." Cambridge Univ. Press, Cambridge.
Somolinos, A. (1977). Stability of Lurie-type functional equations. *J. Differential Equations* **26**, 191–199.

Strauss, A. (1970). On a Perturbed Volterra Integral Equation. *J. Math. Anal. Appl.* **30**, 564-575.

Tan, D. H. (1987). Two fixed point theorems of Krasnosels'kii type. *Rev. Roumanian Math. Pure Appl.* **32**, 397-400.

Taylor, Angus E. and Mann, W. Robert (1983). "Advanced Calculus, Third ed." Wiley, New York.

Tricomi, F. G. (1985). "Integral Equations." Dover, New York.

Windsor, Alistair (2009). A contraction mapping proof of the smooth dependence on parameters of solutions to Volterra integral equations. Preprint.

Vladimirescu, Cristian and Avramescu, Cezar (2006). "Applications of the Fixed Point Method to Ordinary Differential and Integral Equations on Noncompact Intervals." Universitari Press, Craiova, Romania.

Volterra, V. (1913). "Lecons sur les équations intégrales et les équations intégro-differentielles." Collection Borel, Paris.

Volterra, V. (1928). Sur la théorie mathématique des phénomès héréditaires. *J. Math. Pur. Appl.* **7**, 249–298.

Volterra, V. (1931). "Lecons sur la théorie mathématique de la lutte pour la vie." Paris.

Volterra, V. (1959). "Theory of Functionals and of Integral and Integro-differential Equations." Dover, New York.

Yoshizawa, T. (1963). Asymptotic behavior of solutions of a system of differential equations. *Contrib. Differential Equations* **1**, 371–387.

Yoshizawa, T. (1966). "Stability Theory by Liapunov's Second Method." Math. Soc. Japan, Tokyo.

Yoshizawa, T. (1975). "Stability Theory and the Existence of Periodic Solutions and Almost Periodic Solutions." Springer Publ., New York.

Yosida, Kosaku (1991). "Lectures on Differential and Integral Equations." Dover, New York.

Zhang, B. (1995). Asymptotic stability in functional-differential equations by Liapunov functionals. *Trans. Amer. Math. Soc.* **347**, 1375–1382.

Zhang, B. (1997). Asymptotic stability criteria and integrability properties of the resolvent of Volterra and functional equations. *Funkcial. Ekvac.* **40**, 335–351.

Zhang, B. (2001). Formulas of Liapunov functions for systems of linear ordinary and delay differential equations. *Funkcial. Ekvac.* **44**, 253–278.

Zhang, B. (2009). Boundedness and global attractivity of solutions for a system of nonlinear integral equations. *Cubo, A Mathematical Journal* **11**, No. 3, pp. 41-53.

Author Index

Avramescu, C., 202, 203, 213

Barbashin, E. A., 2, 3
Barroso, C. S., 202
Becker, L. C., 26, 34–36, 43, 103, 137, 165
Belmekki, M., 290, 299
Benchohra, M., 290, 299
Boyd, B., 202
Burton, T. A., 3, 28, 103, 110, 124, 164, 165, 167, 190, 193, 198, 202, 203, 206, 209, 211, 213, 239, 242, 257, 267, 280, 283, 287, 288, 290, 294, 295, 298, 303, 319, 324, 328, 340, 344

Cain, G. L., 202
Calvert, B., 202
Cao, Jinde, 206
Coddington, E. A., 304
Conti, R., 169
Corduneanu, C., 41, 42, 167, 177, 204, 209, 215, 326

Dhage, B. C., 202, 213
Driver, R., 293, 305
Dwiggins, D. P., 28

El'sgol'ts, L. E, 212, 293, 295, 305
Eloe, P., 257
Epstein, I. J., 240
Ezzinbi, K., 290

Feller, W., 91, 113
Furumochi, Tetsuo, 202, 242, 333, 339, 340

Garcia-Falset, J., 202, 290, 299
Gopalsamy, K., 293, 319
Graef, J. R., 3
Graves, L. M., 37
Grimmer, R., 287, 291
Grippenburg, G., 167, 204, 215, 326
Grossman, S., 35

Haddock, J. R., 344
Hartman, P., 46, 167, 169, 207, 208
Hatvani, L., 257
Henry, D., 3
Hering, R., 319
Hewitt, E., 190, 306
Hoa, L. H., 202
Huang, Qichang, 340
Hurewicz, W., 37

Islam, M. N., 54, 154, 257

James, Glenn, 216
James, Robert, 216

Kato, J., 168, 169, 199
Kirk, Colleen, 211, 213, 216, 241, 290, 303
Krasnoselskii, M. A., 201
Krasovskii, N. N., 3
Krisztin, T., 103, 165
Kuang, Y., 293, 319

AUTHOR INDEX

Lakshmikantham, V., 173
LaSalle, J. P., 206
Leela, S., 173
Lefschetz, S., 206
Levin, J. J., 187, 219
Levinson, N., 219, 304
Li, H. X., 206
Li, Zhixiang, 203, 210, 213, 299
Liu, Yicheng, 203, 210, 213, 299
Londen, S-O, 167, 204, 326
Lurie, A. I., 206
Lyapunov, A. M., 1

Makay, G., 242, 287, 288
Mann, W. Robert, 207
Massera, J. L., 240
Melville, Herman, 216
Miller, R. K., 3, 28, 29, 35, 114, 163, 167, 177, 204, 252

Nashed, M. Z., 202
Natanson, I. P., 332
Neugebauer, J. T., 54, 154
Ntouyas, S. K., 213, 290, 299

Olmstead, W. E., 216, 241
Olmsted, John M. H., 37

Padmavally, Komarath, 214, 219, 241
Park, Sehie, 202, 209
Perron, O., 53, 139
Purnaras, Ioannis K., 213

Qian, Chuanxi, 3

Reynolds, D. W., 214, 219
Ritt, J. F., 8, 52, 204
Roberts, Catherine A., 216

Sansone, G., 169
Schaefer, H., 169
Schmitt, K., 202
Sehgal, V. M., 202
Singh, S. P., 202

Smart, D. R., 169, 201, 207, 208
Somolinos, A., 3, 206
Staffans, O. J., 167, 204, 326
Strauss, A., 54, 168, 169
Stromberg, K., 190, 306

Tan, D. H., 202
Taylor, Angus E., 207
Teixeira, E. V., 202
Tricomi, F. G., 236, 237

Vladimirescu, C., 202, 203, 213
Volterra, V., 219

Windsor, A., 27
Wong, J. S. W., 202

Yoshizawa, T., 1, 3, 168, 273, 328
Yosida, Kosaku, 237

Zhang, Binggen, 293
Zhang, Bo, 3, 137, 190, 202, 203, 324

Subject Index

$a' \in L^{2^n}$ implies $x \in L^{2^n}$, 83

Adam and Eve Counterpart, 124
Adam and Eve theorem, 22
admissible, 326
Ascoli-Arzela theorem, 170

Becker's resolvent, 25
bound
    a priori, 237
    a priori, 280
    uniform, 338
bounded solution, 306

compact map, 207
complete, 21
Conti-Wintner theorem, 169
contraction, 21
    large, 210
    separate, 210, 299
contraction mapping principle, 21
convexity, 299

equicontinuous, 170
equilibrium
    near, 322, 324
existence, 27, 33, 178

fixed point
    Krasnoselskii-Schaefer, 213, 289
    Schaefer, 169, 237
Fixed points, 21
Functions

    equicontinuous, 179
    initial, 174–176
fundamental matrix solution, 40
Fundamental theorem, 54

Gronwall's inequality, 31

heredity, 290, 299

identity
    L^1-approximate, 60
    L^2-approximate, 79
    L^p-approximate, 53, 60, 78, 79
    approximate, 53, 73, 77, 91, 92
    asymptotic, 53, 56
implicit function, 208
Inequalities
    Gronwall's, 31
initial function, 174
initial interval, 174
integral equations
    first kind, 237
    second kind, 237

Jensen's inequality, 332

Krasnoselskii's fixed point theorem, 202
Krasnoselskii's theorem, 201
Krasnoselskii-Schaefer theorem, 290

Liapunov function, 1
Liapunov functional, 4
Lipschitz condition, 176

SUBJECT INDEX

metric space, 20
mildly unbounded, 168

Notation
    $V(t,x(\cdot))$, 17
    Q, 58
    Y, 58
    BC, 53
    P, 58

partial
    integral equations, 340
perfect functional, 64
periodic
    asymptotically, 58, 59
periodic solution, 57, 261, 267, 275,
    296, 306, 317
    asymptotically, 98, 99
    partial differential equation,
    299
Perron's theorem, 54
principal matrix solution, 35, 41
problem of Lurie, 204

Razumikhin technique, 85
research problems, 206
resolvent, 24, 41, 53

Solutions
    unique, 176
stable
    L^N, 84
    asymptotically, 264, 272, 273,
    328, 330, 333
superposition
    principle of, 39

uniformly bounded, 169

variation of parameters, 24, 44, 53
variation of parameters (nonlinear),
    164

wedge, 330
weighted norm, 27, 30, 48, 56, 177

Young's inequality, 190

www.ingramcontent.com/pod-product-compliance
Lightning Source LLC
Chambersburg PA
CBHW060820170526
45158CB00001B/41